White Rose Maths Edition

Year 2C

A Guide to Teaching for Mastery

Series Editor: Tony Staneff
Lead author: Josh Lury

Contents

Introduction to the author team

Power Maths arises from the work of maths mastery experts who are committed to proving that, given the right mastery mindset and approach, **everyone can do maths**. Based on robust research and best practice from around the world, *Power Maths* was developed in partnership with a group of UK teachers to make sure that it not only meets our children's wide-ranging needs but also aligns with the National Curriculum in England.

Power Maths – White Rose Maths edition

This edition of *Power Maths* has been developed and updated by:

Tony Staneff, Series Editor and Author

Vice Principal at Trinity Academy, Halifax, Tony also leads a team of mastery experts who help schools across the UK to develop teaching for mastery via nationally recognised CPD courses, problem-solving and reasoning resources, schemes of work, assessment materials and other tools.

Josh Lury, Lead Author

Josh is a specialist maths teacher, author and maths consultant with a passion for innovative and effective maths education.

The first edition of *Power Maths* was developed by a team of experienced authors, including:

- **Tony Staneff and Josh Lury**

- **Trinity Academy Halifax** (Michael Gosling CEO, Emily Fox, Kate Henshall, Rebecca Holland, Stephanie Kirk, Stephen Monaghan and Rachel Webster)

- **David Board, Belle Cottingham, Jonathan East, Tim Handley, Derek Huby, Neil Jarrett, Stephen Monaghan, Beth Smith, Tim Weal, Paul Wrangles** – skilled maths teachers and mastery experts

- **Cherri Moseley** – a maths author, former teacher and professional development provider

- **Professors Liu Jian and Zhang Dan**, Series Consultants and authors, and their team of mastery expert authors: **Wei Huinv, Huang Lihua, Zhu Dejiang, Zhu Yuhong, Hou Huiying, Yin Lili, Zhang Jing, Zhou Da and Liu Qimeng**

  Used by over 20 million children, Professor Liu Jian's textbook programme is one of the most popular in China. He and his author team are highly experienced in intelligent practice and in embedding key maths concepts using a C-P-A approach.

- **A group of 15 teachers and maths co-ordinators**

  We consulted our teacher group throughout the development of *Power Maths* to ensure we are meeting their real needs in the classroom.

What is *Power Maths*?

Created especially for UK primary schools, and aligned with the new National Curriculum, *Power Maths* is a whole-class, textbook-based mastery resource that empowers every child to understand and succeed. *Power Maths* rejects the notion that some people simply 'can't do' maths. Instead, it develops growth mindsets and encourages hard work, practice and a willingness to see mistakes as learning tools.

Best practice consistently shows that mastery of small, cumulative steps builds a solid foundation of deep mathematical understanding. *Power Maths* combines interactive teaching tools, high-quality textbooks and continuing professional development (CPD) to help you equip children with a deep and long-lasting understanding. Based on extensive evidence, and developed in partnership with practising teachers, *Power Maths* ensures that it meets the needs of children in the UK.

Power Maths and Mastery

Power Maths makes mastery practical and achievable by providing the structures, pathways, content, tools and support you need to make it happen in your classroom.

To develop mastery in maths children must be enabled to acquire a deep understanding of maths concepts, structures and procedures, step by step. Complex mathematical concepts are built on simpler conceptual components and when children understand every step in the learning sequence, maths becomes transparent and makes logical sense. Interactive lessons establish deep understanding in small steps, as well as effortless fluency in key facts such as tables and number bonds. The whole class works on the same content and no child is left behind.

Power Maths

- Builds every concept in small, progressive steps
- Is built with interactive, whole-class teaching in mind
- Provides the tools you need to develop growth mindsets
- Helps you check understanding and ensure that every child is keeping up
- Establishes core elements such as intelligent practice and reflection

The *Power Maths* approach

Everyone can!

Founded on the conviction that every child can achieve, *Power Maths* enables children to build number fluency, confidence and understanding, step by step.

Child-centred learning

Children master concepts one step at a time in lessons that embrace a concrete-pictorial-abstract (C-P-A) approach, avoid overload, build on prior learning and help them see patterns and connections. Same-day intervention ensures sustained progress.

Continuing professional development

Embedded teacher support and development offer every teacher the opportunity to continually improve their subject knowledge and manage whole-class teaching for mastery.

Whole-class teaching

An interactive, whole-class teaching model encourages thinking and precise mathematical language and allows children to deepen their understanding as far as they can.

What's different in the new edition?

If you have previously used the first editions of *Power Maths*, you might be interested to know how this edition is different. All of the improvements described below are based on feedback from *Power Maths* customers.

Changes to units and the progression

⚡ The order of units has been slightly adjusted, creating closer alignment between adjacent year groups, which will be useful for mixed age teaching.

⚡ The flow of lessons has been improved within units to optimise the pace of the progression and build in more recap where needed. For key topics, the sequence of lessons gives more opportunities to build up a solid base of understanding. Other units have fewer lessons than before, where appropriate, making it possible to fit in all the content.

⚡ Overall, the lessons put more focus on the most essential content for that year, with less time given to non-statutory content.

⚡ The progression of lessons matches the steps in the new White Rose Maths schemes of learning.

Lesson resources

⚡ There is a Quick recap for each lesson in the Teacher Guide, which offers an alternative lesson starter to the Power Up for cases where you feel it would be more beneficial to surface prerequisite learning than general number fluency.

⚡ In the **Discover** and **Share** sections there is now more of a progression from 1 a) to 1 b). Whereas before, 1 b) was mainly designed as a separate question, now 1 a) leads directly into 1 b). This means that there is an improved whole-class flow, and also an opportunity to focus on the logic and skills in more detail. As a teacher, you will be using 1 a) to lead the class into the thinking, then 1 b) to mould that thinking into the core new learning of the lesson.

⚡ In the **Share** section, for KS1 in particular, the number of different models and representations has been reduced, to support the clarity of thinking prompted by the flow from 1 a) into 1 b).

⚡ More fluency questions have been built into the guided and independent practice.

⚡ Pupil pages are as easy as possible for children to access independently. The pages are less full where this supports greater focus on key ideas and instructions. Also, more freedom is offered around answer format, with fewer boxes scaffolding children's responses; squared paper backgrounds are used in the Practice Books where appropriate. Artwork has also been revisited to ensure the highest standards of accessibility.

New components

480 Individual Practice Games are available in *ActiveLearn* for practising key facts and skills in Years 1 to 6. These are designed in an arcade style, to feel like fun games that children would choose to play outside school. They can be accessed via the Pupil World for homework or additional practice in school – and children can earn rewards. There are Support, Core and Extend levels to allocate, with Activity Reporting available for the teacher. There is a Quick Guide on *ActiveLearn* and you can use the Help area for support in setting up child accounts.

There is also a new set of lesson video resources on the Professional Development tile, designed for in-school training in 10- to 20-minute bursts. For each part of the *Power Maths* lesson sequence, there is a slide deck with embedded video, which will facilitate discussions about how you can take your *Power Maths* teaching to the next level.

Your *Power Maths* resources

To help you teach for mastery, *Power Maths* comprises a variety of high-quality resources.

Pupil Textbooks

Discover, **Share** and **Think together** sections promote discussion and introduce mathematical ideas logically, so that children understand more easily.

Using a Concrete-Pictorial-Abstract approach, clear mathematical models help children to make connections and grasp concepts.

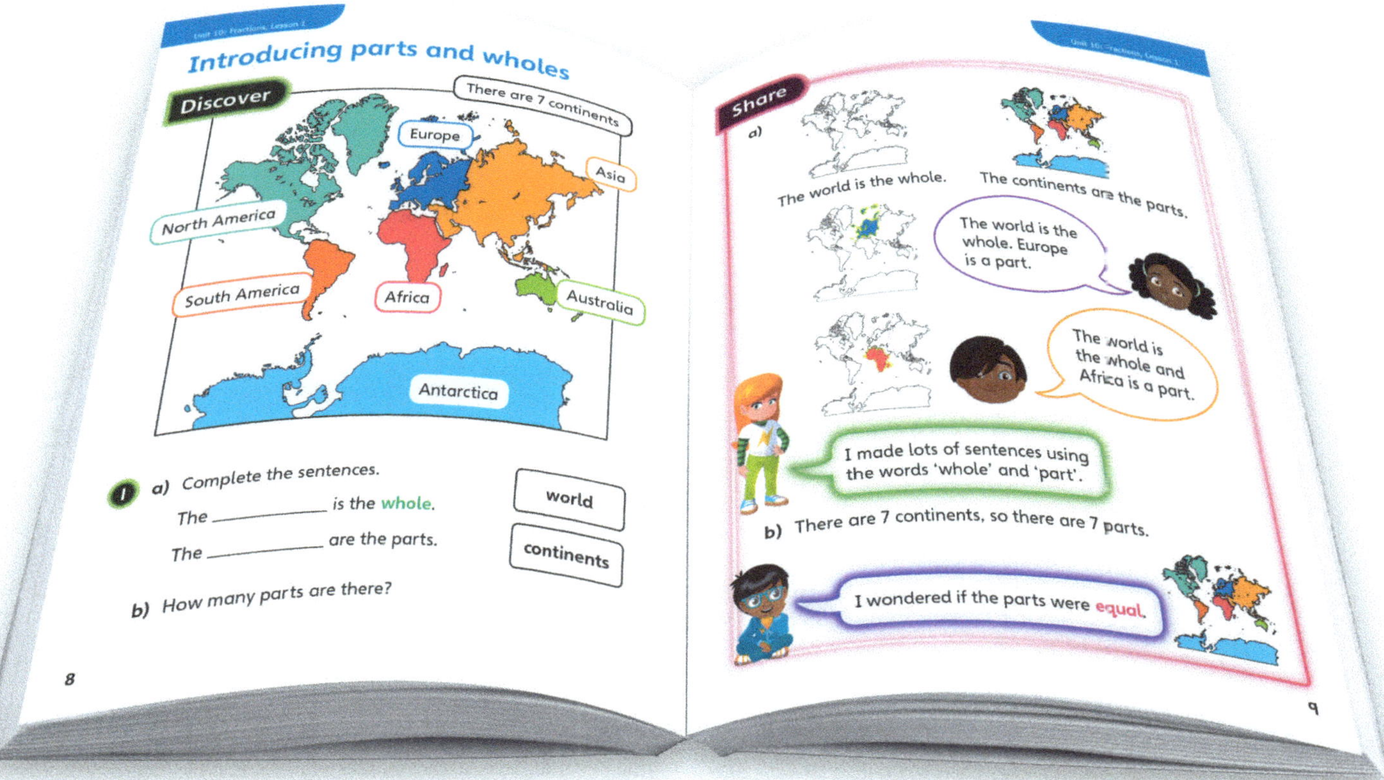

Appealing scenarios stimulate curiosity, helping children to identify the maths problem and discover patterns and relationships for themselves.

Friendly, supportive characters help children develop a growth mindset by prompting them to think, reason and reflect.

The coherent *Power Maths* lesson structure carries through into the vibrant, high-quality textbooks. Setting out the core learning objectives for each class, the lesson structure follows a carefully mapped journey through the curriculum and supports children on their journey to deeper understanding.

Pupil Practice Books

The Practice Books offer just the right amount of intelligent practice for children to complete independently in the final section of each lesson.

Practice questions are finely tuned to move children forward in their thinking and to reveal misconceptions.

The practice questions are for everyone – each question varies one small element to move children on in their thinking.

Calculations are connected so that children think about the underlying concept.

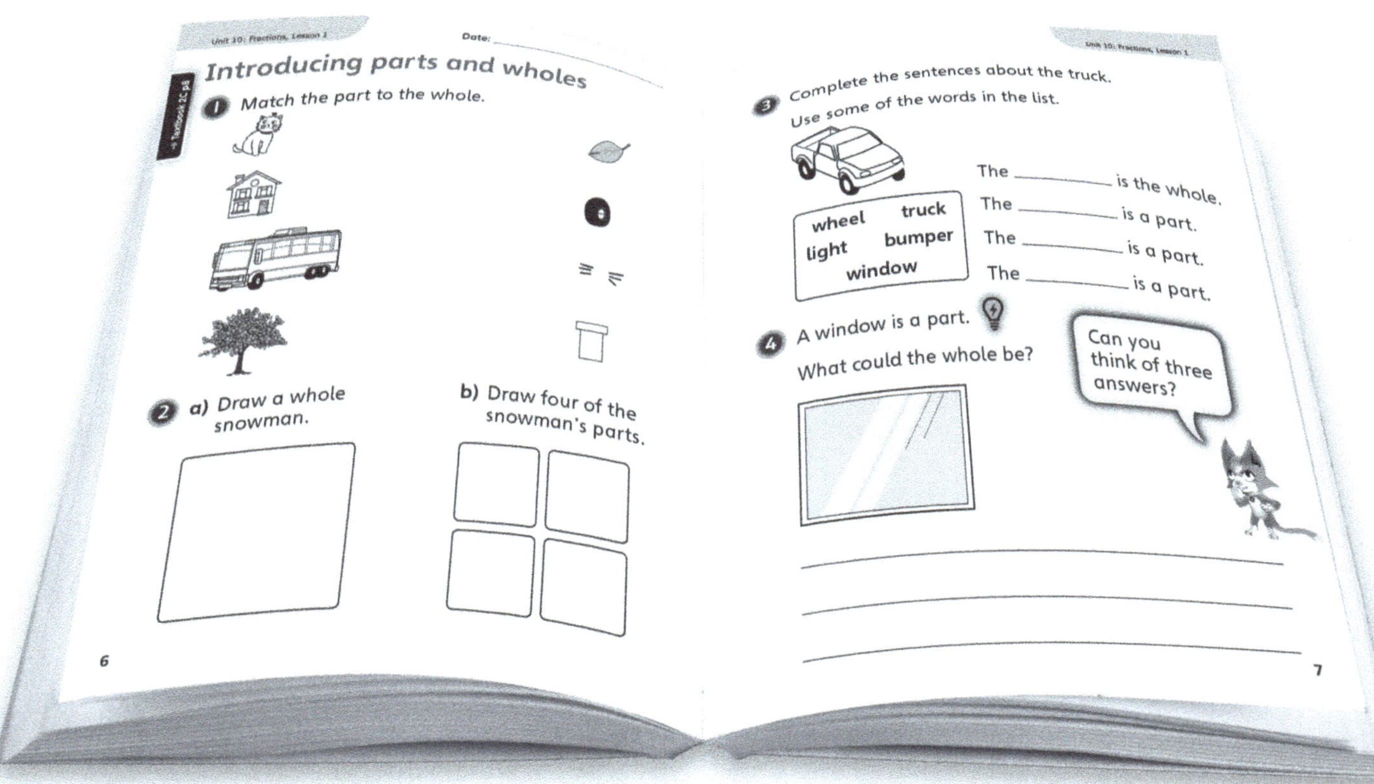

Challenge questions allow children to delve deeper into a concept.

The *Power Maths* characters support and encourage children to think and work in different ways.

Think differently questions encourage children to use reasoning as well as their mathematical knowledge to reach a solution.

Reflect questions reveal the depth of each child's understanding before they move on.

Online subscription

The online subscription will give you access to additional resources and answers from the Textbook and Practice Book.

eTextbooks

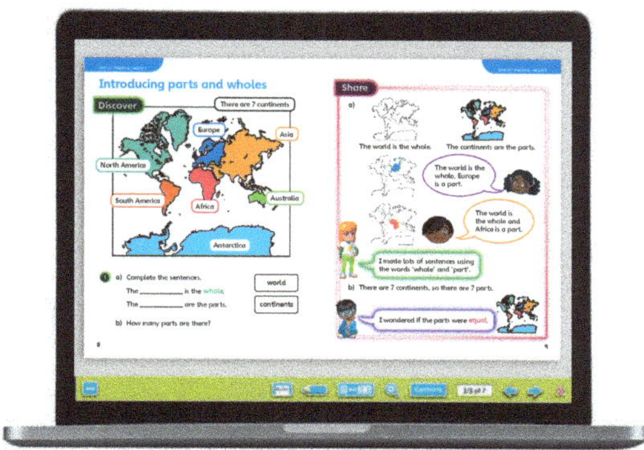

Digital versions of *Power Maths* Textbooks allow class groups to share and discuss questions, solutions and strategies. They allow you to project key structures and representations at the front of the class, to ensure all children are focusing on the same concept.

Teaching tools

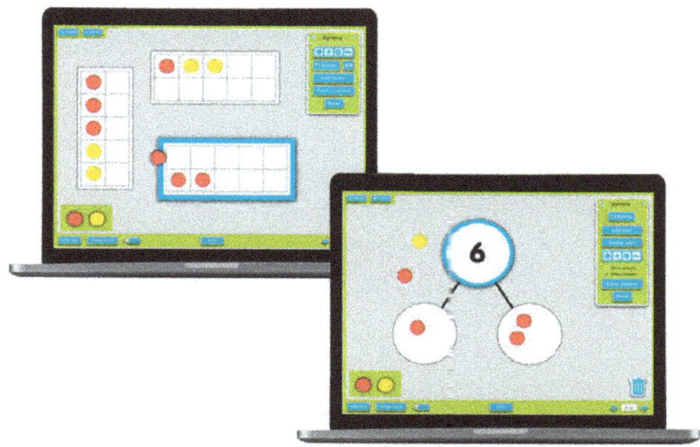

Here you will find interactive versions of key *Power Maths* structures and representations.

Power Ups

Use this series of daily activities to promote and check number fluency.

Online versions of Teacher Guide pages

PDF pages give support at both unit and lesson levels. You will also find help with key strategies and templates for tracking progress.

Unit videos

Watch the professional development videos at the start of each unit to help you teach with confidence. The videos explore common misconceptions in the unit, and include intervention suggestions as well as suggestions on what to look out for when assessing mastery in your students.

End of unit Strengthen and Deepen materials

The Strengthen activity at the end of every unit addresses a key misconception and can be used to support children who need it. The Deepen activities are designed to be low ceiling/high threshold and will challenge those children who can understand more deeply. These resources will help you ensure that every child understands and will help you keep the class moving forward together. These printable activities provide an optional resource bank for use after the assessment stage.

Individual Practice Games

These enjoyable games can be used at home or at school to embed key number skills (see page 6).

Professional Development videos and slides

These slides and videos of *Power Maths* lessons can be used for ongoing training in short bursts or to support new staff (see page 6).

The *Power Maths* teaching model

At the heart of *Power Maths* is a clearly structured teaching and learning process that helps you make certain that every child masters each maths concept securely and deeply. For each year group, the curriculum is broken down into core concepts, taught in units. A unit divides into smaller learning steps – lessons. Step by step, strong foundations of cumulative knowledge and understanding are built.

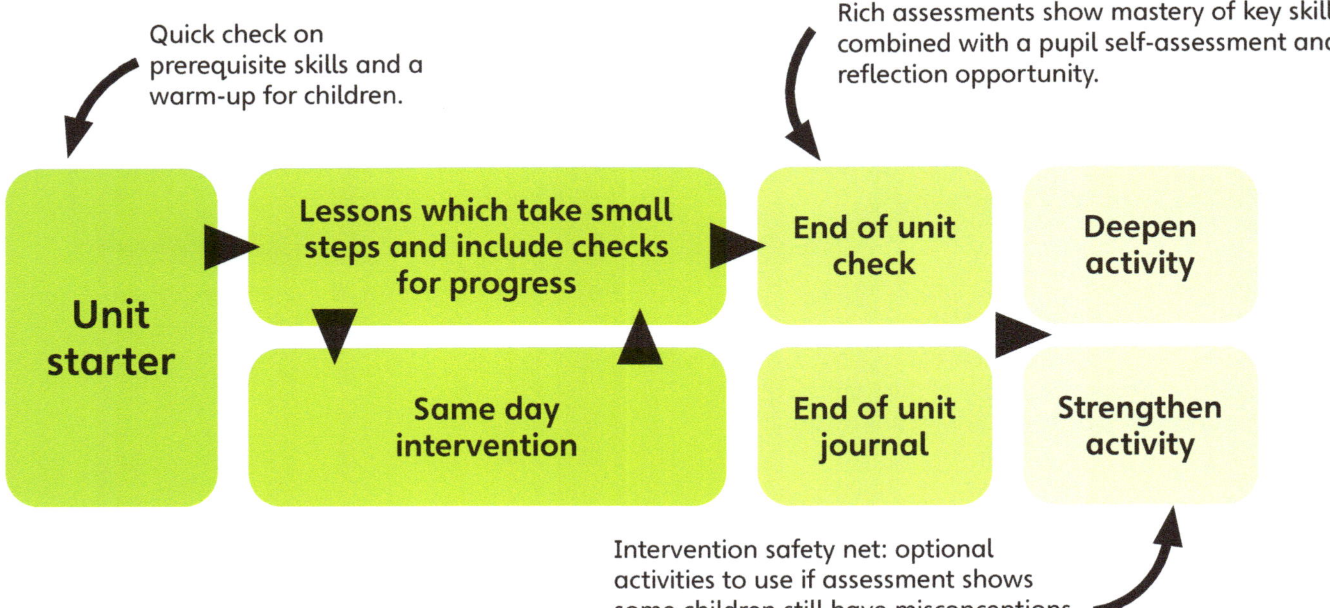

Unit starter

Each unit begins with a unit starter, which introduces the learning context along with key mathematical vocabulary and structures and representations.

- The Textbooks include a check on readiness and a warm-up task for children to complete.
- Your Teacher Guide gives support right from the start on important structures and representations, mathematical language, common misconceptions and intervention strategies.
- Unit-specific videos develop your subject knowledge and insights so you feel confident and fully equipped to teach each new unit. These are available via the online subscription.

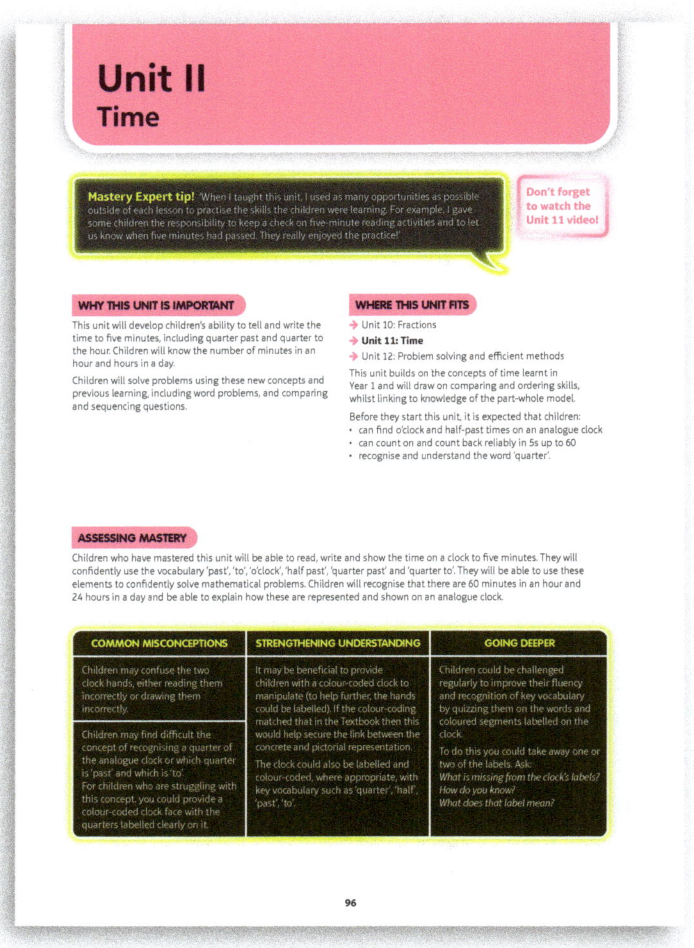

Lesson

Once a unit has been introduced, it is time to start teaching the series of lessons.

- Each lesson is scaffolded with Textbook and Practice Book activities and begins with a Power Up activity (available via online subscription) or the Quick recap activity in the Teacher Guide (see page 15).

- *Power Maths* identifies lesson by lesson what concepts are to be taught.

- Your Teacher Guide offers lots of support for you to get the most from every child in every lesson As well as highlighting key points, tricky areas and how to handle them, you will also find question prompts to check on understanding and clarification on why particular activities and questions are used.

Same-day intervention

Same-day interventions are vital in order to keep the class progressing together. This can be during the lesson as well as afterwards (see page 28). Therefore, *Power Maths* provides plenty of support throughout the journey.

- Intervention is focused on keeping up now, not catching up later, so interventions should happen as soon as they are needed.

- Practice section questions are designed to bring misconceptions to the surface, allowing you to identify these easily as you circulate during independent practice time.

- Child-friendly assessment questions in the Teacher Guide help you identify easily which children need to strengthen their understanding.

End of unit check and journal

For each unit, the End of unit check in the Textbook lets you see which children have mastered the key concepts, which children have not and where their misconceptions lie. The Practice Books also include an End of unit journal in which children can reflect on what they have learned. Each unit also offers Strengthen and Deepen activities, available via the online subscription.

The Teacher Guide offers different ways of managing the End of unit assessments as well as giving support with handling misconceptions.

The End of unit check presents multiple-choice questions. Children think about their answer, decide on a solution and explain their choice.

The End of unit journal is an opportunity for children to test out their learning and reflect on how they feel about it. Tackling the 'journal' problem reveals whether a child understands the concept deeply enough to move on to the next unit.

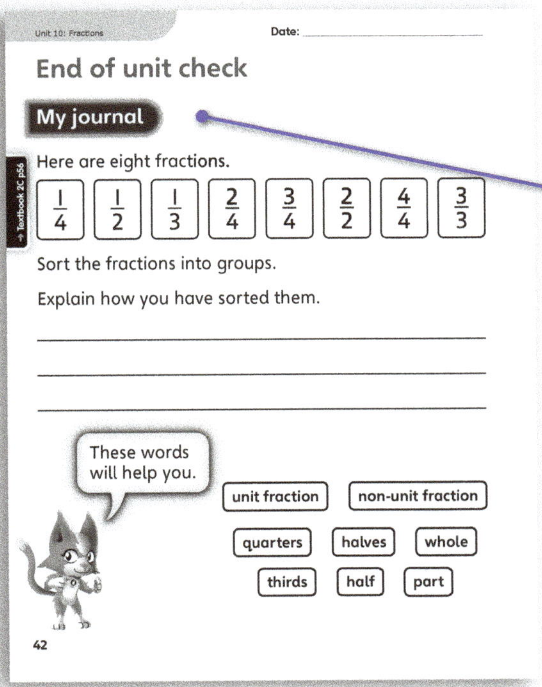

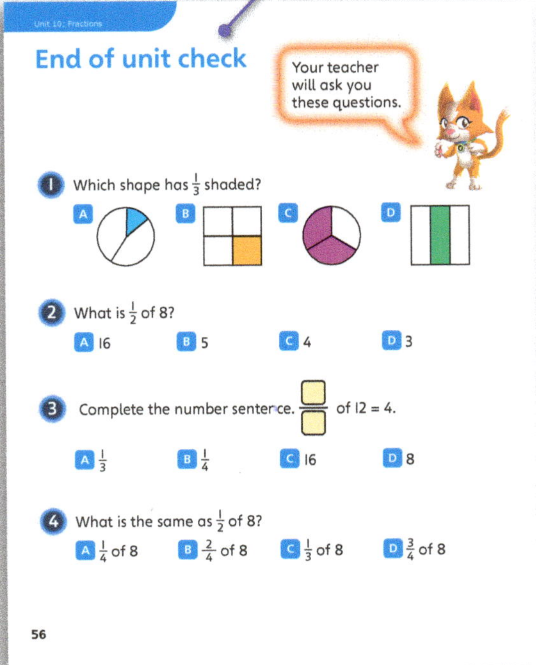

The *Power Maths* lesson sequence

At the heart of *Power Maths* is a unique lesson sequence designed to empower children to understand core concepts and grow in confidence. Embracing the National Centre for Excellence in the Teaching of Mathematics' (NCETM's) definition of mastery, the sequence guides and shapes every *Power Maths* lesson you teach.

Flexibility is built into the *Power Maths* programme so there is no one-to-one mapping of lessons and concepts and you can pace your teaching according to your class. While some children will need to spend longer on a particular concept (through interventions or additional lessons), others will reach deeper levels of understanding. However, it is important that the class moves forward together through the termly schedules.

Power Up  5 minutes

Each lesson begins with a Power Up activity (available via the online subscription) which supports fluency in key number facts.

The whole-class approach depends on fluency, so the Power Up is a powerful and essential activity.

The Quick recap is an alternative starter, for when you think some or all children would benefit more from revisiting pre-requisite work (see page 15).

TOP TIP

If the class is struggling with the task, revisit it later and check understanding.

Power Ups reinforce the two key things that are essential for success: times-tables and number bonds.

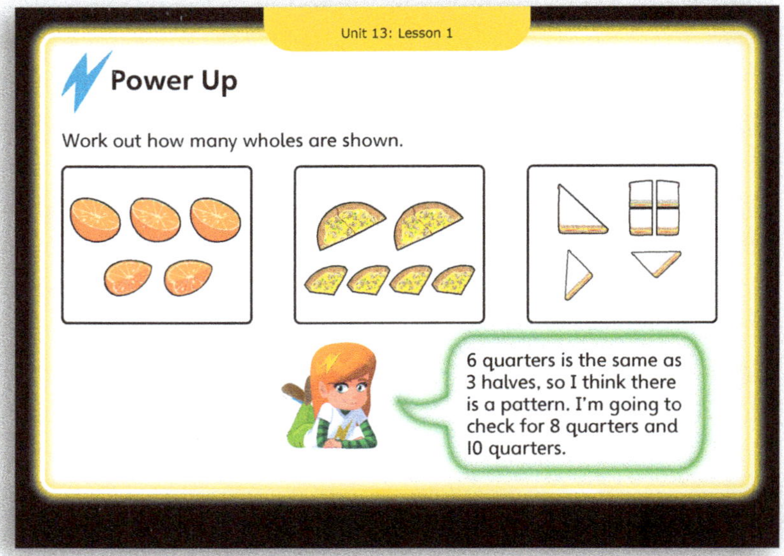

Discover  10 minutes

A practical, real-life problem arouses curiosity. Children find the maths through story telling.

TOP TIP

Discover works best when run at tables, in pairs with concrete objects.

Question ❶ a) tackles the key concept and question ❶ b) digs a little deeper. Children have time to explore, play and discuss possible strategies.

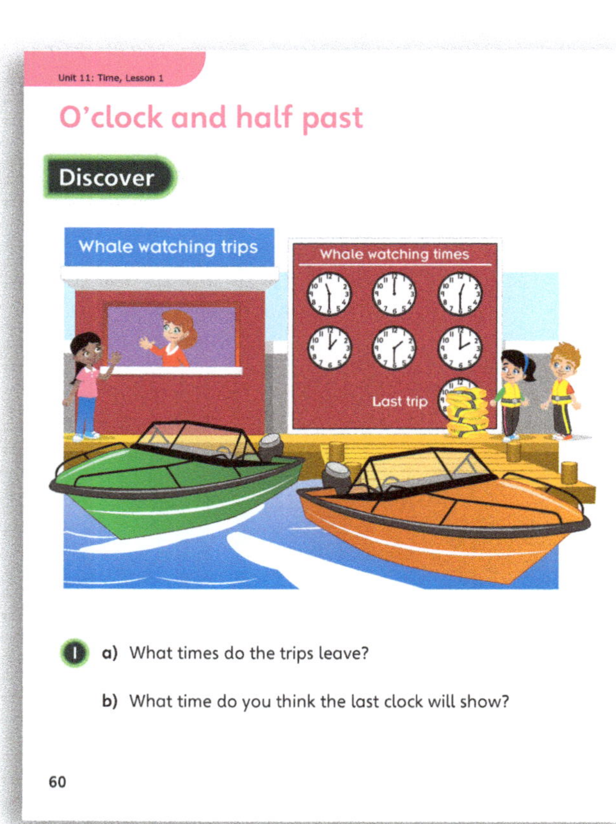

Share ⏱ 10 minutes

Teacher-led, this interactive section follows the **Discover** activity and highlights the variety of methods that can be used to solve a single problem.

TOP TIP

You can use the carpet area if you have this. Pairs sharing a textbook is a great format for **Share**!

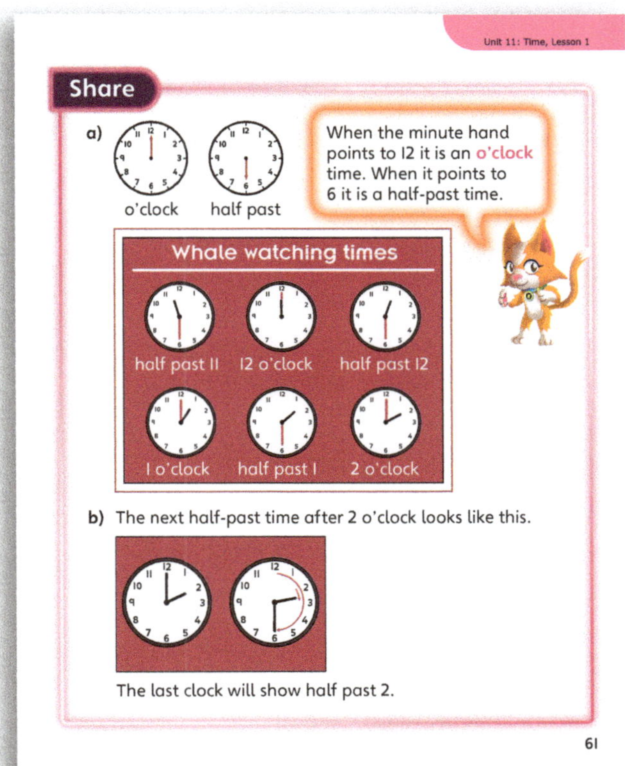

Your Teacher Guide gives target questions for children. The or line toolkit provides interactive structures and representations to link concrete and pictorial to abstract concepts.

Bring children to the front to share and celebrate their solutions and strategies.

Think together

⏱ 10 minutes

Children work in groups on the carpet or at tables, using their textbooks or eBooks.

TOP TIP

Make sure children have mini whiteboards or pads to write on if they are not at their tables.

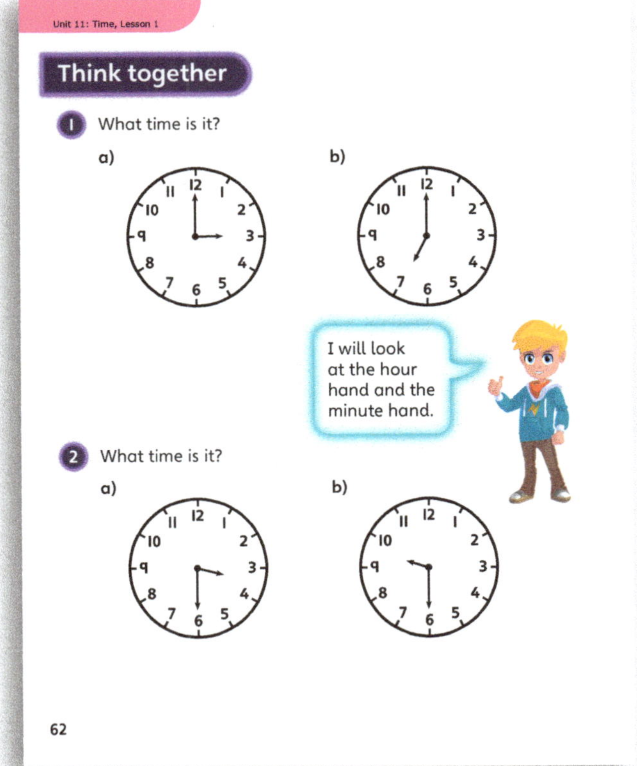

Using the Teacher Guide, model question 1 for your class.

Question 2 is less structured. Children will need to think together in their groups, then discuss their methods and solutions as a class.

Question 3 – the openness of the **Challenge** question helps to check depth of understanding.

Practice   ⏱ 15 minutes

Using their Practice Books, children work independently while you circulate and check on progress.

Questions follow small steps of progression to deepen learning.

TOP TIP
Some children could work separately with a teacher or assistant.

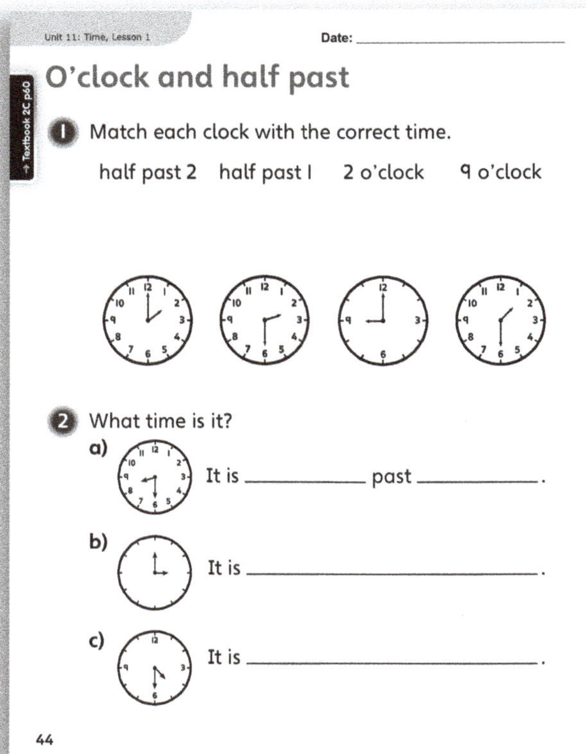

Are some children struggling? If so, work with them as a group, using mathematical structures and representations to support understanding as necessary.

There are no set routines: for real understanding, children need to think about the problem in different ways.

Reflect   ⏱ 5 minutes

'Spot the mistake' questions are great for checking misconceptions.

The **Reflect** section is your opportunity to check how deeply children understand the target concept.

The Practice Books use various approaches to check that children have fully understood each concept.

Looking like they understand is not enough! It is essential that children can show they have grasped the concept.

Using the *Power Maths* Teacher Guide

Think of your Teacher Guides as *Power Maths* handbooks that will guide, support and inspire your day-to-day teaching. Clear and concise, and illustrated with helpful examples, your Teacher Guides will help you make the best possible use of every individual lesson. They also provide wrap-around professional development, enhancing your own subject knowledge and helping you to grow in confidence about moving your children forward together.

There is a Teacher Guide per year group for every term, with unit and lesson level guidance and support.

Never feel stuck! You will find ideas for introducing every unit and lesson and questions to encourage teacher reflection before and after each lesson.

Tips and advice on key elements such as C-P-A approaches, misconceptions, language, modelling growth mindsets and same day intervention.

Annotations for every Textbook and Practice Book page, providing prompts for key questions to ask to expose understanding and explanations as to why key questions have been chosen.

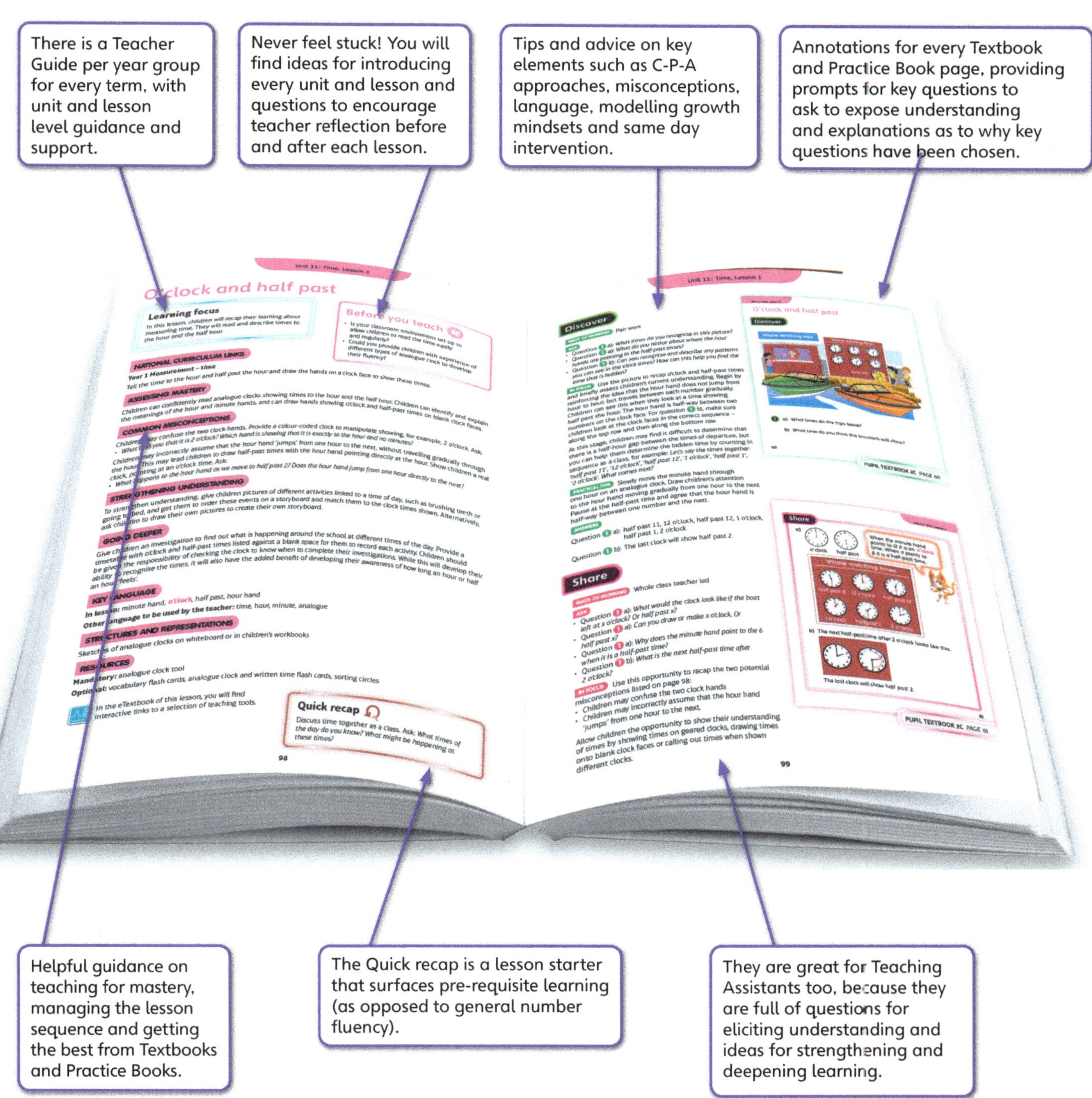

Helpful guidance on teaching for mastery, managing the lesson sequence and getting the best from Textbooks and Practice Books.

The Quick recap is a lesson starter that surfaces pre-requisite learning (as opposed to general number fluency).

They are great for Teaching Assistants too, because they are full of questions for eliciting understanding and ideas for strengthening and deepening learning.

At the end of each unit, your Teacher Guide helps you identify who has fully grasped the concept, who has not and how to move every child forward. This is covered later in the Assessment strategies section.

Power Maths Year 2, yearly overview

Textbook	Strand	Unit		Number of lessons
Textbook A / Practice Book A (Term 1)	Number – number and place value	1	Numbers to 100	17
	Number – addition and subtraction	2	Addition and subtraction (1)	13
	Number – addition and subtraction	3	Addition and subtraction (2)	12
	Geometry – properties of shape	4	Properties of shapes	12
Textbook B / Practice Book B (Term 2)	Measurement	5	Money	10
	Number – multiplication and division	6	Multiplication and division (1)	8
	Number – multiplication and division	7	Multiplication and division (2)	10
	Measurement	8	Length and height	5
	Measurement	9	Mass, capacity and temperature	8
Textbook C / Practice Book C (Term 3)	Number – fractions	10	Fractions	12
	Measurement	11	Time	5
	Number – addition and subtraction	12	Problem solving and efficient methods	11
	Geometry – position and direction	13	Position and direction	5
	Statistics	14	Statistics	7

Power Maths Year 2, Textbook 2C (Term 3) overview

Strand	Unit	Unit title	Lesson number	Lesson title	NC Objective 1	NC Objective 2
Number – fractions	10	Fractions	1	Introducing parts and wholes	Recognise, find and name a quarter as 1 of 4 equal parts of an object, shape or quantity	Recognise, find and name a half as one of two equal parts of an object, shape or quantity (Year 1)
Number – fractions	10	Fractions	2	Equal and unequal parts	Recognise, find and name a half as one of two equal parts of an object, shape or quantity (Year 1)	
Number – fractions	10	Fractions	3	Recognise a half	Recognise, find and name a half as one of two equal parts of an object, shape or quantity (Year 1)	
Number – fractions	10	Fractions	4	Find a half	Recognise, find and name a half as one of two equal parts of an object, shape or quantity (Year 1)	
Number – fractions	10	Fractions	5	Recognise a quarter	Recognise, find and name a half as one of two equal parts of an object, shape or quantity (Year 1)	Recognise, find, name and write fractions $\frac{1}{3}$, $\frac{1}{4}$, $\frac{2}{4}$ and $\frac{3}{4}$ of a length, shape, set of objects or quantity
Number – fractions	10	Fractions	6	Find a quarter	Recognise, find and name a half as one of two equal parts of an object, shape or quantity (Year 1)	Recognise, find, name and write fractions $\frac{1}{3}$, $\frac{1}{4}$, $\frac{2}{4}$ and $\frac{3}{4}$ of a length, shape, set of objects or quantity
Number – fractions	10	Fractions	7	Thirds	Recognise, find, name and write fractions $\frac{1}{3}$, $\frac{1}{4}$, $\frac{2}{4}$ and $\frac{3}{4}$ of a length, shape, set of objects or quantity	
Number – fractions	10	Fractions	8	Find the whole	Recognise, find, name and write fractions $\frac{1}{3}$, $\frac{1}{4}$, $\frac{2}{4}$ and $\frac{3}{4}$ of a length, shape, set of objects or quantity	
Number – fractions	10	Fractions	9	Unit and non-unit fractions	Write simple fractions for example, $\frac{1}{2}$ of 6 = 3 and recognise the equivalence of $\frac{2}{4}$ and $\frac{1}{2}$	
Number – fractions	10	Fractions	10	Recognise the equivalence of a half and two quarters	Write simple fractions for example, $\frac{1}{2}$ of 6 = 3 and recognise the equivalence of $\frac{2}{4}$ and $\frac{1}{2}$	
Number – fractions	10	Fractions	11	Recognise three quarters	Recognise, find, name and write fractions $\frac{1}{3}$, $\frac{1}{4}$, $\frac{2}{4}$ and $\frac{3}{4}$ of a length, shape, set of objects or quantity	
Number – fractions	10	Fractions	12	Count in fractions up to a whole	Non-statutory guidance: Pupils should count in fractions up to 10, starting from any number and using the $\frac{1}{2}$ and $\frac{2}{4}$ equivalence on the number line (for example, $1\frac{1}{2}$, $1\frac{2}{4}$ (or $1\frac{1}{2}$), $1\frac{3}{4}$, 2)	

Strand	Unit	Unit title	Lesson number	Lesson title	NC Objective 1	NC Objective 2
Measurement	11	Time	1	O'clock and half past	Tell the time to the hour and half past the hour and draw the hands on a clock face to show these times (Year 1)	
Measurement	11	Time	2	Quarter past and quarter to	Tell and write the time to five minutes, including quarter past/to the hour and draw the hands on a clock face to show these times	
Measurement	11	Time	3	Tell the time to 5 minutes	Tell and write the time to five minutes, including quarter past/to the hour and draw the hands on a clock face to show these times	
Measurement	11	Time	4	Minutes in an hour	Know the number of minutes in an hour and the number of hours in a day	
Measurement	11	Time	5	Hours in a day	Know the number of minutes in an hour and the number of hours in a day	
Number – addition and subtraction	12	Problem solving and efficient methods	1	My way, your way!	Use place value and number facts to solve problems	Recognise and use the inverse relationship between addition and subtraction and use this to check calculations and solve missing number problems
Number – addition and subtraction	12	Problem solving and efficient methods	2	Use number facts	Use place value and number facts to solve problems	
Number – addition and subtraction	12	Problem solving and efficient methods	3	Use a 100 square	Use place value and number facts to solve problems	Recognise and use the inverse relationship between addition and subtraction and use this to check calculations and solve missing number problems
Number – addition and subtraction	12	Problem solving and efficient methods	4	Getting started	Use place value and number facts to solve problems	Solve problems with addition and subtraction: using concrete objects and pictorial representations, including those involving numbers, quantities and measures
Number – addition and subtraction	12	Problem solving and efficient methods	5	Missing numbers	Recognise and use the inverse relationship between addition and subtraction and use this to check calculations and solve missing number problems	

Strand	Unit	Unit title	Lesson number	Lesson title	NC Objective 1	NC Objective 2
Number – addition and subtraction	12	Problem solving and efficient methods	6	Mental addition and subtraction (1)	Use place value and number facts to solve problems	Solve problems with addition and subtraction: applying their increasing knowledge of mental and written methods
Number – addition and subtraction	12	Problem solving and efficient methods	7	Mental addition and subtraction (2)	Use place value and number facts to solve problems	Solve problems with addition and subtraction: applying their increasing knowledge of mental and written methods
Number – addition and subtraction	12	Problem solving and efficient methods	8	Efficient subtraction	Solve problems with addition and subtraction: using concrete objects and pictorial representations, including those involving numbers, quantities and measures	
Number – addition and subtraction	12	Problem solving and efficient methods	9	Solve problems – addition and subtraction	Use place value and number facts to solve problems	Solve problems with addition and subtraction: applying their increasing knowledge of mental and written methods
Number – addition and subtraction	12	Problem solving and efficient methods	10	Solve problems – multiplication and division	Solve problems involving multiplication and division, using materials, arrays, repeated addition, mental methods, and multiplication and division facts, including problems in contexts.	
Number – addition and subtraction	12	Problem solving and efficient methods	11	Solve problems – using the four operations	Use place value and number facts to solve problems	
Geometry – position and direction	13	Position and direction	1	Language of position	Use mathematical vocabulary to describe position, direction and movement, including movement in a straight line and distinguishing between rotation as a turn and in terms of right angles for quarter, half and three-quarter turns (clockwise and anti-clockwise)	
Geometry – position and direction	13	Position and direction	2	Describe movement	Use mathematical vocabulary to describe position, direction and movement, including movement in a straight line and distinguishing between rotation as a turn and in terms of right angles for quarter, half and three-quarter turns (clockwise and anti-clockwise)	
Geometry – position and direction	13	Position and direction	3	Describe turns	Use mathematical vocabulary to describe position, direction and movement, including movement in a straight line and distinguishing between rotation as a turn and in terms of right angles for quarter, half and three-quarter turns (clockwise and anti-clockwise)	

Strand	Unit	Unit title	Lesson number	Lesson title	NC Objective 1	NC Objective 2
Geometry – position and direction	13	Position and direction	4	Describe movement and turns	Use mathematical vocabulary to describe position, direction and movement, including movement in a straight line and distinguishing between rotation as a turn and in terms of right angles for quarter, half and three-quarter turns (clockwise and anti-clockwise)	
Geometry – position and direction	13	Position and direction	5	Make patterns by turning shapes	Use mathematical vocabulary to describe position, direction and movement, including movement in a straight line and distinguishing between rotation as a turn and in terms of right angles for quarter, half and three-quarter turns (clockwise and anti-clockwise)	Order and arrange combinations of mathematical objects in patterns and sequences
Statistics	14	Statistics	1	Make tally charts	Interpret and construct simple pictograms, tally charts, block diagrams and simple tables	
Statistics	14	Statistics	2	Tables	Interpret and construct simple pictograms, tally charts, block diagrams and simple tables	
Statistics	14	Statistics	3	Block diagrams	Interpret and construct simple pictograms, tally charts, block diagrams and simple tables	
Statistics	14	Statistics	4	Draw pictograms (1 to 1)	Interpret and construct simple pictograms, tally charts, block diagrams and simple tables	
Statistics	14	Statistics	5	Interpret pictograms (1 to 1)	Ask and answer simple questions by counting the number of objects in each category and sorting the categories by quantity	Ask and answer questions about totalling and comparing categorical data
Statistics	14	Statistics	6	Draw pictograms (1 to 2, 5 or 10)	Interpret and construct simple pictograms, tally charts, block diagrams and simple tables	
Statistics	14	Statistics	7	Interpret pictograms (1 to 2, 5 or 10)	Ask and answer simple questions by counting the number of objects in each category and sorting the categories by quantity	Ask and answer questions about totalling and comparing categorical data

Mindset: an introduction

Global research and best practice deliver the same message: learning is greatly affected by what learners perceive they can or cannot do. What is more, it is also shaped by what their parents, carers and teachers perceive they can do. Mindset – the thinking that determines our beliefs and behaviours – therefore has a fundamental impact on teaching and learning.

Everyone can!

Power Maths and mastery methods focus on the distinction between 'fixed' and 'growth' mindsets (Dweck, 2007).[1] Those with a fixed mindset believe that their basic qualities (for example, intelligence, talent and ability to learn) are pre-wired or fixed: 'If you have a talent for maths, you will succeed at it. If not, too bad!' By contrast, those with a growth mindset believe that hard work, effort and commitment drive success and that 'smart' is not something you are or are not, but something you become. In short, everyone can do maths!

Key mindset strategies

A growth mindset needs to be actively nurtured and developed. *Power Maths* offers some key strategies for fostering healthy growth mindsets in your classroom.

It is okay to get it wrong

Mistakes are valuable opportunities to re-think and understand more deeply. Learning is richer when children and teachers alike focus on spotting and sharing mistakes as well as solutions.

Praise hard work

Praise is a great motivator, and by focusing on praising effort and learning rather than success, children will be more willing to try harder, take risks and persist for longer.

Mind your language!

The language we use around learners has a profound effect on their mindsets. Make a habit of using growth phrases, such as, 'Everyone can!', 'Mistakes can help you *learn*' and 'Just try for a little longer'. The king of them all is one little word, 'yet'... I can't solve this...yet!' Encourage parents and carers to use the right language too.

Build in opportunities for success

The step-by-small-step approach enables children to enjoy the experience of success. In addition, avoid ability grouping and encourage every child to answer questions and explain or demonstrate their methods to others.

[1] Dweck, C (2007) *The New Psychology of Success*, Ballantine Books: New York

The *Power Maths* characters

The *Power Maths* characters model the traits of growth mindset learners and encourage resilience by prompting and questioning children as they work. Appearing frequently in the Textbooks and Practice Books, they are your allies in teaching and discussion, helping to model methods, alternatives and misconceptions, and to pose questions. They encourage and support your children, too: they are all hardworking, enthusiastic and unafraid of making and talking about mistakes.

Meet the team!

Creative Flo is open-minded and sometimes indecisive. She likes to think differently and come up with a variety of methods or ideas.

Determined Dexter is resolute, resilient and systematic. He concentrates hard, always tries his best and he'll never give up – even though he doesn't always choose the most efficient methods!

Curious Ash is eager, interested and inquisitive, and he loves solving puzzles and problems. Ash asks lots of questions but sometimes gets distracted.

Sparks the Cat

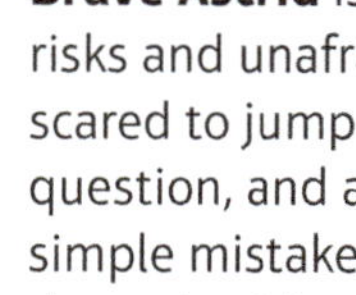

Brave Astrid is confident, willing to take risks and unafraid of failure. She's never scared to jump straight into a problem or question, and although she often makes simple mistakes she's happy to talk them through with others.

Mathematical language

Traditionally, we in the UK have tended to try simplifying mathematical language to make it easier for young children to understand. By contrast, evidence and experience show that by diluting the correct language, we actually mask concepts and meanings for children. We then wonder why they are confused by new and different terminology later down the line! *Power Maths* is not afraid of 'hard' words and avoids placing any barriers between children and their understanding of mathematical concepts. As a result, we need to be deliberate, precise and thorough in building every child's understanding of the language of maths. Throughout the Teacher Guides you will find support and guidance on how to deliver this, as well as individual explanations throughout the pupil Textbooks.

Use the following key strategies to build children's mathematical vocabulary, understanding and confidence.

Precise and consistent

Everyone in the classroom should use the correct mathematical terms in full, every time. For example, refer to 'equal parts', not 'parts'. Used consistently, precise maths language will be a familiar and non-threatening part of children's everyday experience.

Full sentences

Teachers and children alike need to use full sentences to explain or respond. When children use complete sentences, it both reveals their understanding and embeds their knowledge.

Stem sentences

These important sentences help children express mathematical concepts accurately, and are used throughout the *Power Maths* books. Encourage children to repeat them frequently, whether working independently or with others. Examples of stem sentences are:

'4 is a part, 5 is a part, 9 is the whole.'

'There are …. groups. There are …. in each group.'

Key vocabulary

The unit starters highlight essential vocabulary for every lesson. In the Pupil books, characters flag new terminology and the Teacher Guide lists important mathematical language for every unit and lesson. New terms are never introduced without a clear explanation.

Symbolic language

Symbols are used early on so that children quickly become familiar with them and their meaning. Often, the *Power Maths* characters will highlight the connection between language and particular symbols.

The role of talk and discussion

When children learn to talk purposefully together about maths, barriers of fear and anxiety are broken down and they grow in confidence, skills and understanding. Building a healthy culture of 'maths talk' empowers their learning from day one.

Explanation and discussion are integral to the *Power Maths* structure, so by simply following the books your lessons will stimulate structured talk. The following key 'maths talk' strategies will help you strengthen that culture and ensure that every child is included.

Sentences, not words

Encourage children to use full sentences when reasoning, explaining or discussing maths. This helps both speaker and listeners to clarify their own understanding. It also reveals whether or not the speaker truly understands, enabling you to address misconceptions as they arise.

Working together

Working with others in pairs, groups or as a whole class is a great way to support maths talk and discussion. Use different group structures to add variety and challenge. For example, children could take timed turns for talking, work independently alongside a 'discussion buddy', or perhaps play different *Power Maths* character roles within their group.

Think first – then talk

Provide clear opportunities within each lesson for children to think and reflect, so that their talk is purposeful, relevant and focused.

Give every child a voice

Where the 'hands up' model allows only the more confident child to shine, *Power Maths* involves everyone. Make sure that no child dominates and that even the shyest child is encouraged to contribute – and praised when they do.

Assessment strategies

Teaching for mastery demands that you are confident about what each child knows and where their misconceptions lie; therefore, practical and effective assessment is vitally important.

Formative assessment within lessons

The **Think together** section will often reveal any confusions or insecurities; try ironing these out by doing the first **Think together** question as a class. For children who continue to struggle, you or your Teaching Assistant should provide support and enable them to move on.

Performance in practice can be very revealing: check Practice Books and listen out both during and after practice to identify misconceptions.

The **Reflect** section is designed to check on the all-important depth of understanding. Be sure to review how the children performed in this final stage before you teach the next lesson.

End of unit check – Textbook

Each unit concludes with a summative check to help you assess quickly and clearly each child's understanding, fluency, reasoning and problem solving skills. Your Teacher Guide will suggest ideal ways of organising a given activity and offer advice and commentary on what children's responses mean. For example, 'What misconception does this reveal?'; 'How can you reinforce this particular concept?'

For Year 1 and Year 2 children, assess in small, teacher-led groups, giving each child time to think and respond while also consolidating correct mathematical language. Assessment with young children should always be an enjoyable activity, so avoid one-to-one individual assessments, which they may find threatening or scary. If you prefer, the End of unit check can be carried out as a whole-class group using whiteboards and Practice Books.

End of unit check – Practice Book

The Practice Book contains further opportunities for assessment, and can be completed by children independently whilst you are carrying out diagnostic assessment with small groups. Your Teacher Guide will advise you on what to do if children struggle to articulate an explanation – or perhaps encourage you to write down something they have explained well. It will also offer insights into children's answers and their implications for next learning steps. It is split into three main sections, outlined below.

My journal is designed to allow children to show their depth of understanding of the unit. It can also serve as a way of checking that children have grasped key mathematical vocabulary. The question children should answer is first presented in the Textbook in the Think! section. This provides an opportunity for you to discuss the question first as a class to ensure children have understood their task. Children should have some time to think about how they want to answer the question, and you could ask them to talk to a partner about their ideas. Then children should write their answer in their Practice Book, using the word bank provided to help them with vocabulary.

The **Power check** allows pupils to self-assess their level of confidence on the topic by colouring in different smiley faces. You may want to introduce the faces as follows:

Each unit ends with either a Power play or a Power puzzle. This is an activity, puzzle or game that allows children to use their new knowledge in a fun, informal way.

Progress Tests

There are *Power Maths* Progress Tests for each half term and at the end of the year, including an Arithmetic test and Reasoning test in each case. You can enter results in the online markbook to track and analyse results and see the average for all schools' results. The tests use a 6-step scale to show results against age-related expectation.

How to ask diagnostic questions

The diagnostic questions provided in children's Practice Books are carefully structured to identify both understanding and misconceptions (if children answer in a particular way, you will know why). The simple procedure below may be helpful:

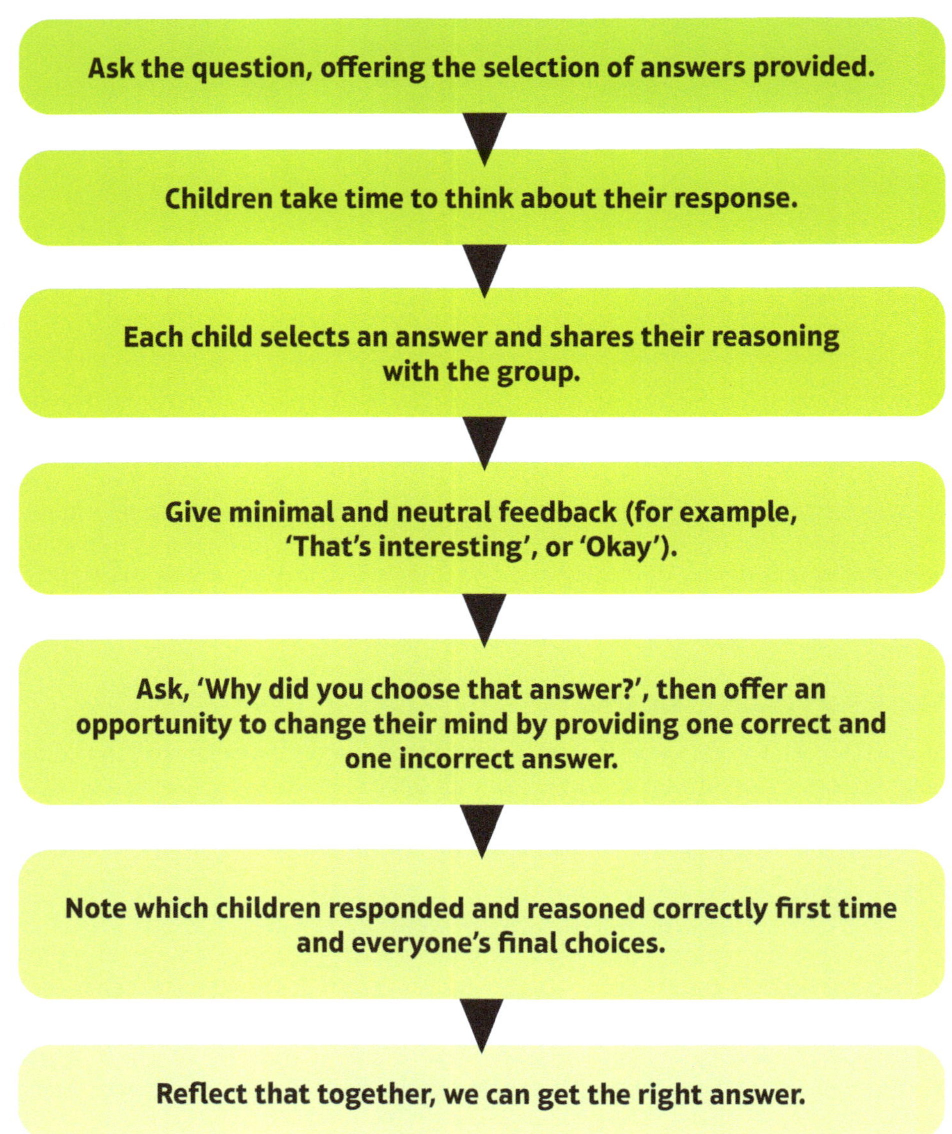

Keeping the class together

Traditionally, children who learn quickly have been accelerated through the curriculum. As a consequence, their learning may be superficial and will lack the many benefits of enabling children to learn with and from each other.

By contrast, *Power Maths'* mastery approach values real understanding and richer, deeper learning above speed. It sees all children learning the same concept in small, cumulative steps, each finding and mastering challenge at their own level. Remember that when you teach for mastery, EVERYONE can do maths! Those who grasp a concept easily have time to explore and understand that concept at a deeper level. The whole class therefore moves through the curriculum at broadly the same pace via individual learning journeys.

For some teachers, the idea that a whole class can move forward together is revolutionary and challenging. However, the evidence of global good practice clearly shows that this approach drives engagement, confidence, motivation and success for all learners, and not just the high flyers. The strategies below will help you keep your class together on their maths journey.

Mix it up

Do not stick to set groups at each table. Every child should be working on the same concept, and mixing up the groupings widens children's opportunities for exploring, discussing and sharing their understanding with others.

Recycling questions

Reuse the Textbook and Practice Book questions with concrete materials to allow children to explore concepts and relationships and deepen their understanding. This strategy is especially useful for reinforcing learning in same-day interventions.

Strengthen at every opportunity

The next lesson in a *Power Maths* sequence always revises and builds on the previous step to help embed learning. These activities provide golden opportunities for individual children to strengthen their learning with the support of Teaching Assistants.

Prepare to be surprised!

Children may grasp a concept quickly or more slowly. The 'fast graspers' won't always be the same individuals, nor does the speed at which a child understands a concept predict their success in maths. Are they struggling or just working more slowly?

Same-day intervention

Since maths competence depends on mastering concepts one by one in a logical progression, it is important that no gaps in understanding are ever left unfilled. Same-day interventions – either within or after a lesson – are a crucial safety net for any child who has not fully made the small step covered that day. In other words, intervention is always about keeping up, not catching up, so that every child has the skills and understanding they need to tackle the next lesson. That means presenting the same problems used in the lesson, with a variety of concrete materials to help children model their solutions.

We offer two intervention strategies below, but you should feel free to choose others if they work better for your class.

Within-lesson intervention

The **Think together** activity will reveal those who are struggling, so when it is time for practice, bring these children together to work with you on the first practice questions. Observe these children carefully, ask questions, encourage them to use concrete models and check that they reach and can demonstrate their understanding.

After-lesson intervention

You might like to use the **Think together** questions to recap the lesson with children who are working behind expectations during assembly time. Teaching Assistants could also work with these children at other convenient points in the school day. Some children may benefit from revisiting work from the same topic in the previous year group. Note also the suggestion for recycling questions from the Textbook and Practice Book with concrete materials on page 26.

The role of practice

Practice plays a pivotal role in the *Power Maths* approach. It takes place in class groups, smaller groups, pairs, and independently, so that children always have the opportunities for thinking as well as the models and support they need to practise meaningfully and with understanding.

Intelligent practice

In *Power Maths*, practice never equates to the simple repetition of a process. Instead we embrace the concept of intelligent practice, in which all children become fluent in maths through varied, frequent and thoughtful practice that deepens and embeds conceptual understanding in a logical, planned sequence. To see the difference, take a look at the following examples.

Traditional practice

- Repetition can be rote – no need for a child to think hard about what they are doing

- Praise may be misplaced

- Does this prove understanding?

Intelligent practice

- Varied methods – concrete, pictorial and abstract

- Equation expressed in different ways, requiring thought and understanding

- Constructive feedback

All practice questions are designed to move children on and reveal misconceptions.

Simple, logical steps build onto earlier learning.

C-P-A runs throughout – different ways of modelling and understanding the same concept.

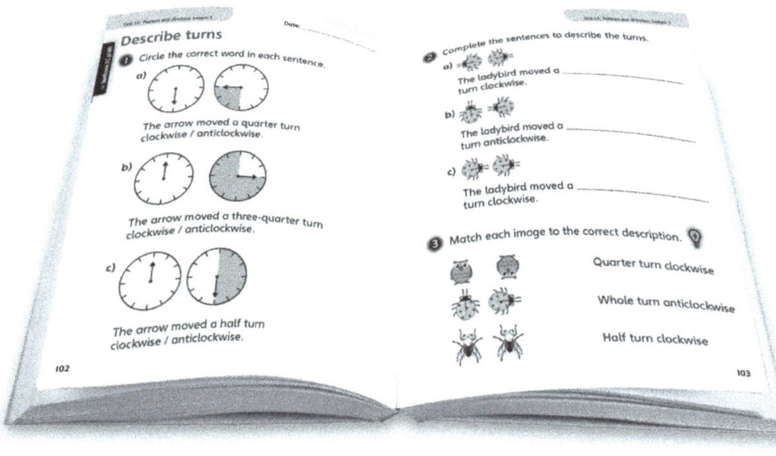

Conceptual variation – children work on different representations of the same maths concept.

Friendly characters offer support and encourage children to try different approaches.

A carefully designed progression

The Practice Books provide just the right amount of intelligent practice for children to complete independently in the final sections of each lesson. It is really important that all children are exposed to the practice questions, and that children are not directed to complete different sections. That is because each question is different and has been designed to challenge children to think about the maths they are doing. The questions become more challenging so children grasping concepts more quickly will start to slow down as they progress. Meanwhile, you have the chance to circulate and spot any misconceptions before they become barriers to further learning.

Homework and the role of parents and carers

While *Power Maths* does not prescribe any particular homework structure, we acknowledge the potential value of practice at home. For example, practising fluency in key facts, such as number bonds and times-tables, is an ideal homework task. You can share the Individual Practice Games for homework (see page 6), or parents and carers could work through uncompleted Practice Book questions with children at either primary stage.

However, it is important to recognise that many parents and carers may themselves lack confidence in maths, and few, if any, will be familiar with mastery methods. A Parents' and Carers' evening that helps them understand the basics of mindsets, mastery and mathematical language is a great way to ensure that children benefit from their homework. It could be a fun opportunity for children to teach their families that everyone can do maths!

Structures and representations

Unlike most other subjects, maths comprises a wide array of abstract concepts – and that is why children and adults so often find it difficult. By taking a concrete-pictorial-abstract (C-P-A) approach, *Power Maths* allows children to tackle concepts in a tangible and more comfortable way.

Non-linear stages

Concrete

Replacing the traditional approach of a teacher working through a problem in front of the class, the concrete stage introduces real objects that children can use to 'do' the maths – any familiar object that a child can manipulate and move to help bring the maths to life. It is important to appreciate, however, that children must always understand the link between models and the objects they represent. For example, children need to first understand that three cakes could be represented by three pretend cakes, and then by three counters or bricks. Frequent practice helps consolidate this essential insight. Although they can be used at any time, good concrete models are an essential first step in understanding.

Pictorial

This stage uses pictorial representations of objects to let children 'see' what particular maths problems look like. It helps them make connections between the concrete and pictorial representations and the abstract maths concept. Children can also create or view a pictorial representation together, enabling discussion and comparisons. The *Power Maths* teaching tools are fantastic for this learning stage, and bar modelling is invaluable for problem solving throughout the primary curriculum.

Abstract

Our ultimate goal is for children to understand abstract mathematical concepts, symbols and notation and of course, some children will reach this stage far more quickly than others. To work with abstract concepts, a child must be comfortable with the meaning of and relationships between concrete, pictorial and abstract models and representations. The C-P-A approach is not linear, and children may need different types of models at different times. However, when a child demonstrates with concrete models and pictorial representations that they have grasped a concept, we can be confident that they are ready to explore or model it with abstract symbols such as numbers and notation.

Use at any time and with any age to support understanding

Variation helps visualisation

Children find it much easier to visualise and grasp concepts if they see them presented in a number of ways, so be prepared to offer and encourage many different representations.

For example, the number six could be represented in various ways:

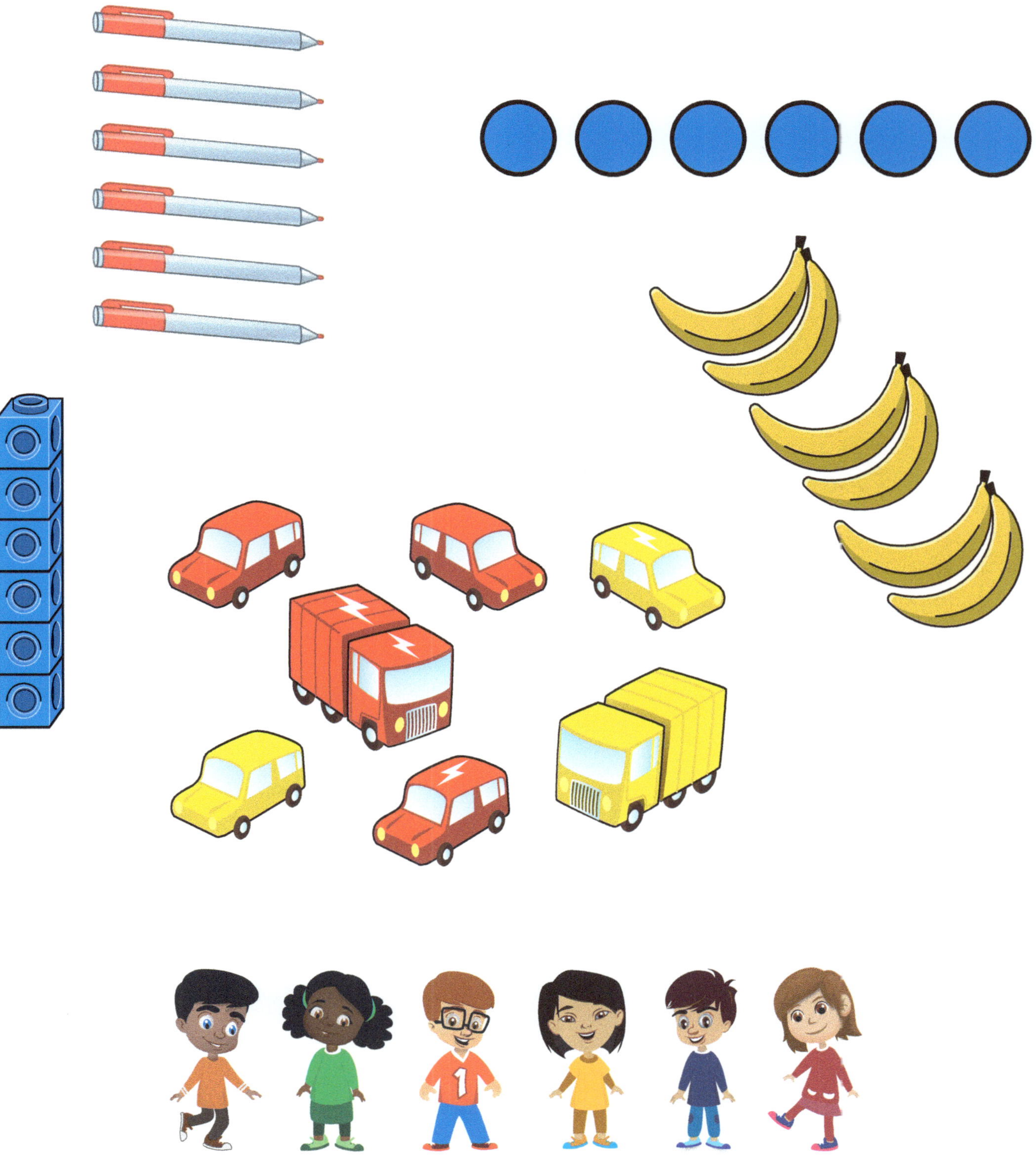

Practical aspects of *Power Maths*

One of the key underlying elements of *Power Maths* is its practical approach, allowing you to make maths real and relevant to your children, no matter their age.

Manipulatives are essential resources for both key stages and *Power Maths* encourages teachers to use these at every opportunity, and to continue the Concrete-Pictorial-Abstract approach right through to Year 6.

The Textbooks and Teacher Guides include lots of opportunities for teaching in a practical way to show children what maths means in real life.

Discover and Share

The **Discover** and **Share** sections of the Textbook give you scope to turn a real-life scenario into a practical and hands-on section of the lesson. Use these sections as inspiration to get active in the classroom. Where appropriate, use the **Discover** contexts as a springboard for your own examples that have particular resonance for your children – and allow them to get their hands dirty trying out the mathematics for themselves.

Unit videos

Every term has one unit video which incorporates real-life classroom sequences.

These videos show you how the reasoning behind mathematics can be carried out in a practical manner by showing real children using various concrete and pictorial methods to come to the solution. You can see how using these practical models, such as part-whole and bar models, helps them to find and articulate their answer.

Mastery tips

Mastery Experts give anecdotal advice on where they have used hands-on and real-life elements to inspire their children.

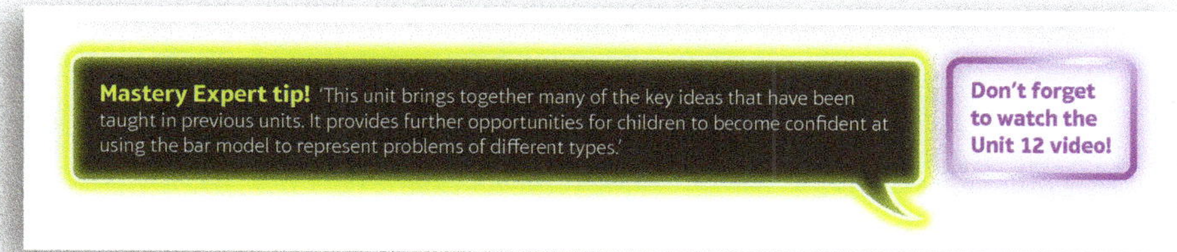

Concrete-Pictorial-Abstract (C-P-A) approach

Each **Share** section uses various methods to explain an answer, helping children to access abstract concepts by using concrete tools, such as counters. Remember, this isn't a linear process, so even children who appear confident using the more abstract method can deepen their knowledge by exploring the concrete representations. Encourage children to use all three methods to really solidify their understanding of a concept.

Pictorial representation – drawing the problem in a logical way that helps children visualise the maths

Concrete representation – using manipulatives to represent the problem. Encourage children to physically use resources to explore the maths.

Abstract representation – using words and calculations to represent the problem.

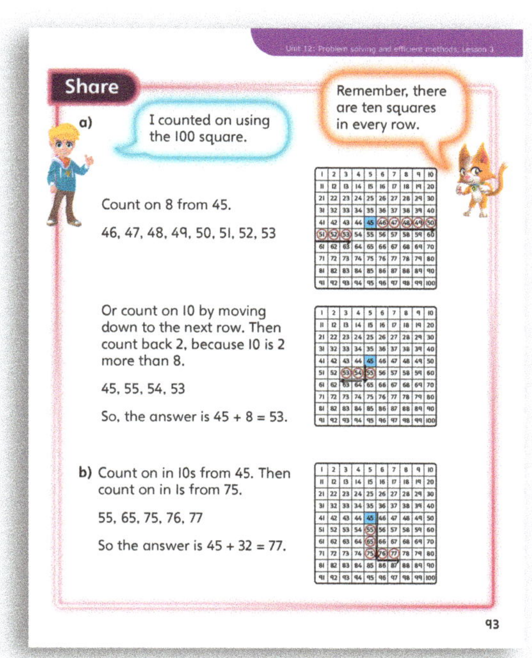

Practical tips

Every lesson suggests how to draw out the practical side of the **Discover** context.

You'll find these in the **Discover** section of the Teacher Guide for each lesson.

PRACTICAL TIPS Play this game together as a class, starting with numbers in the range 0 to 10.

Resources

Every lesson lists the practical resources you will need or might want to use. There is also a summary of all of the resources used throughout the term on page 41 to help you be prepared.

RESOURCES

Mandatory: counters, ten frames, number lines
Optional: number tracks, multilink cubes

Using *Power Maths* flexibly in Key Stage I

Power Maths lessons have a coherent, regular structure that supports you in building up children's understanding in a series of small steps. This is something most classes will need to build up to, rather than running in from a standing start at the beginning of Year 1.

Start by using the Practice Books in small groups

In most Year 1 classes, it won't be realistic for the whole class to complete the Practice Book pages independently at the start of the year, but they will learn to do this gradually. For the Textbooks, children will need to get used to direct teaching and recording answers in their own books. And, of course, this will set them up well for the rest of Primary school.

Small teacher-led groups are likely to be the best approach for independent practice at the beginning of KS1. This format allows you to talk children through the question, discuss their ideas using manipulatives (often there will be manipulatives on the page as a hint), and guide them in representing their answer. (For instance, they can tell you the answer is 5, but they may need help writing 5 or knowing that they should colour in 5 apples.)

Go through the questions one-by-one with the group. You can mark their work/give feedback there and then. As children get used to the materials, the next stage could be for the small group to work through the questions at their own pace. The style of questions in *Power Maths* is quite regular, so children will get better at knowing what they need to do.

To facilitate small group work, you are likely to need some other activities as a carousel. A good way to do this is by turning a question from the Textbook into a game (usually **Think together** question 3 will work well) and teaching this to children before you break into groups. For instance, look at the example below (pages 78–79 in Textbook 1A). You could teach children a game with a part-whole model where one child puts in the whole using counters and the other children have to put in the parts. Or they could try this with beanbags and hoops. Base the practice on the key learning from the lesson.

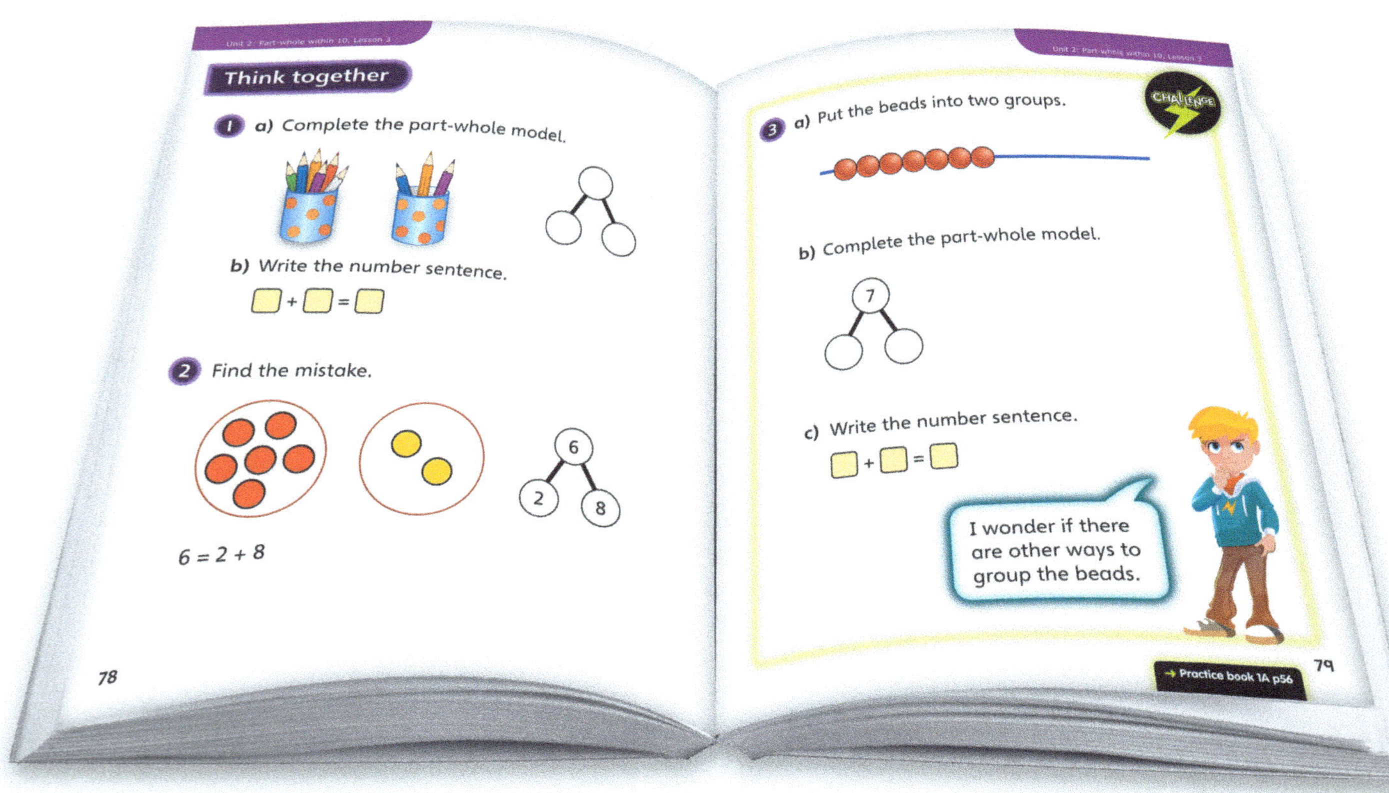

Are there any other ways to use the resources flexibly?

Don't be afraid to bring the **Discover** activity to life! Perhaps you could turn it into a game, or a role play. For instance, if the context is a teddy bear's picnic, you could share out fruit between teddies in the class. Or could you find a toy rocket to launch for the lesson below? (Textbook 1A page 32).

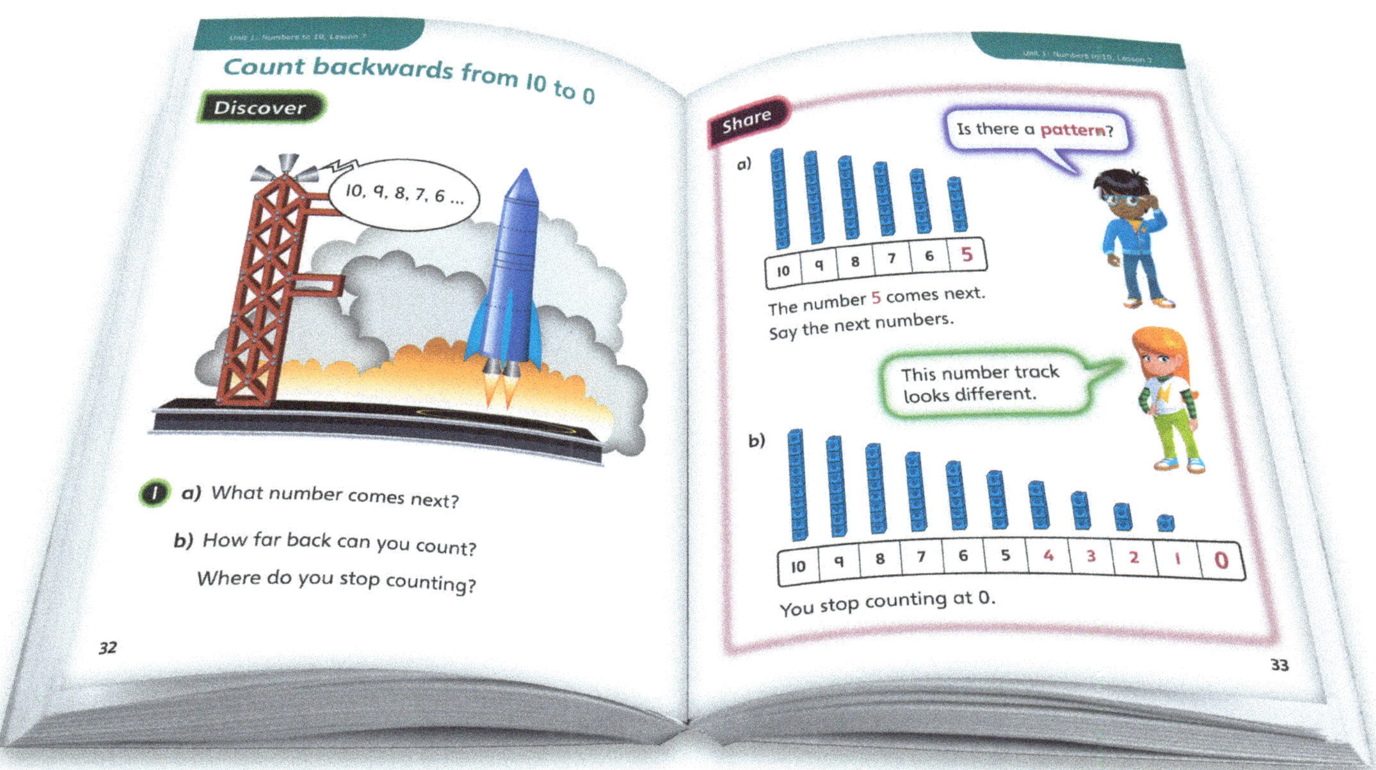

For some lessons you could consider a slightly different approach where you move backwards and forwards between the Textbook and Practice Book. If **Think together** question 1 links well with Practice Book question 1, you could do the **Think together** question together and then let children complete the Practice Book question, then the same for question 2, etc. This works better for some lessons than others, but it is one way of making practice more independent in short bursts, as a way of building up independence.

Don't forget, there isn't a *Power Maths* lesson for every lesson in the year. You can take more time where you need to, so that children's understanding is secure. In Key Stage 1, it will be all the more important to take your time, because children need to get used to the format as well as master the key learning. If using some of the ideas above means that a *Power Maths* lesson actually takes two lessons, e.g. for the first part of the year, then that's fine!

There are some further ideas for using the materials flexibly in the next section.

Working with children below age-related expectation

This section offers advice on using *Power Maths* with children who are significantly behind age-related expectation. Teacher judgement will be crucial in terms of where and why children are struggling, and in choosing the right approach. The suggestions can of course be adapted for children with special educational needs, depending on the specific details of those needs.

General approaches to support children who are struggling

Keeping the pace manageable

Remember, you have more teaching days than *Power Maths* lessons so you can cover a lesson over more than one day, and revisit key learning, to ensure all children are ready to move on. You can use the + and − buttons to adjust the time for each unit in the online planning. The NCETM's Ready-to-Progress criteria can be used to help determine what should be highest priority.

Same-day intervention

You could go over the Textbook pages or revisit the previous year's work if necessary (see Addressing gaps). Remember that same-day intervention can be within the lesson, as well as afterwards (see page 28). As children start their independent practice, you can work with those who found the first part of the lesson difficult, checking understanding using manipulatives.

Fluency sessions

Fit in as much practice as you can for number bonds and times-tables, etc., at other times of the day. If you can, plan a short 'maths meeting' for this in the afternoon. You might choose to use a Power Up you haven't used already.

Addressing gaps

Use material from the same topic in the previous year to consolidate or address gaps in learning, e.g. Textbook pages and Strengthen activities. The End of unit check will help gauge children's understanding.

Pre-teaching

Find a 5- to 10-minute slot before the lesson to work with the children you feel would benefit. The afternoon before the lesson can work well, because it gives children time to think in between. Recap previous work on the topic (addressing any gaps you're aware of) and do some fluency practice, targeting number facts etc. that will help children access the learning.

Focusing on the key concepts

If children are a long way behind, it can be helpful to take a step back and think about the key concepts for children to engage with, not just the fine detail of the objective for that year group (e.g. addition with a specific number of columns). Bearing that in mind, how could children advance their understanding of the topic?

Providing extra support within the lesson

Support in the Teacher Guide

First of all, use the Strengthen support in the Teacher Guide for guided and independent work in each lesson, and share this with Teaching Assistants, where relevant. As you read through the lesson content and corresponding Teacher Guide pages before the lesson, ask yourself what key idea or nugget of understanding is at the heart of the lesson. If children are struggling, this should help you decide what's essential for all children before they move on.

Annotating pages

You can annotate questions to provide extra scaffolding or hints if you need to, but aim to build up children's ability to access questions independently wherever you can. Children tend to get used to the style of the *Power Maths* questions over time.

Quick recap as lesson starter

The Quick recap for each lesson in the Teacher Guide is an alternative starter activity to the Power Up. You might choose to use this with some or all children if you feel they will need support accessing the main lesson.

Consolidation questions

If you think some children would benefit from additional questions at the same level before moving on, write one or two similar questions on the board. (This shouldn't be at the expense of reasoning and problem-solving opportunities: take longer over the lesson if you need to.)

Hard copy Textbooks

The Textbooks help children focus in more easily on the mathematical representations, read the text more comfortably, and revisit work from a previous lesson that you are building on, as well as giving children ownership of their learning journey. In main lessons, it can work well to use the e-Textbook for Discover and give out the books when discussing the methods in the Share section.

Reading support

It's important that all children are exposed to problem solving and reasoning questions, which often involves reading. For whole-class work you can read questions together. For independent practice you could consider annotating pages to help children see what the question is asking, and stem sentences to help structure their answer. A general focus on specific mathematical language and vocabulary will help children access the questions. You could consider pairing weaker readers with stronger readers, or read questions as a group if those who need support are on the same table.

Providing extra depth and challenge with *Power Maths*

Just as prescribed in the National Curriculum, the goal of *Power Maths* is never to accelerate through a topic but rather to gain a clear, deep and broad understanding. Here are some suggestions to help ensure all children are appropriately challenged as you work with the resources.

Overall approaches

First of all, remember that the materials are designed to help you keep the class together, allowing all children to master a concept while those who grasp it quickly have time to explore it in more depth. Use the Deepen support in the Teacher Guide (see below) to challenge children who work through the questions quickly. Here are some questions and ideas to encourage breadth and depth during specific parts of the lesson, or at any time (where no part of the lesson sequence is specified):

- **Discover**: 'Can you demonstrate your solution another way?'

- **Share**: Make sure every child is encouraged to give answers and engage with the discussion, not just the most confident.

- **Think together**: 'Can you model your answers using concrete materials? Can you explain your solution to a partner?'

- Practice: Allow all children to work through the full set of questions, so that they benefit from the logical sequence.

- **Reflect**: 'Is there another way of working out the answer? And another way?'
  'Have you found all the solutions?'
  'Is that always true?'
  'What's different between this question and that question? And what's the same?'

Note that the **Challenge** questions are designed so that all children can access and attempt them, if they have worked through the steps leading up to them. There may be some children in a given lesson who don't manage to do the **Challenge**, but it is not supposed to be a distinct task for a subset of the class. When you look through the lesson materials before teaching, think about what each question is specifically asking, and compare this with the key learning point for the lesson. This will help you decide which questions you feel it's essential for all children to answer, before moving on. You can at least aim for all children to try the **Challenge**!

Deepen activities and support

The Teacher Guide provides valuable support for each stage of the lesson. This includes Deepen tips for the guided and independent practice sections, which will help you provide extra stretch and challenge within your lesson, without having to organise additional tasks. If you have a Teaching Assistant, they can also make use of this advice. There are also suggestions for the lesson as a whole in the 'Going Deeper' section on the first page of the Teacher Guide section for that lesson. Every class is different, so you can always go a bit further in the direction indicated, if appropriate, and build on the suggestions given.

There is a Deepen activity for each unit. These are designed to follow on from the End of unit check, stretching children who have a firm understanding of the key learning from the unit. Children can work on them independently, which makes it easier for the teacher to facilitate the Strengthen activity for children who need extra support. Deepen activities could also be introduced earlier in the unit if the necessary work has been covered. The Deepen activities are on *ActiveLearn* on the Planning page for each unit, and also on the Resources page).

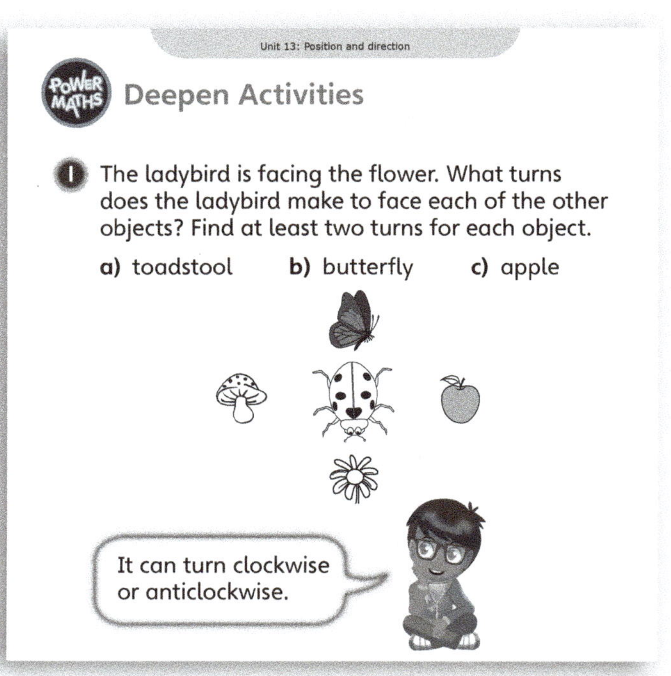

Using the questions flexibly to provide extra challenge

Sometimes you may want to write an extra question on the board or provide this on paper. You can usually do this by tweaking the lesson materials. The questions are designed to form a carefully structured sequence that builds understanding step by step, but, with careful thought about the purpose of each question, you can use the materials flexibly where you need to. Sometimes you might feel that children would benefit from another similar question for consolidation before moving on to the next one, or you might feel that they would benefit from a harder example in the same style. It should be quick and easy to generate 'more of the same' type questions where this is the case.

When you see a question like this one (from Unit 3, Lesson 10), it's easy to make harder examples to do afterwards if you need an extra challenge. For example, you could do something similar using multiples of 10, using 2-digit numbers with 1-digit numbers, or using 2-digit numbers. And, of course, more difficult examples will be 'across 10'.

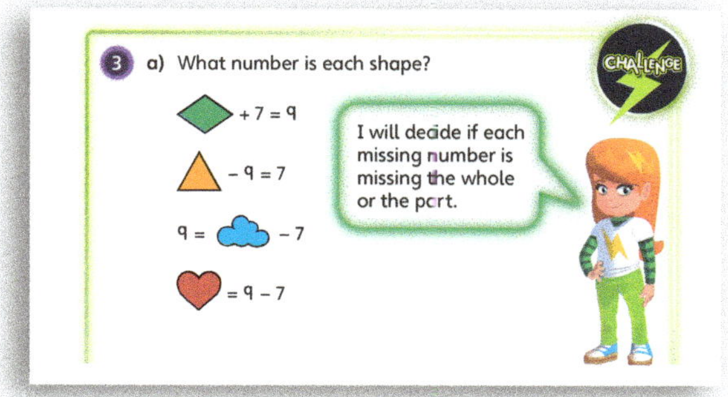

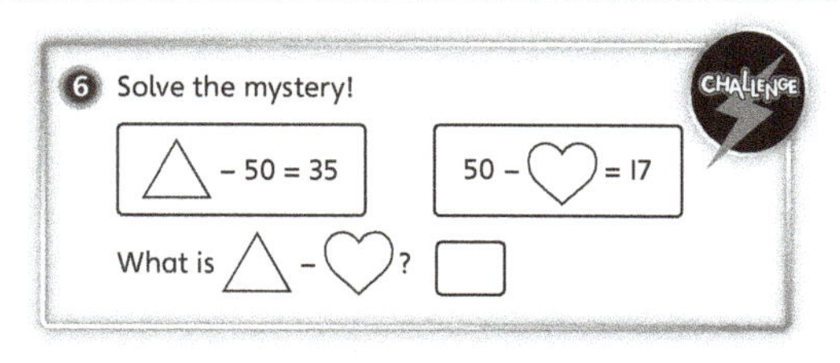

For this example (from Unit 3, Lesson 10), you could ask children to make up their own question(s) for a partner to solve. (In fact, for any of these examples you could ask early finishers to create their own question for a partner.)

Here's an example (from Unit 3, Lesson 12) where some of the sums and differences feature as questions in the lesson, but others don't. Clearly there are plenty of extra two-step problems you could ask using the same context. Children could work out the difference between Kasim's total and Amy and Ben's joint total. Or, if Kasim gives Ben 10 marbles, do they each have as many as Amy? A trickier one might be, how many more marbles does Kat need to have as many as Ben, Amy and Kasim combined.

Besides creating additional questions, you should be able to find a question in the lesson that you can adapt into a game or open-ended investigation, if this helps to keep everyone engaged. It could simply be that, instead of answering 5 + 6 etc. on the page, they could build a robot with 5 cubes and 6 cylinders.

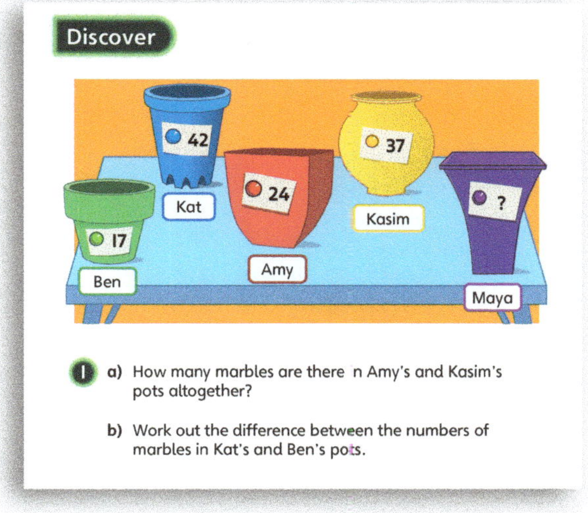

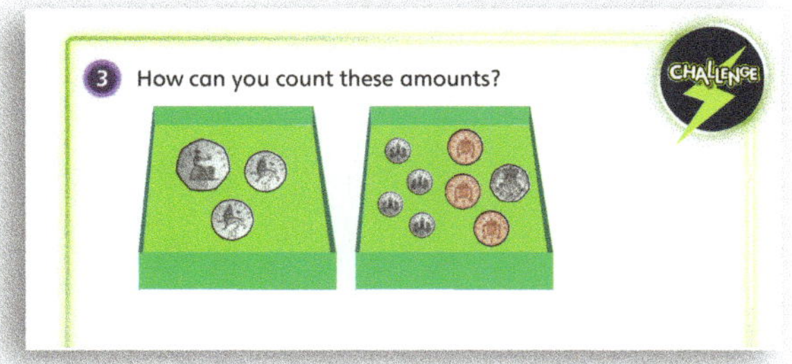

With a question like this (from Unit 5, Lesson 1), an extra challenge could be for children to play a game where they each take a handful of coins and then find the difference between the totals, or perhaps one child makes a total and their partner has to match it with different coins.

See the bullets above for some general ideas that will help with 'opening out' questions in the books, e.g. 'can you find all the solutions?' type questions.

Other suggestions

Another way of stretching children is through mixed ability pairs, or via other opportunities for children to explain their understanding in their own way. This is a good way of encouraging children to go deeper into the learning, rather than, for instance, tackling questions that are computationally more challenging but conceptually equivalent in level.

Using *Power Maths* with mixed age classes

Overall approaches

There are many variables between schools that would make it inadvisable to recommend a one-size-fits-all approach to mixed age teaching with *Power Maths*. These include how year groups are merged, availability of Teaching Assistants, experience and preference of teaching staff, range in pupil attainment across years, classroom space and layout, level of flexibility around timetables, and overall organisational structure (whether the school is part of a trust).

Some schools will find it best to timetable separate maths lessons for the different year groups. Others will aim to teach the class together as much as possible using the mixed age planning support on *ActiveLearn* (see the lesson exemplars for ways of organising lessons with strong/medium/weak correlation between year groups). There will also be ways of adapting these general approaches. For example, offset lessons where Year A start their lesson with the teacher, while Year B work independently on the practice from the previous lesson, and then start the next lesson with the teacher while Year A work independently; or teachers may choose to base their provision around the lesson from one year group and tweak the content up/down for the other group.

Key strategies for mixed age teaching

The mixed age teaching webinar on *ActiveLearn* provides advice on all aspects of mixed age teaching, including more detail on the ideas below.

Developing independence over time

Investing time in building up children's independence will pay off in the medium term.

Clear rationale

If someone asked, 'Why did you teach both Unit 3 and 4 in the same lesson/separate lessons?', what would your answer be?

Designing a lesson

1. Identify the core learning for each group
2. Identify any number skills necessary to access the core
3. Consider the flow of concepts and how one core leads to the other

Challenging all children

The questions are designed to build understanding step by step, but with careful thought about the purpose of each question you can tweak them to increase the challenge.

Multiple years combined

With more than two years together, teachers will inevitably need to use the resources flexibly if delivering a single lesson.

Enjoy the positives!

Comparison deepens understanding and there will be lots of opportunities for children, as well as misconceptions to explore. There is also in-built pre-teaching and the chance to build up a concept from its foundations. For teachers there is double the material to draw on! Mixed age teachers require a strong understanding of the progression of ideas across year groups, which is highly valuable for all teachers. Also, it is necessary to engage deeply with the lesson to see how to use the materials flexibly – this is recommended for all teachers and will help you bring your lesson to life!

List of practical resources

Year 2C Mandatory resources

100 squares (laminated)	**Unit 12** Lessons 3, 7
2D shapes (paper or card)	**Unit 10** Lesson 9
2D shapes (printed, for children to fold into parts)	**Unit 10** Lesson 2
2D shapes (printed, to be folded into equal and unequal parts)	**Unit 10** Lesson 3
2D shapes (printed, to fold and cut)	**Unit 10** Lesson 5
2D shapes (variety)	**Unit 13** Lesson 5
Analogue clock tool	**Unit 11** Lessons 1, 2, 3, 4, 5
Bar models (blank)	**Unit 12** Lesson 9
Base 10 equipment	**Unit 12** Lessons 2, 5, 6
Circles of card (split into quarters)	**Unit 10** Lesson 5
Coins	**Unit 12** Lesson 1
Coloured rods	**Unit 12** Lesson 11
Coloured rods or pre-cut strips of card (to make the bar model)	**Unit 12** Lesson 10
Colouring pencils	**Unit 10** Lesson 10
Counters	**Unit 10** Lessons 5, 6, 10, 11, 12 **Unit 14** Lessons 1, 5
Cubes or counters (for sorting)	**Unit 10** Lesson 8
Fraction (large '$\frac{1}{2}$' printed, with numerator and denominator labelled)	**Unit 10** Lesson 3

Interlocking cubes	**Unit 14** Lesson 3
Multilink cubes	**Unit 10** Lessons 2, 3, 6, 10, 11, 12
Multilink cubes and counters (to represent objects)	**Unit 10** Lesson 4
Number cards 1–9	**Unit 12** Lesson 4
Number lines	**Unit 14** Lessons 1, 5, 6
Number lines (1–100)	**Unit 12** Lessons 3, 7
Number lines (blank)	**Unit 12** Lessons 1, 8
Number lines (completed)	**Unit 12** Lesson 1
Number sentence scaffolds	**Unit 12** Lesson 4
Objects (classroom) and shapes (to split, fold or cut in half)	**Unit 10** Lesson 3
Paper (for folding)	**Unit 10** Lesson 10
Part-whole models (blank)	**Unit 12** Lesson 2
Pictogram templates	**Unit 14** Lesson 6
Pictogram templates (outline)	**Unit 14** Lesson 4
Pictograms (extra)	**Unit 14** Lesson 5
Rulers	**Unit 10** Lesson 5 **Unit 14** Lessons 1, 3, 4 **Unit 14** Lessons 1, 3, 4
Sorting hoops	**Unit 10** Lessons 4, 6
Squared paper	**Unit 10** Lessons 4, 5

Year 2C Optional resources

Resource	Lesson
100 squares	**Unit 12** Lesson 8
2D or 3D shapes	**Unit 13** Lesson 1
3D shapes	**Unit 14** Lesson 2
Analogue clock and flashcards (with written times)	**Unit 11** Lessons 1, 2, 3
Animal pictures or toy animals	**Unit 13** Lesson 3
Arrow cards	**Unit 13** Lesson 2
Bar models (completed)	**Unit 12** Lessons 1, 5
Bar models (showing the four operations)	**Unit 12** Lesson 11
Base 10 equipment	**Unit 11** Lesson 4 **Unit 12** Lessons 3, 7, 11
Bead strings	**Unit 12** Lessons 5, 6
Cake (to be cut into unequal and equal parts)	**Unit 10** Lesson 2
Card (yellow and blue)	**Unit 10** Lesson 9
Clock faces (laminated pictures)	**Unit 11** Lesson 5
Clocks	**Unit 13** Lesson 3
Coloured pencils	**Unit 12** Lesson 2
Coloured rods	**Unit 12** Lesson 1
Coloured rods (to create bar models) or strips of paper	**Unit 12** Lessons 5, 9
Counters	**Unit 10** Lesson 9 **Unit 12** Lessons 3, 7, 8 **Unit 13** Lesson 4 **Unit 14** Lesson 4
Counters (or cubes)	**Unit 10** Lesson 7 **Unit 12** Lesson 11
Counting materials (e.g. number lines, 100 squares, multiplication grids (2s, 5s, 10s))	**Unit 14** Lesson 7
Cubes	**Unit 12** Lesson 4
Cubes (or counters, or other objects for counting)	**Unit 10** Lesson 1
Dry-wipe pens	**Unit 13** Lessons 2, 4
Flag templates (to shade in)	**Unit 10** Lesson 7
Fractions (visual representations)	**Unit 10** Lesson 7
Grids (laminated copies)	**Unit 13** Lesson 4
Grids and objects (selection, including 2D and 3D shapes, to model the problems)	**Unit 13** Lesson 2
Laminated copies of the problems	**Unit 13** Lesson 2
Metre sticks	**Unit 10** Lesson 10
Mirrors	**Unit 10** Lesson 3
Money	**Unit 12** Lesson 5
Multilink cubes	**Unit 10** Lesson 9
Number bonds (within 20)	**Unit 12** Lesson 4
Number lines	**Unit 12** Lessons 5, 6
Number lines (counting up in 5s)	**Unit 14** Lesson 1
Objects (classroom, containing multiple parts, e.g. pencil cases)	**Unit 10** Lesson 1

Resource	Lesson
Objects (classroom, e.g. pots and paintbrushes, or pencils and pencil cases)	**Unit 10** Lesson 11
Objects (classroom, groups of, e.g. pencils and erasers)	**Unit 10** Lesson 4
Objects (classroom, to be used to demonstrate $\frac{1}{2}$ and $\frac{2}{4}$ – e.g. pencil pots)	**Unit 10** Lesson 10
Objects (easy to split equally, e.g. packs of pens or pencils)	**Unit 10** Lesson 2
Objects (hard to split into equal pieces, e.g. bananas)	**Unit 10** Lesson 2
Objects (multiple, groups of, e.g. pencils, sweets or erasers)	**Unit 10** Lesson 6
Objects (suitable to use on the grids)	**Unit 13** Lesson 4
Objects (to carry out rotations)	**Unit 13** Lesson 3
Objects (to cut into different parts, e.g. sandwiches, cakes, oranges or card shapes)	**Unit 10** Lesson 12
Paper plates	**Unit 10** Lesson 11
Paper squares or rectangles	**Unit 10** Lesson 7
Paper strips to fold or cut	**Unit 10** Lesson 7
Part-whole diagrams (laminated)	**Unit 12** Lesson 9
Part-whole models (blank)	**Unit 12** Lessons 3, 5
Pictures of activities (24, linked to each hour on the clock)	**Unit 11** Lesson 5
Place value counters	**Unit 12** Lesson 5
Reading book baskets or file dividers	**Unit 10** Lesson 10
Resource trays	**Unit 10** Lesson 10
Rope	**Unit 11** Lesson 2
Rulers	**Unit 14** Lesson 2
Sandwiches (square, to cut into parts)	**Unit 10** Lesson 5
Small items or outdoor apparatus to model positioning practically	**Unit 13** Lesson 1
Sorting circles	**Unit 11** Lesson 1
Squared paper (1 cm or 2 cm – same size as interlocking cubes)	**Unit 14** Lesson 3
Starting point prompts	**Unit 12** Lesson 4
Storage box	**Unit 14** Lesson 2
Times-table multiplication facts	**Unit 12** Lesson 10
Toys (for counting)	**Unit 14** Lesson 2
Toys (programmable)	**Unit 13** Lessons 3, 4
Toys (soft, for role play)	**Unit 11** Lesson 5
Tracing paper	**Unit 10** Lessons 3, 5
Vocabulary flashcards	**Unit 11** Lessons 1, 2, 3
Weekday timeline (noting hours and weekdays)	**Unit 11** Lesson 5

Getting started with *Power Maths*

As you prepare to put *Power Maths* into action, you might find the tips and advice below helpful.

STEP 1: Train up!

A practical, up-front full day professional development course will give you and your team a brilliant head-start as you begin your *Power Maths* journey. You will learn more about the ethos, how it works and why.

STEP 2: Check out the progression

Take a look at the yearly and termly overviews. Next take a look at the unit overview for the unit you are about to teach in your Teacher Guide, remembering that you can match your lessons and pacing to match your class.

STEP 3: Explore the context

Take a little time to look at the context for this unit: what are the implications for the unit ahead? (Think about key language, common misunderstandings and intervention strategies, for example.) If you have the online subscription, don't forget to watch the corresponding unit video.

STEP 4: Prepare for your first lesson

Familiarise yourself with the objectives, essential questions to ask and the resources you will need. The Teacher Guide offers tips, ideas and guidance on individual lessons to help you anticipate children's misconceptions and challenge those who are ready to think more deeply.

STEP 5: Teach and reflect

Deliver your lesson — and enjoy!

Afterwards, reflect on how it went… Did you cover all five stages? Does the lesson need more time? How could you improve it?

Mastery Expert tip! 'There were plenty of physical resources in the classroom that I could use to model equals in fractions, which allowed children to make links between what they saw on the page and real-life objects. Every time a lesson showed an image or shape split into equal parts, I would show children a physical object that also represents it, such as the window, an egg box or a pack of pencils.'

Don't forget to watch the Unit 10 video!

WHY THIS UNIT IS IMPORTANT

This unit builds on children's knowledge of equal groups, which they have come across in Unit 6. This unit also exposes children to equal parts in a range of contexts, including shape and numbers.

Within this unit, children will be introduced to fraction-specific key language, such as numerator and denominator, and will be able to explain what each word means in context.

At first, children will deal with unit fractions where the numerator is always one, focusing on halves, quarters and thirds. Children will then move on to non-unit fractions and learn about the equivalence between them, particularly between $\frac{1}{2}$ and $\frac{2}{4}$. Children will practise counting to 1 whole in quarters and halves.

WHERE THIS UNIT FITS

→ Unit 9: Mass, capacity and temperature

→ **Unit 10: Fractions**

→ Unit 11: Time

This unit builds on children's knowledge of sharing and grouping in division, asking children to divide a whole into equal parts and learn that the equal parts have given names. Children also learn to halve shapes by folding them or cutting them in two. Children can find a fraction of an amount using the previous strategy of sharing objects into equal groups but can now name these parts, for example by saying that $\frac{1}{2}$ of 6 is 3.

Before they start this unit, it is expected that children:

- know how to split an amount into equal parts by sharing or grouping
- understand that the same whole can have a different number of equal parts (building upon Unit 6)
- know what the ÷ sign means.

ASSESSING MASTERY

Children who have mastered this unit will be able to explain what each part of a fraction represents. They will be able to relate fractions to different contexts where there are equal parts and will be able to find a unit and non-unit fraction of a number or shape. Children will be able to recognise equivalent fractions and explain the relationship between the numerator and denominator in equivalent fractions. They will also be able to find a whole from a given fraction.

COMMON MISCONCEPTIONS	STRENGTHENING UNDERSTANDING	GOING DEEPER
Children may think that more parts always means a greater amount.	Fold a strip of paper into equal parts, label the parts and then fold them into equal parts again. Count up how many equal pieces there are now and count the whole again.	Show children a shape that has been halved and then each half halved again in a different way. Discuss whether each of the parts still represents $\frac{1}{4}$, even though they may be different shapes.
Children may think that it is only possible to count in whole numbers and not in fractions.	Demonstrate that there are lots of ways of making one half of a square. Give children pieces of squared paper and encourage them to investigate this.	Give children a whole number, such as 12, and ask them to make as many different equal parts as they can. Investigate which whole numbers allow children to make the most equal parts.

UNIT STARTER PAGES

Use these pages to introduce the unit. Use the characters to discuss concepts and phrases that children have heard before. For example, children were introduced to the concept of equal parts when learning about division.

STRUCTURES AND REPRESENTATIONS

Fraction strip: This model will help children recognise patterns when interpreting and comparing fractions and will help them to find equivalence with quarters and halves.

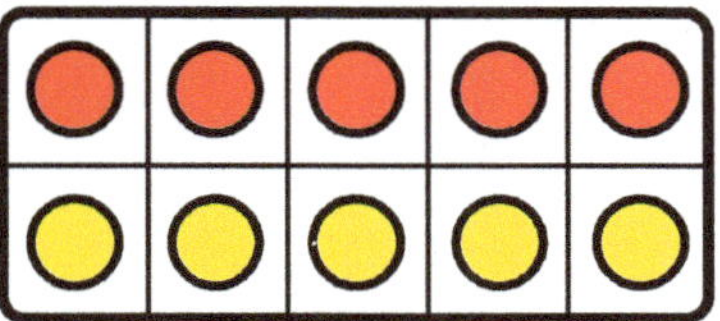

Ten frame: This model is used to represent doubles and halves and to support finding half of a given number.

KEY LANGUAGE

There is some key language that children will need to know as part of the learning in this unit:

→ fraction
→ half ($\frac{1}{2}$), quarter ($\frac{1}{4}$), third ($\frac{1}{3}$)
→ whole
→ part, equal parts
→ numerator, denominator
→ unit fraction, non-unit fraction
→ equivalent
→ three quarters ($\frac{3}{4}$)
→ equal
→ divided by (÷)
→ share

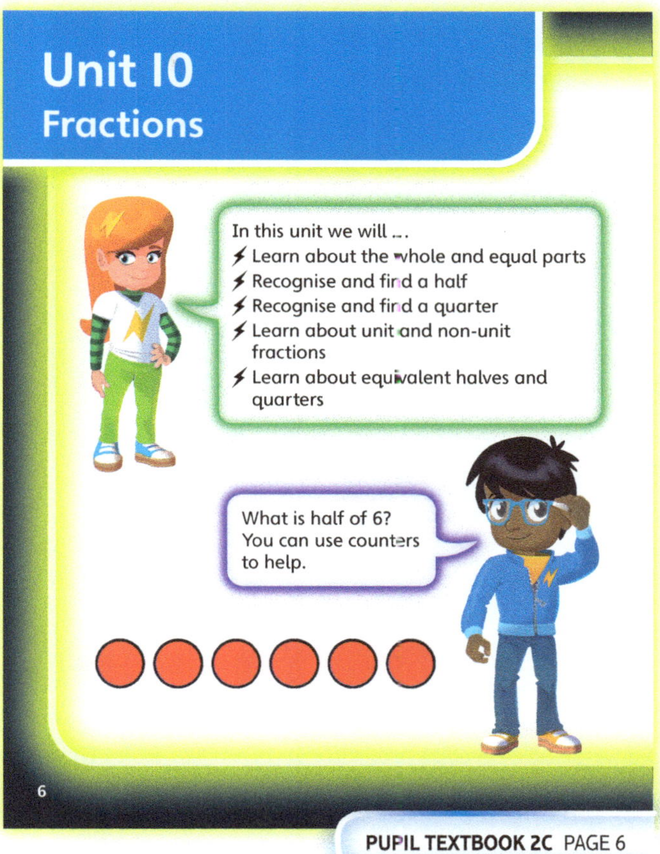

PUPIL TEXTBOOK 2C PAGE 6

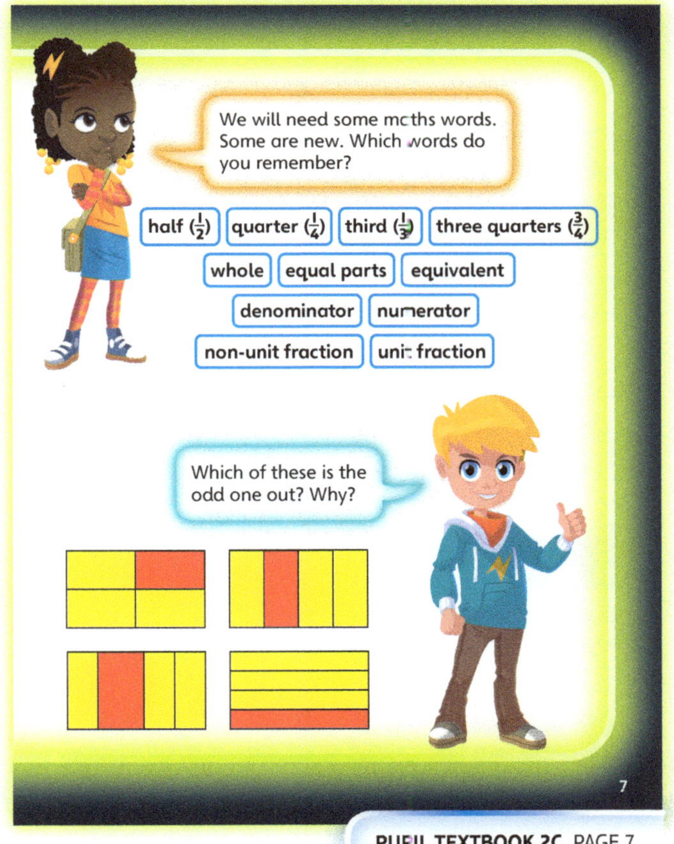

PUPIL TEXTBOOK 2C PAGE 7

Introducing parts and wholes

Learning focus

In this lesson, children will talk about the difference between a whole and a part in different contexts. Children will match parts to the correct wholes and fill in sentence scaffolds to match parts and wholes.

Before you teach 

- Are children confident with the words 'whole' and 'part' in relation to number?
- What curriculum links could make this lesson more tangible?

NATIONAL CURRICULUM LINKS

Year 1 Number – fractions

Recognise, find and name a half as one of two equal parts of an object, shape or quantity.

ASSESSING MASTERY

Children can imagine what the whole could be if they are given one part. Children understand that there can be many wholes and many parts within each whole and they can find multiple parts of one whole, understanding each time that the whole has not changed.

COMMON MISCONCEPTIONS

Children may believe that a whole needs to be something very big or that a part has to be something very small. Show children lots of different examples, such as wholes that are different shapes and sizes and parts that are big as well as small. Ask:

- *Does a whole or part have to be a certain size? Does the size matter?*

STRENGTHENING UNDERSTANDING

A number of objects in the classroom can represent a whole and can be taken apart into different parts. For example, a pencil case is a whole and holds a number of small objects that can be different parts. Ask children to find their own physical examples of wholes and parts in the classroom, guiding them towards understanding that the part is always smaller than the whole and the whole is always bigger than the part.

GOING DEEPER

Challenge children to zoom in on or zoom out of a whole or a part so that each object becomes both a whole and a part. For example, a book can be a part of a book shelf (the whole), and it can also be the whole, with a page being a part.

KEY LANGUAGE

In lesson: whole, equal, part, how many?, different

Other language to be used by the teacher: split

STRUCTURES AND REPRESENTATIONS

Part-whole model, sentence scaffolds involving the words 'whole' and 'part'

RESOURCES

Optional: objects in the classroom such as pencil cases that contain multiple parts; cubes, counters or other objects for counting

 In the eTextbook of this lesson, you will find interactive links to a selection of teaching tools.

Quick recap 

Provide children with 10 counters for them to split into two parts. Ask: *What are the parts? What is the whole?*

Discover

WAYS OF WORKING Pair work

ASK

- Question **1** a): *What is the whole?*
- Question **1** b): *Is the number of parts the same as the number of continents? Could you split up the whole in any other way?*

IN FOCUS The world map should be familiar to children so it is a good starting point for working out parts. If children struggle with continent names, use familiar places such as their home and school. This allows children themselves to also be part of the whole.

PRACTICAL TIPS Children can make a whole from a given number of cubes or similar objects that can be counted. Ask: *How many parts?* Agree that each cube is a part and all the cubes together are the whole. Count each of the cubes.

Revisit the part-whole model. Can children remember how to partition a number into two parts, and write the whole and the two parts in the model? Can they connect this with partitioning the world map into continents and realise that, in the same way, when the different parts are added together they make the whole?

ANSWERS

Question **1** a): The world is the whole. The continents are the parts.

Question **1** b): There are 7 continents, so there are 7 parts.

PUPIL TEXTBOOK 2C PAGE 8

Share

WAYS OF WORKING Whole class teacher led

ASK

- Question **1** a): *Does the whole change because it has different parts?*
- Question **1** a): *Can the whole have lots of different parts?*
- Question **1** b): *What important word does Ash say? Do you think the parts are equal?*

IN FOCUS Astrid says that lots of different sentences can be made using the words whole and parts. How many different sentences can children make using these words? Now bring children's attention to what Ash says and see whether they can spot the important word, 'equal'. Refer to the images to encourage children to compare the different parts and see whether they are equal.

PUPIL TEXTBOOK 2C PAGE 9

Think together

WAYS OF WORKING Whole class teacher led (I do, We do, You do)

ASK

- Questions ❶, ❷ and ❸: *Does the whole change when you identify a different part?*
- Question ❷: *Can you think of any more parts?*

IN FOCUS In question ❶, it is important that children identify that the United Kingdom is made up of England, Scotland, Wales and Northern Ireland. They may not have encountered this before, as there are few other countries in the world that unite together to form a conglomerate country. In question ❷, you can draw out that the whole (the elephant) remains the same, regardless of which part you identify.

Question ❸ arranges the parts and wholes in different orders so that the first sentence is not necessarily the sentence about the whole. This requires children to look closely and work out whether each object is a part or a whole.

STRENGTHEN Ask children to think of different parts for each object in question ❸. Can they name other parts of the duck, boat and pizza?

DEEPEN Encourage children to take a whole and create a new sentence where that whole is now a part of a new whole. For example, the elephant is the part and the zoo is the whole.

ASSESSMENT CHECKPOINT When reading the sentence scaffolds, can children instantly identify the whole or do they have to work it out? Are children confident in their ability to label each object a whole or a part, even when the order of the sentence scaffold has changed?

ANSWERS

Question ❶ a): The United Kingdom is the whole.

Question ❶ b): Scotland, England, Wales or Northern Ireland are each a part. (Accept any one answer.)

Question ❷ a): The elephant is the whole.

Question ❷ b): The trunk, ear, tail, tusk or eye are each a part. (Accept any one answer.)

Question ❸ a): The duck is the whole. The beak is a part.

Question ❸ b): The sail is a part. The boat is the whole.

Question ❸ c): The pizza is the whole. The cheese is a part.

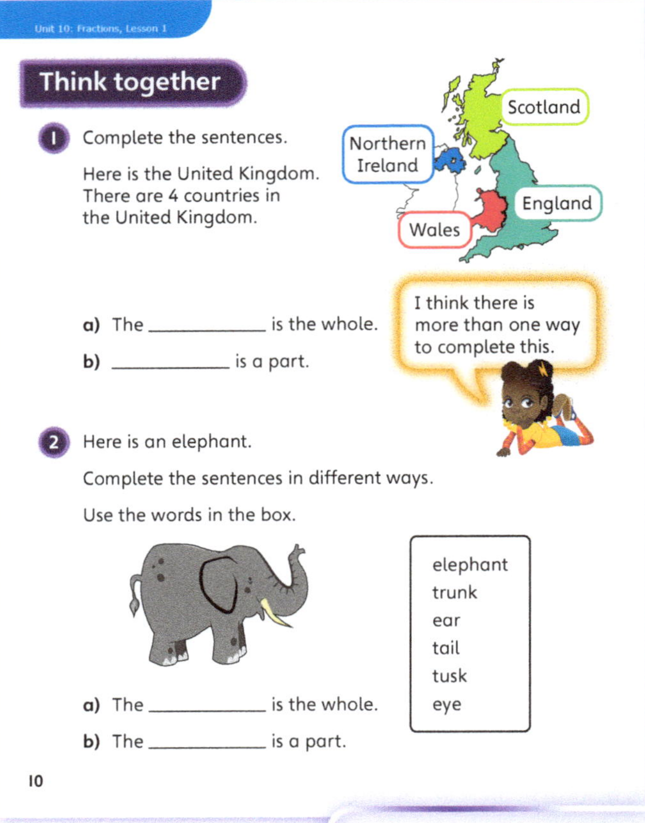

PUPIL TEXTBOOK 2C PAGE 10

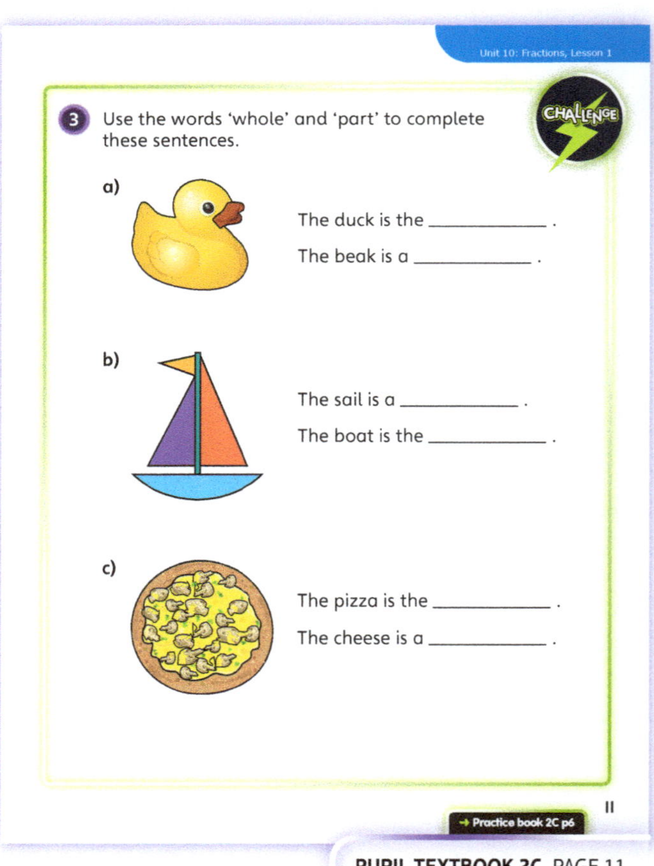

PUPIL TEXTBOOK 2C PAGE 11

Practice

WAYS OF WORKING Independent thinking

IN FOCUS In questions ❶ and ❷, children use images to represent parts and wholes. Sentence scaffolds are introduced in question ❸ with the words 'part' and 'whole'. Children identify the word that could correctly complete each sentence.

STRENGTHEN Ask children to use the sentence scaffolds they have heard in the lesson to explain which objects in question ❶ they are linking together and whether each object is the whole or a part. Ask children to write the whole and the part in different orders when answering question ❺ so that they have to identify each part for themselves rather than relying on the order in which parts and wholes usually appear.

DEEPEN Take the objects in question ❺ and ask children what the whole would be if the flower or the swing was a part. For example, the flower would be a part if a field was the whole. Expand this further by discussing the fact that a petal would be a part of both the flower and the field.

THINK DIFFERENTLY Question ❹ is the first question to present children with a part rather than the whole. It requires children to imagine the whole rather than identify a part.

ASSESSMENT CHECKPOINT Do children correctly choose the truck in question ❸ as the whole? Do they make up an inaccurate set of statements because they are just trying to fill the blank, such as 'The light is the whole, the window is a part'? Do children think of the window in question ❹ as a part or do they try to follow the pattern of previous questions and think of it as a whole? Do they answer with legitimate wholes or do they try to think of possible parts of the window?

ANSWERS Answers for the **Practice** part of the lesson can be found in the *Power Maths* online subscription.

Reflect

WAYS OF WORKING Independent thinking

IN FOCUS Children choose objects found in the classroom to make sentences using the words part and whole, with the freedom to choose an object of their choice.

ASSESSMENT CHECKPOINT Do children put their objects in the correct order when describing them as part or whole in relation to each other? Do children understand that a part must be smaller than the whole they have chosen? Equally, if they decide that an object is a part, do they understand that the whole must be bigger?

ANSWERS Answers for the **Reflect** part of the lesson can be found in the *Power Maths* online subscription.

After the lesson ⏸

- Were children confident with the relationship between a part and a whole across a range of contexts?
- Could children think of both a part and a whole from a given object?
- Did children show any awareness of parts being unequal or equal, such as commenting on the relative size of different parts of the same whole?

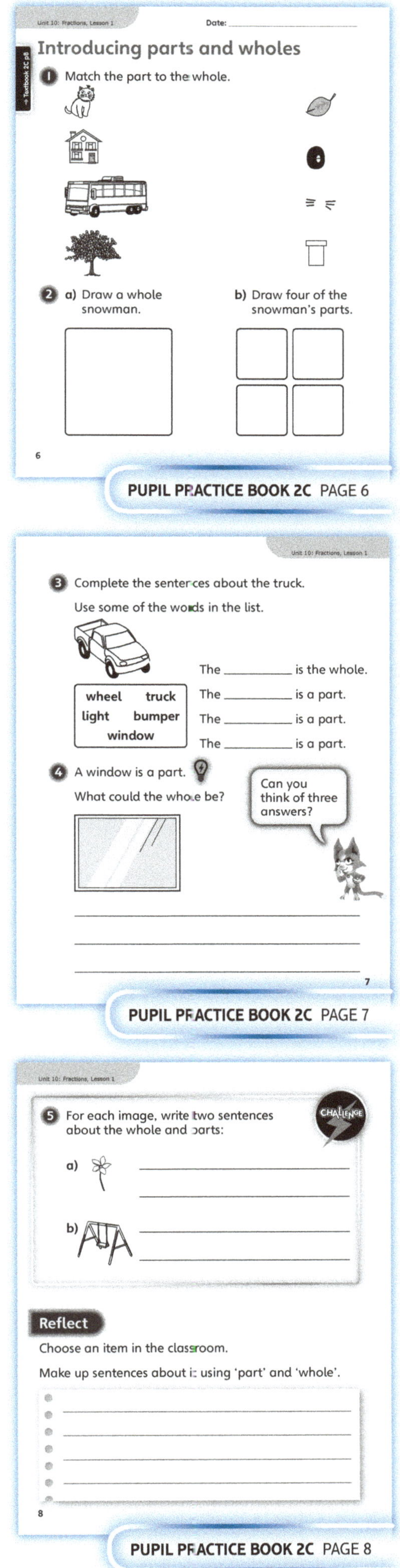

PUPIL PRACTICE BOOK 2C PAGE 6

PUPIL PRACTICE BOOK 2C PAGE 7

PUPIL PRACTICE BOOK 2C PAGE 8

Equal and unequal parts

Learning focus

In this lesson, children will identify equal parts of a whole in different contexts including shape, quantity, volume and money.

Before you teach 

- Where might children have come across the concept of fair before? Did they discuss it when learning division?
- Are children confident with what the word equal means when talking about numbers from previous units?

NATIONAL CURRICULUM LINKS

Year 1 Number – fractions

Recognise, find and name a half as one of two equal parts of an object, shape or quantity.

ASSESSING MASTERY

Children can associate the act of splitting a whole into equal parts with being fair and understand that the parts need to be the same. Children can recognise when parts of a variety of objects, shapes and volumes are unequal, can explain why and can rearrange unequal parts to make them equal parts.

COMMON MISCONCEPTIONS

When there is the same number of objects in each part but each object has a different value, children may believe that the parts are equal. For example, if there are three parts, each of which is one different value of coin, children may say that these parts are equal because there is one coin in each part. However, the coins have different values, which makes them unequal parts. Ask:

- *What is each group worth?*

STRENGTHENING UNDERSTANDING

Use real-life objects in the classroom to split equally between children. Ask children to think of times when a whole has to be split into equal parts to be fair, such as sharing out a pack of pens amongst the class or a packet of sweets with other children.

GOING DEEPER

When splitting an object into equal parts where the parts are not clear (perhaps because they are different shapes), ask children how they might check that all the parts are equal. For example, how would children cut a banana into equal pieces when the ends of the banana are different shapes to the middle? How would they check that the parts are equal?

KEY LANGUAGE

In lesson: part, equal, unequal, fair, fairly, differently, divide, odd one out, **equal parts**

Other language to be used by the teacher: whole, split, exactly

STRUCTURES AND REPRESENTATIONS

2D shapes

RESOURCES

Mandatory: multilink cubes, printed 2D shapes for children to fold into parts

Optional: a cake to cut into unequal and equal parts, objects that are easy to split equally (such as packs of pens or pencils), objects that are less easy to split into equal pieces (such as a banana)

 In the eTextbook of this lesson, you will find interactive links to a selection of teaching tools.

Quick recap 

Challenge children to list as many different ways as they can of sorting 8 counters into equal parts.

Discover

WAYS OF WORKING Pair work

ASK

- Question ❶ a): *Why is it not fair? Who has more? Who has less?*
- Question ❶ b): *If you divided the cake for four children, would the slices be the same as the boy's slice or the girl's slice?*

IN FOCUS Explain that each child has a piece of cake but that one child does not look happy. Guide children to conclude that it is not fair because the children have different amounts of cake.

PRACTICAL TIPS Bring in a cake or use circular pieces of paper to represent the cake in the scenario.

ANSWERS

Question ❶ a): It is not fair as one of the parts is bigger than the other. They could have cut the cake into 2 equal parts (halves).

Question ❶ b): The cake has been cut into 4 equal parts. Each child will get the same size piece.

PUPIL TEXTBOOK 2C PAGE 12

Share

WAYS OF WORKING Whole class teacher led

ASK

- Question ❶ a): *Do the pieces of cake look the same?*
- Question ❶ b): *Do these pieces of cake look the same as each other or different to each other?*
- Questions ❶ a) and b): *Are any of the parts bigger than the others?*

IN FOCUS Dexter explains that the pieces of cake in question ❶ a) are not equal. The next picture then shows the whole split into 2 equal parts to be shared equally between two children. The picture in question ❶ b) shows the whole split into 4 equal parts to be shared equally between four children.

PUPIL TEXTBOOK 2C PAGE 13

Think together

WAYS OF WORKING Whole class teacher led (I do, We do, You do)

ASK

- Question ❶: *Do all of the parts look the same or different? What makes them equal?*
- Question ❶: *How many parts has each been split into?*
- Question ❶: *Have all of the shapes been split?*

IN FOCUS Question ❶ shows a variety of pictures and contexts that have been divided. Children need to work out which shapes have parts that are all the same. Question ❷ requires children to explain which picture is the odd one out based on the criterion of having equal parts.

STRENGTHEN In question ❶, B and C are the same shape split differently so children can compare them if they are unsure. You could print the shapes out so that children can fold them into the constituent parts to check whether the parts are the same or not.

DEEPEN Question ❸ is a **Challenge** question, so provide children with physical resources to model it. For example, if one multilink cube represents one child, cubes can be put into three unequal piles to represent the children in the boats.

ASSESSMENT CHECKPOINT In question ❶, do children understand the idea of equal parts in shapes, volumes, money, and so on? Does F confuse them because each hand contains the same number of coins but they have different values? Do children want to use rulers to check whether E has been split into parts that have the same length and width?

ANSWERS

Question ❶: B and E show equal parts.

Question ❷: B has 2 plates with an equal number of 3 apples on each plate. C has 2 rabbits, each with 1 carrot. A is not equal as one train carriage has 2 children and the other has 3 children. A is the odd one out.

Question ❸: If one child is moved from boat C into boat B, there will be 2 children in each boat, giving equal groups of children. If not all the boats are used, then there could be 2 boats with 3 children in each.

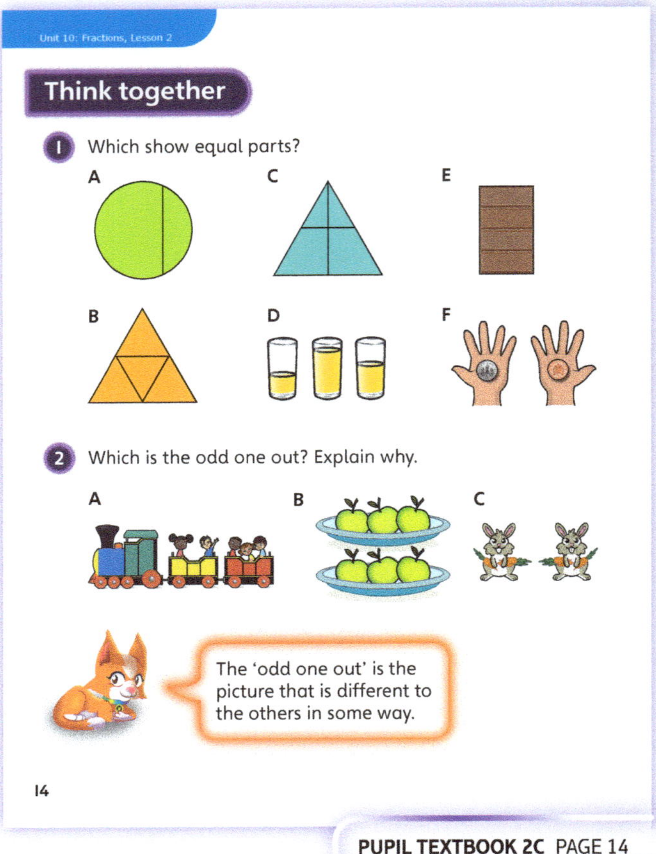

PUPIL TEXTBOOK 2C PAGE 14

PUPIL TEXTBOOK 2C PAGE 15

Practice

WAYS OF WORKING Independent thinking

IN FOCUS Question **1** requires children to count how many equal parts each whole has been split into. Question **2** asks children to decide whether the food has been split into equal or unequal parts.

STRENGTHEN When children are answering question **3**, ask them to explain what the whole is and how many equal groups they will split the whole into. Ask how they would share out 12 cubes if they were using a sharing strategy. When children have worked out that there should be 3 cubes in each group, can they recreate the unequal arrangement in question **3** and rearrange them in order to have 3 cubes in each group? Can they arrange them into groups of 6 cubes?

DEEPEN Question **4** requires children to identify which shapes show equal parts and which do not. Ask: *How did you know? Can you prove it to me?* Children can fold pieces of paper into equal and unequal parts to show their thinking.

THINK DIFFERENTLY Question **3** asks children to rearrange unequal parts into equal parts. This can be modelled using multilink cubes, as suggested by Flo. Look for the methods children are using to do this. The best method here is to use sharing (putting one cube on each plate at a time, and repeating until all cubes are gone). Grouping is not as helpful here as you do not know how many cubes need to be in each 'group', but instead you know the number of groups (4).

ASSESSMENT CHECKPOINT Use question **1** to assess whether children can accurately count a given number of equal parts. Use question **3** to assess whether children can use sharing to rearrange unequal groups to make equal groups.

ANSWERS Answers for the **Practice** part of the lesson can be found in the *Power Maths* online subscription.

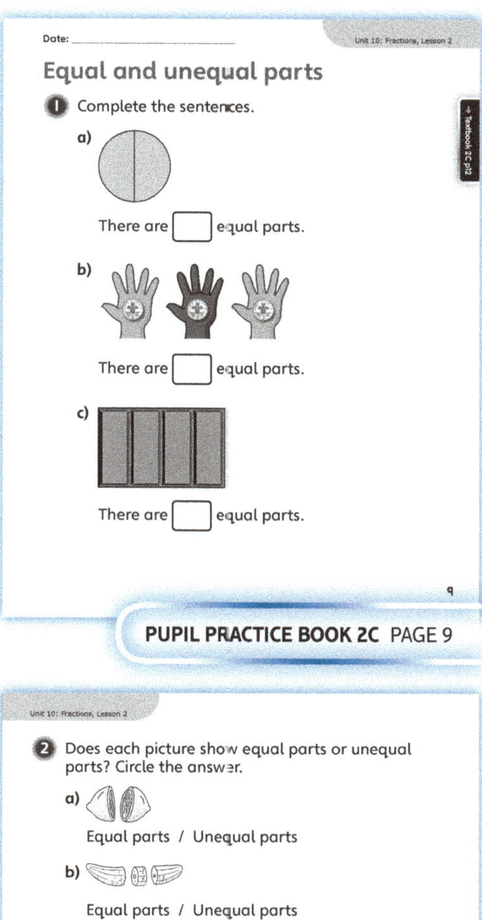

PUPIL PRACTICE BOOK 2C PAGE 9

PUPIL PRACTICE BOOK 2C PAGE 10

PUPIL PRACTICE BOOK 2C PAGE 11

Reflect

WAYS OF WORKING Independent thinking

IN FOCUS Children are asked to make both equal and unequal parts by folding pieces of paper. They should identify how many equal parts they want to fold the paper into and how they can achieve this.

ASSESSMENT CHECKPOINT Can children give a reason why one part is more or less than another? Do they identify a way to fold the paper equally, such as by measuring it?

ANSWERS Answers for the **Reflect** part of the lesson can be found in the *Power Maths* online subscription.

After the lesson

- Were children confident when splitting a range of objects into equal parts?
- Could children spot when parts were unequal and suggest how to arrange them to make them equal?

Recognise a half

Learning focus

In this lesson, children will identify which objects have been split into two equal parts. They will be introduced to $\frac{1}{2}$ as a written fraction with a numerator and a denominator.

Before you teach 

- Do children know the meaning of the word 'half' in the context of halving and doubling numbers?
- Could you create a classroom display explaining $\frac{1}{2}$ for children to refer to?

NATIONAL CURRICULUM LINKS

Year 1 Number – fractions

Recognise, find and name a half as one of two equal parts of an object, shape or quantity.

ASSESSING MASTERY

Children can explain what the 1 and 2 represent in one half written as a fraction and can understand that it represents one whole divided into two equal parts in a range of contexts. Children can link finding one half of a shape to finding a line of symmetry.

COMMON MISCONCEPTIONS

Children may associate a shape that has been split into two equal parts with $\frac{1}{2}$ when it actually represents $\frac{2}{2}$. Show children that each part in the whole represents $\frac{1}{2}$. For example, you could label each part of a shape one half and label the whole shape one whole. Ask:

- *Where can you see $\frac{1}{2}$? What is the whole?*

STRENGTHENING UNDERSTANDING

When splitting a shape into two equal parts, have a mirror available so children can check either part. In addition, encourage children to experiment with folding cut-out shapes in half. Provide children with tracing paper or baking paper so they can trace one half of the shape and turn it over to check that the other half is the same.

GOING DEEPER

There are many contexts in which to find one half, such as half a distance to a place, half the weight of an object, half the height of a person, half a cup of water. Give children the opportunity to explore finding one half in as many contexts as possible so that they do not think that finding a half is only ever related to shapes, numbers or numbers of objects.

KEY LANGUAGE

In lesson: $\frac{1}{2}$, **denominator**, **numerator**, equal parts, half, halves, half-way

Other language to be used by the teacher: whole, symmetrical, the same, line of symmetry, middle

STRUCTURES AND REPRESENTATIONS

2D shapes, fraction strip, fraction number line

RESOURCES

Mandatory: classroom objects and shapes to split, fold or cut in half, 2D shapes printed out to be folded into equal and unequal parts, multilink cubes, a large '$\frac{1}{2}$' printed, with numerator and denominator labelled

Optional: mirrors, tracing paper

 In the eTextbook of this lesson, you will find interactive links to a selection of teaching tools.

Quick recap 

Gradually fill a jug with water. Ask children to shout 'Stop!' as a class when they think the jug is half full.

Discover

WAYS OF WORKING Pair work

ASK

- Question **1** a): *How could the paper show one half?*
- Question **1** b): *What is special about this number? What does the 1 mean? What does the 2 mean?*

IN FOCUS Question **1** a) prompts children to recognise and identify one half of a shape. Children then use this to consider the properties of one half and to describe $\frac{1}{2}$ as a fractional number in question **1** b).

PRACTICAL TIPS Provide children with a variety of objects and shapes that they can split, fold or cut in half. Confirm that there must be exactly two parts and that they must both be equal.

ANSWERS

Question **1** a): The paper should be folded into 2 equal parts, known as 'halves'.

Question **1** b): The denominator is the total number of equal parts. The denominator of $\frac{1}{2}$ is 2. There are 2 equal parts.

The numerator is how many of the equal parts we have. The numerator of $\frac{1}{2}$ is 1. It is one of 2 equal parts.

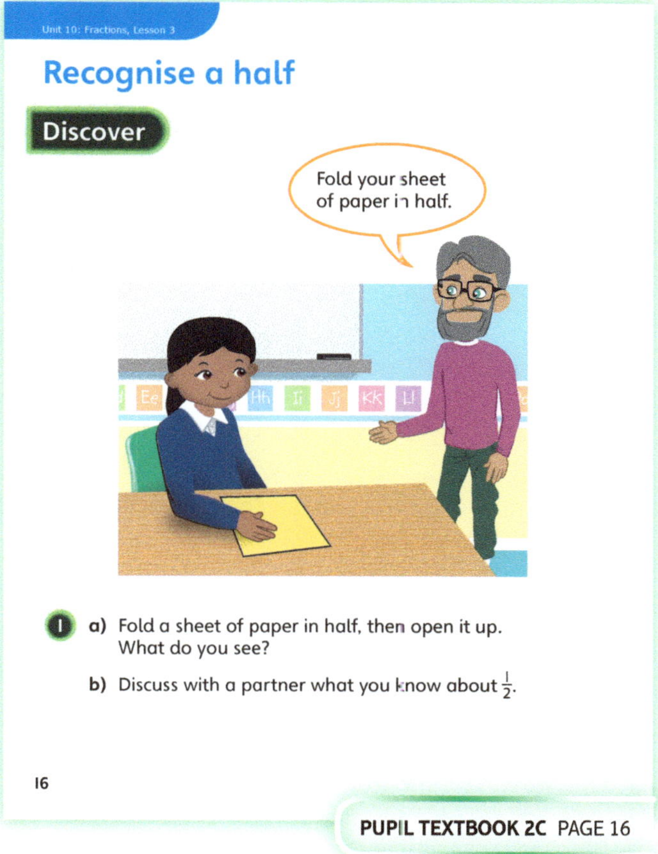

PUPIL TEXTBOOK 2C PAGE 16

Share

WAYS OF WORKING Whole class teacher led

ASK

- Question **1** a): *Where can you see two equal parts in these diagrams?*
- Question **1** b): *What does the 1 represent in $\frac{1}{2}$? What does the 2 represent in $\frac{1}{2}$?*

IN FOCUS Explain that the fraction bar in question **1** b) is like the line in the ÷ sign and means divide by. Encourage children to label each part of their paper '$\frac{1}{2}$' to avoid confusion with the whole. Each half of the piece of paper is $\frac{1}{2}$, whereas the piece of paper is a whole.

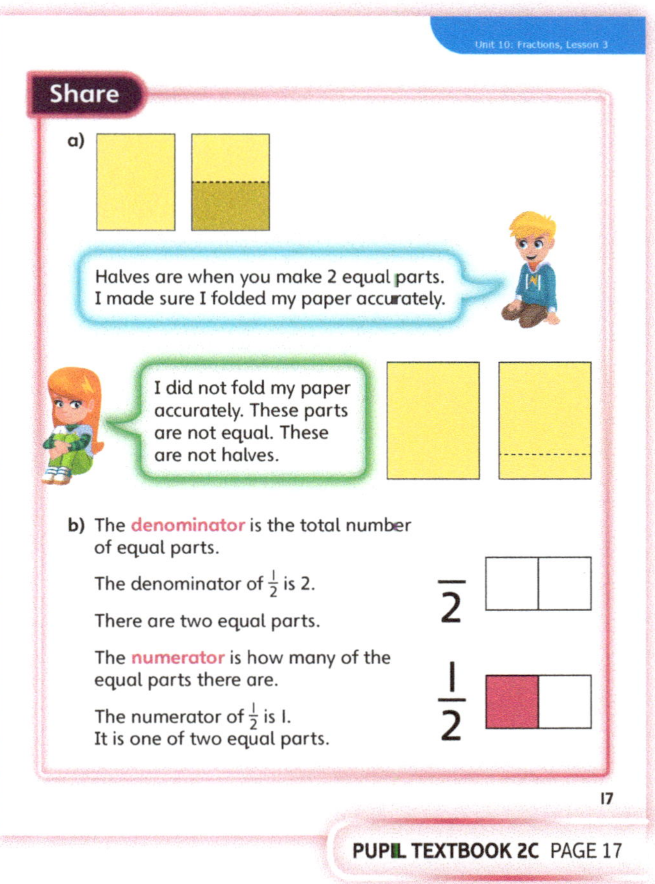

PUPIL TEXTBOOK 2C PAGE 17

Think together

WAYS OF WORKING Whole class teacher led (I do, We do, You do)

ASK

- Question **1**: *How do we know that these all show halves?*
- Question **3**: *What does half-way mean?*

IN FOCUS Question **1** asks children to apply what they have learnt by accurately folding squares in several different ways to make halves. Remind them that the whole needs to be folded into two parts that are equal. Agree that if the two parts are not equal, then they do not show halves.

STRENGTHEN Print out the circles in question **2** and place them in front of each child. Ask children to split them physically by using a mirror to check and draw where the two halves meet, by folding them, or by using tracing paper to trace one half and check that the other side is the same.

DEEPEN Look together at the diagram in question **3** a). Explain that the race track is 12 metres long. Can children begin to make sensible estimates about how many metres each runner has run?

ASSESSMENT CHECKPOINT What method do children use to halve the shapes in question **2**? Are they able to find multiple ways of splitting the circle in half? In question **3**, do children appear confident that a half can be represented on a line and that the middle of the line is half-way?

ANSWERS

Question **1**: Practical activity. Check children can accurately fold squares of paper in half, as in the examples.

Question **2**: Answers may vary but children should draw 3 circles, split each circle into 2 equal parts and shade 1 of the parts. For example:

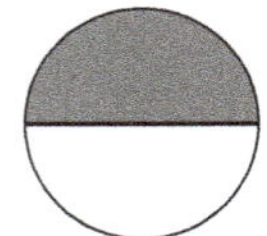 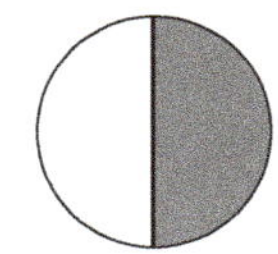 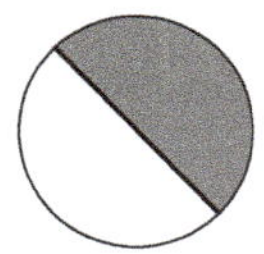

Question **3** a): B
         b): B

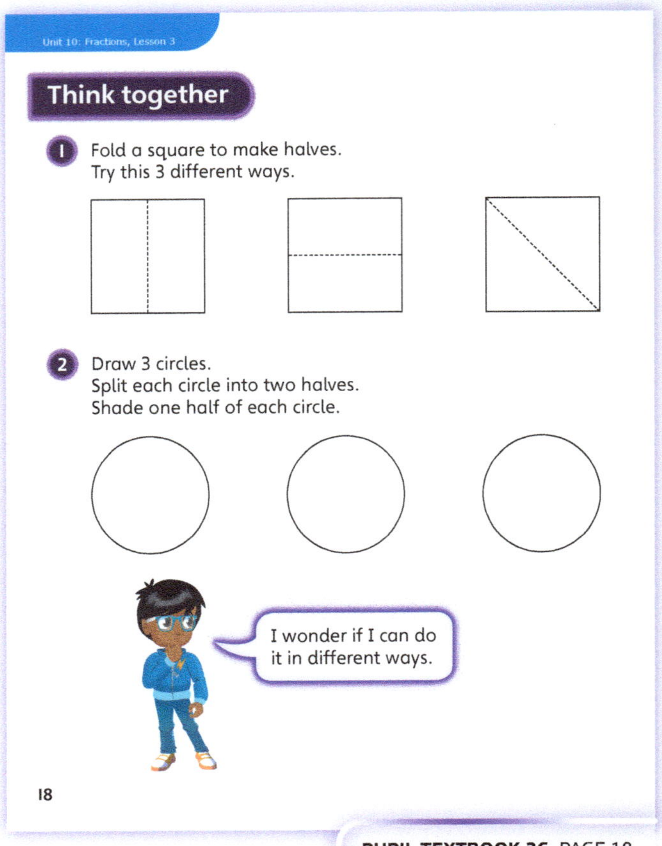

PUPIL TEXTBOOK 2C PAGE 18

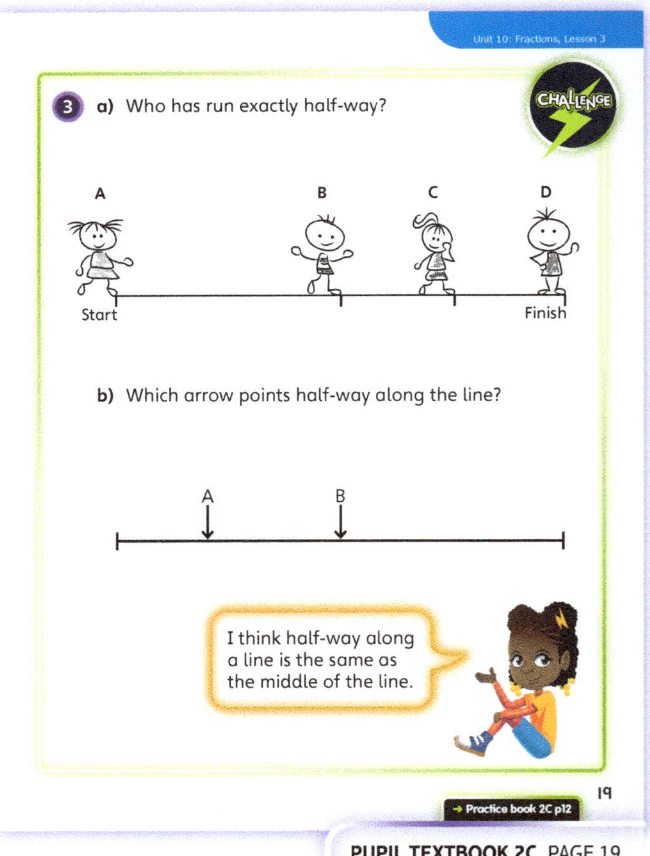

PUPIL TEXTBOOK 2C PAGE 19

Practice

WAYS OF WORKING Independent thinking

IN FOCUS Question **2** gives children a selection of different shapes and asks them to shade in $\frac{1}{2}$ of each one. Question **3** presents children with squares and rectangles that do not include a line to indicate halves. Children must draw the lines themselves.

STRENGTHEN Copy and cut out the shapes in question **4** and give them to children. Provide tracing paper for children to check whether each part is the same by rotating the tracing paper to fit one half over the other. Point out that they need to be a perfect match.

DEEPEN Fractions on a number line are introduced in question **5**. Children should notice that $\frac{1}{2}$ is always half-away along the number line: it is in the middle. Ask: *What number would go at the start of this number line?* [0] *What number would go at the end?* [1] Agree that 1 is the same as one whole.

ASSESSMENT CHECKPOINT Use question **4** to assess whether children can accurately recognise a half and can explain how they know when a shape does not show halves. Use question **5** to assess whether children are becoming confident with the concept of $\frac{1}{2}$ as a fractional number.

ANSWERS Answers for the **Practice** part of the lesson can be found in the *Power Maths* online subscription.

Reflect

WAYS OF WORKING Pair work

IN FOCUS Children are asked to work with a partner in order to both make and recognise halves. They do this by folding strips of paper in half and checking that the resulting two parts are equal.

ASSESSMENT CHECKPOINT Do children instantly recall that halves are two equal parts and can they identify whether or not the strip of paper has been correctly folded to show this?

ANSWERS Answers for the **Reflect** part of the lesson can be found in the *Power Maths* online subscription.

After the lesson ⏸

- Could children explain what 1 and 2 represent in $\frac{1}{2}$?
- Did children understand that there is more than one way to split regular shapes?
- Did children relate any part of the lesson to halving numbers?

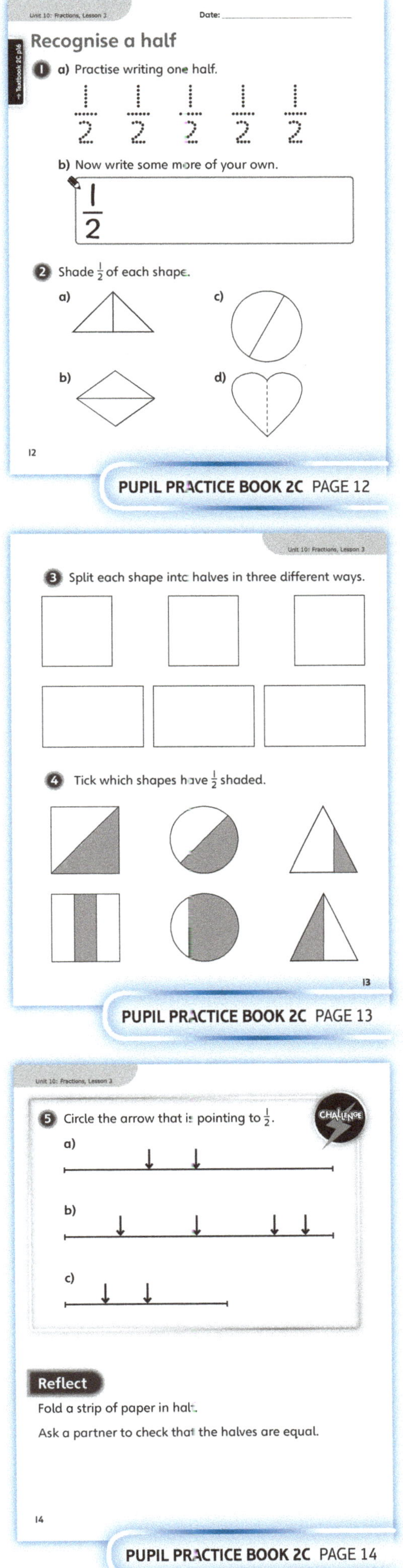

PUPIL PRACTICE BOOK 2C PAGE 12

PUPIL PRACTICE BOOK 2C PAGE 13

PUPIL PRACTICE BOOK 2C PAGE 14

Find a half

Learning focus

In this lesson, children will find one half of different amounts of objects, shapes and numbers.

Before you teach 

- How will you build on children's previous strategies of recognising a half?
- Do children recall doubles to double 10?

NATIONAL CURRICULUM LINKS

Year 1 Number – fractions

Recognise, find and name a half as one of two equal parts of an object, shape or quantity.

ASSESSING MASTERY

Children can find one half of numbers of objects up to 20. Children can fill in sentence scaffolds stating that '$\frac{1}{2}$ of ☐ is ☐', sometimes filling in the $\frac{1}{2}$ for themselves too. Children should be able to also link this with dividing by 2 and times-table facts.

COMMON MISCONCEPTIONS

Children may struggle to split a group of objects in half when the objects are scattered on the page rather than arranged in neat rows. Encourage children to develop a strategy, such as first counting how many objects there are in the whole, representing each object with cubes and rearranging them into neat groups, or circling one object and leaving one object alternately. Children may lose track of the original whole number and may not check that their two groups are equal. Ask:
- *How many objects have you circled or crossed out? How many have you not circled or crossed out? Are the numbers the same?*

STRENGTHENING UNDERSTANDING

When finding half of a number, encourage children to represent the number using physical resources, such as cubes. When splitting the physical resources into halves by adding one to each stack, children will find that their stacks of cubes will be the same height at the end.

GOING DEEPER

Some children may be able to spot that half of an odd number is possible if you include $\frac{1}{2}$ in the answer. For example, half of 7 is $3\frac{1}{2}$. Guide children to spot this in contexts such as splitting seven squares in half, and to understand the difference between this and splitting seven children in half.

KEY LANGUAGE

In lesson: half, equal, total, double, share

Other language to be used by the teacher: numerator, denominator, odd, even, fair

STRUCTURES AND REPRESENTATIONS

Arrays, ten frames

RESOURCES

Mandatory: squared paper, multilink cubes and counters to represent objects, sorting hoops

Optional: groups of classroom objects, such as pencils and erasers

 In the eTextbook of this lesson, you will find interactive links to a selection of teaching tools.

Quick recap 

Challenge children to draw any picture that accurately shows halves.

Discover

 Pair work

ASK

- Question **1** a): *How many players are there?*
- Question **1** a): *What strategy could you use to split these players into two teams?*

IN FOCUS Discuss the sports context and consider why the coach wants the players to split into 2 teams. Emphasise the word 'equal' and ask children to explain what this means. Agree why it is necessary to have the same number of players on each team, referring back to previous learning about equal parts being fair. Do children know half of 10 already as a number fact or do they have to use a sharing technique to work it out?

PRACTICAL TIPS Recreate the **Discover** scenario by asking 10 children to stand at the front of the room. Take suggestions from the rest of the class about how they can be arranged to make 2 equal teams. Ask: *How many children will be in each of our teams?*

ANSWERS

Question **1** a): There are 10 players in total, so to make 2 equal teams, each team needs 5 players.

Question **1** b): $\frac{1}{2}$ of 10 is 5.

PUPIL TEXTBOOK 2C PAGE 20

Share

 Whole class teacher led

ASK

- Question **1** a): *What does it mean to have equal teams? Are these 2 teams equal? How do you know?*
- Question **1** b): *How does Astrid's double fact help you?*

IN FOCUS This part of the lesson gives children a pictorial representation of 10 players split into 2 teams. There is the same number of players in both teams. Splitting 10 into 2 groups is modelled with counters in 2 colours in a ten frame. Astrid mentions the related double fact. Can children make this link for themselves or do you have to explain it to them?

PUPIL TEXTBOOK 2C PAGE 21

Think together

WAYS OF WORKING Whole class teacher led (I do, We do, You do)

ASK

- Question **1**: *How many groups do you have to share between if you are finding $\frac{1}{2}$?*
- Question **2**: *Are the numbers of sweets in each box equal?*

IN FOCUS Question **1** provides ten frames as visual representations of numbers for children to find $\frac{1}{2}$ of. Question **2** shows 14 sweets which need to be shared equally into 2 boxes. Encourage children to cover 1 sweet each time a sweet is shared. Can children also see how to halve the amounts pictorially rather than sharing objects out? Are they able to identify the double that Ash refers to, which could help them find the answer to question **2**?

STRENGTHEN Encourage children to use a wide variety of physical resources to demonstrate halving. Question **2** requires children to halve a pictorial representation of 14 sweets, so ask children whether they can also do this physically with a range of other objects such as 14 counters or cubes. This will reinforce the number fact in a range of contexts.

DEEPEN Question **3** requires children to identify which of the numbers from 1 to 20 can be split into 2 halves. Do children instinctively know that, for example, 4 can be halved again but that 3 cannot be halved again because 4 is even and 3 is odd? Encourage children to explore scenarios where an odd number of objects could be halved. Ask: *What would you do if 2 children wanted to eat 3 apples? Is there any way to share the apples fairly?*

ASSESSMENT CHECKPOINT Use question **1** to assess whether children can use ten frames to model how to halve even numbers. Are children then able to apply this in question **3** to show why it is not possible for odd numbers?

ANSWERS

Question **1** a): $\frac{1}{2}$ of 4 is 2.

Question **1** b): $\frac{1}{2}$ of 6 is 3.

Question **1** c): $\frac{1}{2}$ of 8 is 4.

Question **2**: Double 7 is 14, so half of 14 is 7.

Question **3** a): Children should notice that with 7 counters (or any odd number of counters), they cannot make equal groups or halves – there is always one left over.

Question **3** b): 2, 4, 6, 8, 10, 12, 14, 16, 18, 20 Children should notice that these are all even numbers.

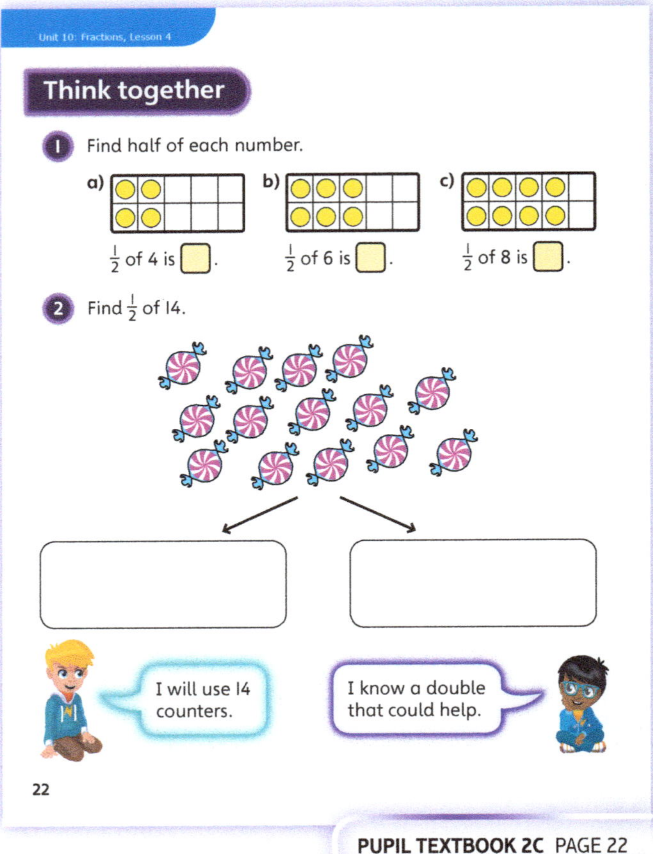

PUPIL TEXTBOOK 2C PAGE 22

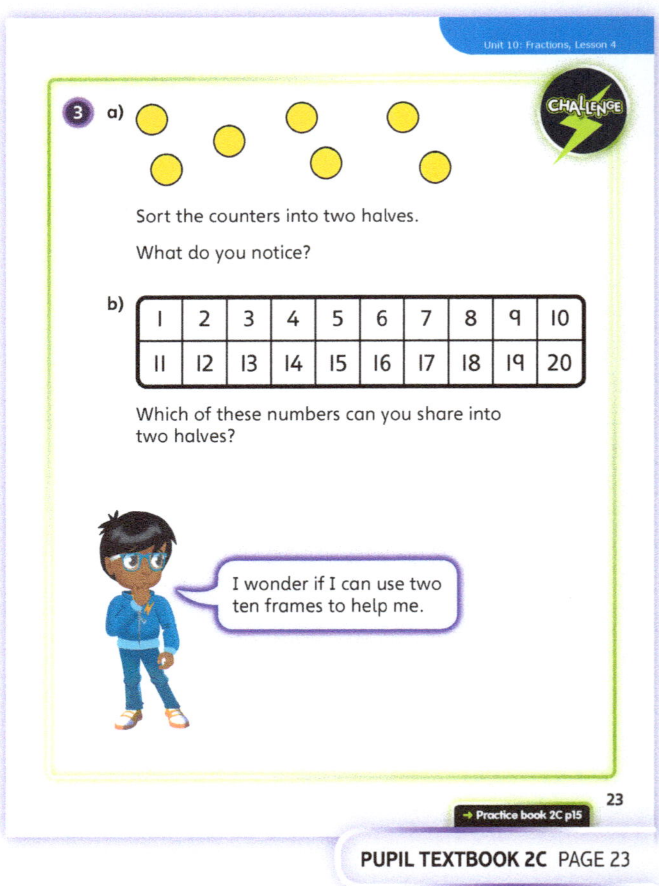

PUPIL TEXTBOOK 2C PAGE 23

Practice

WAYS OF WORKING Independent thinking

IN FOCUS Questions ① to ④ ask children to find half the number of objects or squares and record it in the sentence scaffolds. Question ⑤ presents an amount as one half of an unknown whole rather than presenting the whole itself. This differs to what children have seen so far in this lesson.

STRENGTHEN Ask children to cut the arrays in question ③ out of squared paper. Children could fold them in different ways to find a half. Encourage children to rearrange the squares in question ③ a) so that they look more like the array in question ③ b), guiding them to realise that the array in question ③ a) is half the array in question ③ b).

DEEPEN Provide children with more questions similar to question ⑤, giving one half rather than one whole, and ask children about their strategy each time. Do children realise that they are doubling the given number? Ask: *Is it possible to halve all numbers? Is it possible to double all numbers? Can all numbers be both a half and a whole?*

ASSESSMENT CHECKPOINT In question ⑤, are children able to take what they have learnt about halving and apply it to the process of doubling a half to find a whole? Do they count on from the half or can they recall a previously learnt fact to identify the whole?

ANSWERS Answers for the **Practice** part of the lesson can be found in the *Power Maths* online subscription.

Reflect

WAYS OF WORKING Independent thinking or pair work

IN FOCUS This question provides an odd number for children to share and asks them to explain why they cannot do this equally. Give children the opportunity to draw it or model it using physical resources.

ASSESSMENT CHECKPOINT Look for words such as odd, even, fair or equal in children's answers. Are children aware there will be an odd one out so it is not possible to split the sweets into two equal parts? Do they attempt to split the leftover sweet in half?

ANSWERS Answers for the **Reflect** part of the lesson can be found in the *Power Maths* online subscription.

After the lesson ⏸

- Did children show any awareness of the link between the numbers in this lesson and the 2 times-table and related division facts?
- Did children struggle to find a whole from a half?
- Were children confident that odd numbers cannot be halved because there is always one left over?

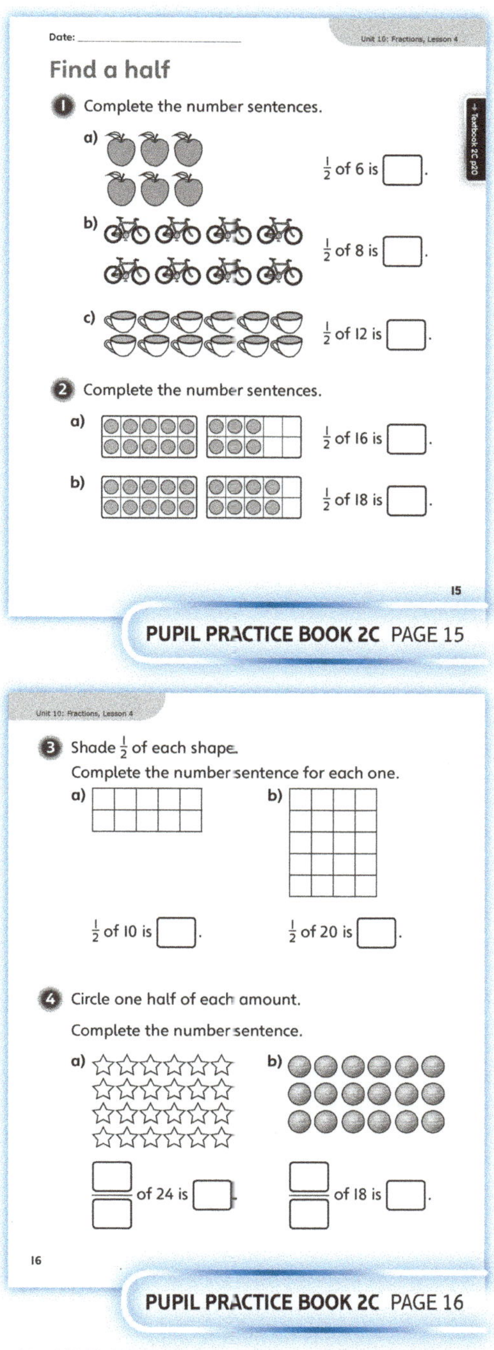

PUPIL PRACTICE BOOK 2C PAGE 15

PUPIL PRACTICE BOOK 2C PAGE 16

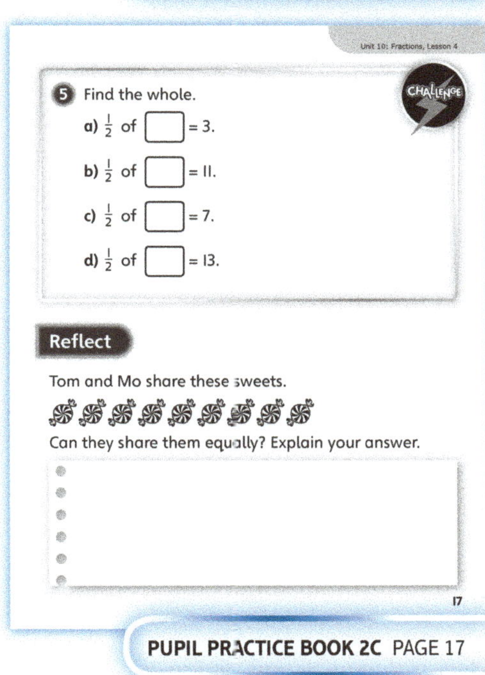

PUPIL PRACTICE BOOK 2C PAGE 17

Recognise a quarter

Learning focus

In this lesson, children will recognise shapes that have been split into four equal parts and will identify one quarter. They will split given shapes into quarters and will recognise which shapes do not show quarters.

Before you teach 

- How will you recap the meaning of numerator and denominator before starting this lesson?
- Could you change your classroom display of fractions to show $\frac{1}{4}$ and clearly display a representation of four equal parts for children to refer to?

NATIONAL CURRICULUM LINKS

Year 1 Number – fractions

Recognise, find and name a quarter as 1 of 4 equal parts of an object, shape or quantity.

Year 2 Number – fractions

Recognise, find, name and write fractions $\frac{1}{3}$, $\frac{1}{4}$, $\frac{2}{4}$ and $\frac{3}{4}$ of a length, shape, set of objects or quantity.

ASSESSING MASTERY

Children can recognise different shapes that have been correctly or incorrectly split into four equal parts and shade $\frac{1}{4}$ of different shapes. Children can split given shapes into quarters and identify one quarter.

COMMON MISCONCEPTIONS

Sometimes children may believe that a shape cannot be split into four equal parts if the equal parts have different orientations to the original shape. Drawing around the parts on tracing paper and turning the tracing paper will help. Ask:
- *Does it matter which way round the part is or is it still an equal part?*
- *Does it matter if the quarters look different?*

STRENGTHENING UNDERSTANDING

Give children pieces of paper to fold into quarters. Demonstrate that there are two ways of doing this: children can fold the piece of paper in half and then half again or they can fold the piece of paper in half, unfold it so that it is flat and then fold it in half again in a different way.

GOING DEEPER

Help children recognise that $\frac{1}{4}$ is half of $\frac{1}{2}$ by showing them two identical squares and explaining that, when put together, these two squares form *half* of an original shape. Ask children to shade in $\frac{1}{4}$ of the original shape. Do they realise that one square needs to be shaded because the original shape would have been made up of four squares?

KEY LANGUAGE

In lesson: $\frac{1}{4}$, equal parts, unequal parts, quarters, compare, split, square

Other language to be used by the teacher: whole, half

STRUCTURES AND REPRESENTATIONS

2D shapes, fraction strips

RESOURCES

Mandatory: counters, squared paper, rulers, printed 2D shapes to fold and cut, circles of card split into quarters

Optional: tracing paper, square sandwiches to cut into parts

 In the eTextbook of this lesson, you will find interactive links to a selection of teaching tools.

Quick recap 

Challenge children to draw a picture that accurately shows four parts.

Discover

WAYS OF WORKING Pair work

ASK

• Question ❶ a): *How many parts has each sandwich been cut into? Do all the parts looks the same?*
• Question ❶ b): *What shape could your equal parts be?*

IN FOCUS This question introduces quarters as four equal parts of a shape. Children should identify that all of the sandwiches have been cut into four parts but that not all parts of each sandwich are the same. Do children recognise that Kat's and Kasim's sandwiches have been split into quarters in different but equally correct ways? Ask: *Which sandwich would you like to share with three friends? Why? Which part of Lena's sandwich would you choose? Why?*

PRACTICAL TIPS Children can recreate the sandwiches by folding or cutting up large squares of squared paper. Encourage children to find other ways of splitting the paper, such as into four columns. Discuss if the parts are equal or not. Ask four children to each choose a part. Ask: *Which part did you choose? Why? Is it fair?*

ANSWERS

Question ❶ a): Kat and Kasim's sandwiches have each been cut into 4 equal parts. Lena's sandwich has been cut into 4 unequal parts, so it is not in equal parts.

Question ❶ b):  

Share

WAYS OF WORKING Whole class teacher led

ASK

• Question ❶ a): *What does unequal mean?*
• Question ❶ b): *Are there any other ways to split a square into quarters? How do you know?*

IN FOCUS Give children square pieces of paper to cut up to help them understand. Children cut each square into four parts as shown. Ask: *How can we find out if the parts of each sandwich are equal?* Model laying the parts over each other to see if they are the same. Confirm that a shape that has been cut into four equal parts, shows quarters.

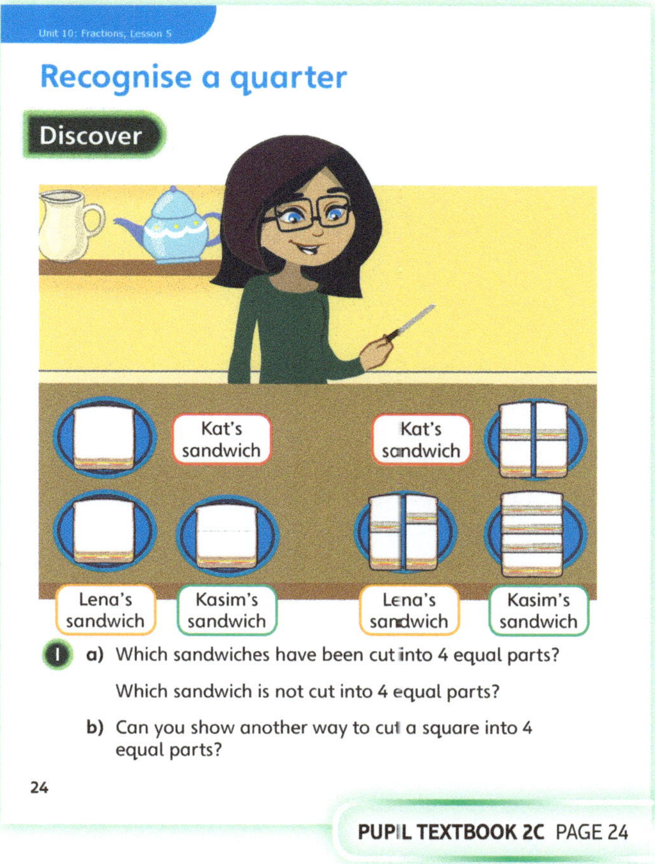

PUPIL TEXTBOOK 2C PAGE 24

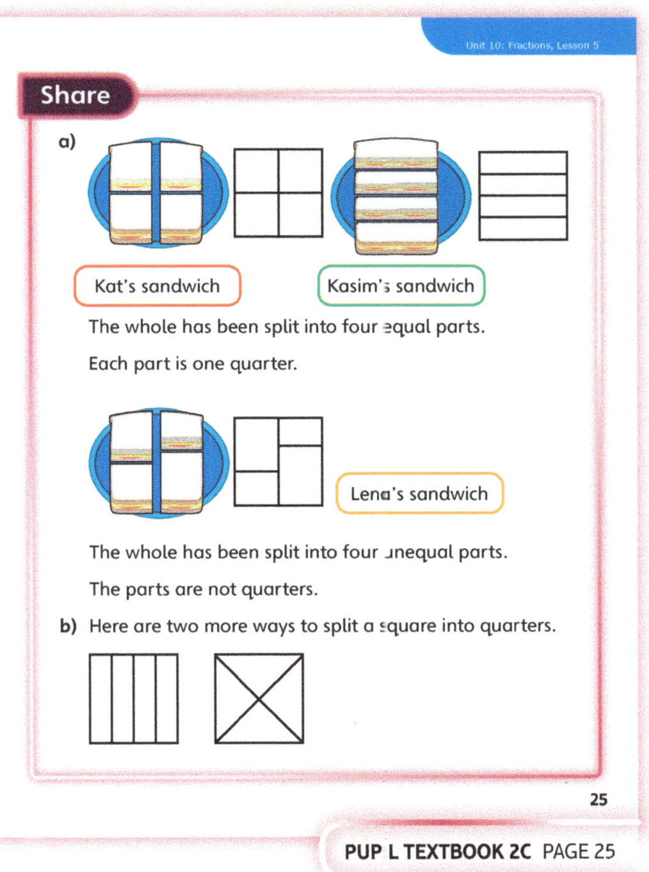

PUPIL TEXTBOOK 2C PAGE 25

Think together

WAYS OF WORKING Whole class teacher led (I do, We do, You do)

ASK

- Question **1**: *What is the numerator? What is the denominator? How do you know?*
- Question **3**: *Do all of these strips show quarters? How do you know?*

IN FOCUS Questions **1** and **2** show different shapes that have each been split into 4 parts in different ways. Children should recognise quarters of different shapes by seeing them as 4 equal parts rather than being distracted by the shape of the whole or the parts. Make the link between a quarter as 1 of 4 equal parts and the fractional number $\frac{1}{4}$.

Question **3** has four fraction strips that each show quarters. However, each strip is a different length, so the quarters of each strip are different sizes. Use this to explore the misconception that all shapes representing quarters must look the same.

STRENGTHEN When answering question **2**, children should use rulers to accurately copy the shapes and confirm that all 4 parts of each shape are equal before they shade 1 part in. Reinforce the fact that if the 4 parts are unequal, then any one of those parts does not accurately represent 1 quarter.

DEEPEN Question **3** asks children to compare fraction strips of different lengths. Ask children to make their own copies of these fraction strips and use these to play a matching game. They cut out all the quarters and challenge a partner to identify all the equal parts in order to piece the four fraction strips back together.

ASSESSMENT CHECKPOINT In question **1**, do children confidently write and describe $\frac{1}{4}$ as a fractional number? Can they make the link between the visual representation and the written number? In question **3**, do children understand that as all 4 parts of each strip are equal, they all accurately represent quarters?

ANSWERS

Question **1**: Check children can write $\frac{1}{4}$ with confidence. The number below the line (denominator) shows how many equal parts the whole is cut into (4). The number above the line (numerator) shows how many of those equal parts have been shaded (1).

Question **2** a): Children should shade 1 out of the 4 rectangles.

Question **2** b): Children should shade 1 out of the 4 rectangles.

Question **3**: Each strip has been cut into 4 equal parts and each of the 4 parts is $\frac{1}{4}$. The strips are different lengths, but if each strip is treated as 1 whole, then they have all been cut into 4 equal parts, or quarters.

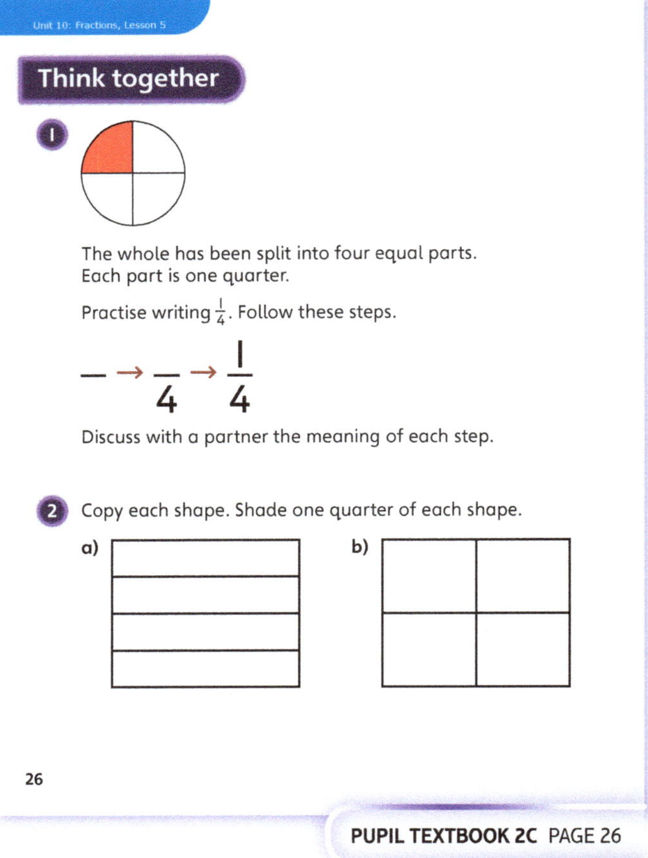

PUPIL TEXTBOOK 2C PAGE 26

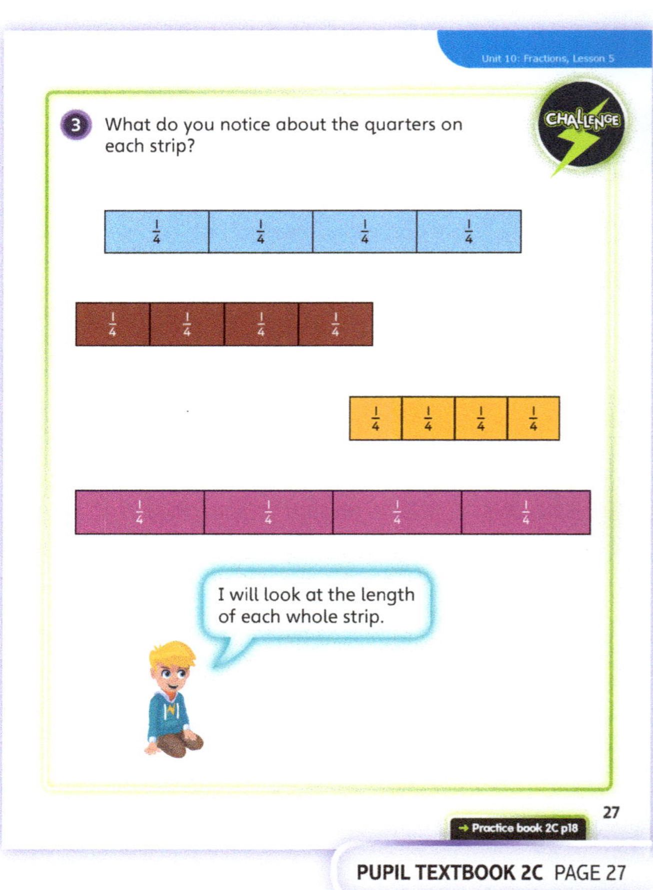

PUPIL TEXTBOOK 2C PAGE 27

Practice

WAYS OF WORKING Independent thinking

IN FOCUS Question **2** asks children to shade $\frac{1}{4}$ of each shape, in a variety of orientations. Question **4** requires children to say which of the given shapes show $\frac{1}{4}$ and which do not. Where children identify the shapes that do not show $\frac{1}{4}$, are they able to state what fraction some of these shapes do show?

STRENGTHEN Demonstrate how to split one of the blank diagonally orientated squares in question **6** into triangles by drawing a horizontal line and vertical line. Give children the opportunity to explore and discuss this.

DEEPEN To challenge children who are confident answering question **3**, ask them to trace the shape and to find as many ways as they can to split it into quarters. Ask: *Have you found all the possible ways? How do you know? Do all the quarters in one shape have to look exactly the same?*

ASSESSMENT CHECKPOINT Are children confident that each part of each shape in question **2** shows $\frac{1}{4}$ even though the quarters look different? When splitting the strips into quarters in question **5**, do children accurately draw four equal parts? Do they split each rectangle in the same way or do they explore different possibilities?

ANSWERS Answers for the **Practice** part of the lesson can be found in the *Power Maths* online subscription.

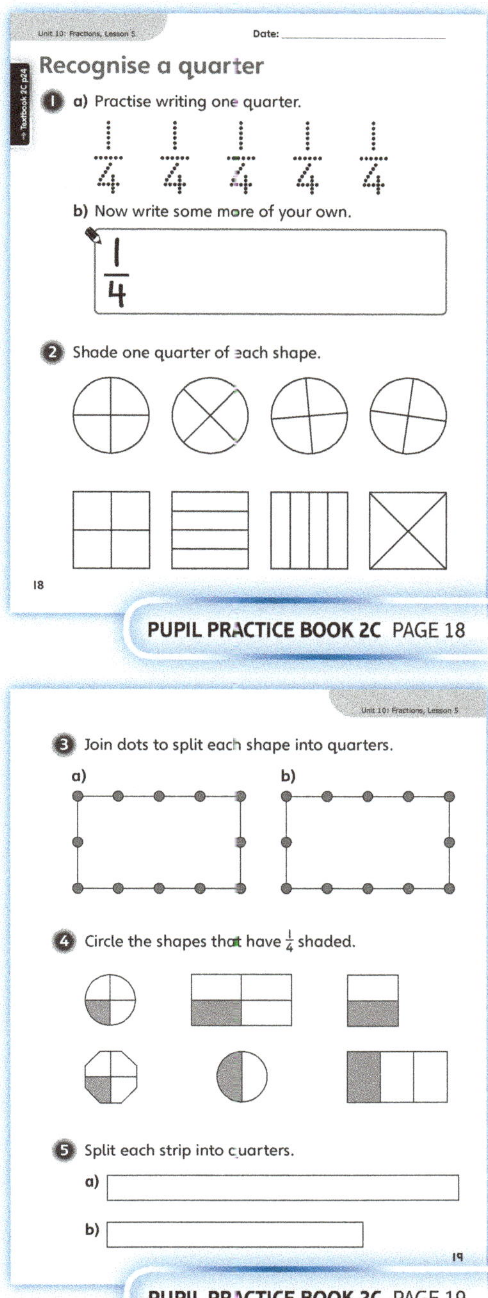

PUPIL PRACTICE BOOK 2C PAGE 18

PUPIL PRACTICE BOOK 2C PAGE 19

Reflect

WAYS OF WORKING Independent thinking

IN FOCUS This part of the lesson gives children the opportunity to demonstrate their understanding of recognising quarters by drawing their own visual representation.

ASSESSMENT CHECKPOINT Do children choose an appropriate model for their drawing and are they able to accurately split their shape into four equal parts?

ANSWERS Answers for the **Reflect** part of the lesson can be found in the *Power Maths* online subscription.

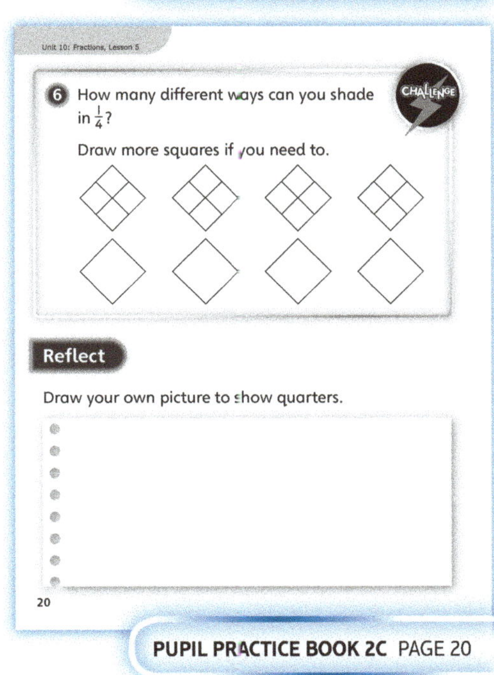

PUPIL PRACTICE BOOK 2C PAGE 20

After the lesson 

- Were children able to relate their understanding of the role of 1 and 2 in $\frac{1}{2}$ to the role of 1 and 4 in $\frac{1}{4}$?
- Did children know that, when finding $\frac{1}{4}$ of a shape, they could fold it in half and then fold in half again?

Find a quarter

Learning focus

In this lesson, children will find one quarter of different amounts by sharing them into four equal groups.

Before you teach 

- Do children understand that $\frac{1}{4}$ means one of four equal parts?
- Are children confident in sharing a number of objects into two groups to find $\frac{1}{2}$?
- How can you encourage children to say 'a quarter' rather than 'a fourth'?

NATIONAL CURRICULUM LINKS

Year 1 Number – fractions

Recognise, find and name a quarter as one of four equal parts of an object, shape or quantity.

Year 2 Number – fractions

Recognise, find, name and write fractions $\frac{1}{3}$, $\frac{1}{4}$, $\frac{2}{4}$ and $\frac{3}{4}$ of a length, shape, set of objects or quantity.

ASSESSING MASTERY

Children can use sharing to split a number of objects into four equal groups and can recognise when a number cannot be divided into four equal groups and how much greater it needs to be in order to be divided into four. Children can work out the value of a whole based on knowledge of the value of one quarter.

COMMON MISCONCEPTIONS

So far, children have mainly seen quarters of objects that have been split into four parts and may associate $\frac{1}{4}$ as being one object split into four parts. They may find it confusing to be presented with multiple objects that either need to be split into quarters or are one quarter of a whole. Relate the number of objects to being one group of four groups. Ask:
- *How many groups do you share these objects into in order to split them into quarters?*

STRENGTHENING UNDERSTANDING

Relate sharing between four to sharing between two and ask: *What are the similarities between these?* Do children see that dividing by four could be achieved by dividing into two groups, as when finding one half, and then further dividing each of those groups into two more groups? Model this by sharing out different amounts of counters and reinforcing the link between $\frac{1}{2}$ and $\frac{1}{4}$.

GOING DEEPER

Some children might spot that quartering certain numbers will give a whole number and one additional fraction. For example, when sharing out 13 multilink cubes, there will be one leftover cube. Children may say that the leftover cube can be split into quarters to share out among the other four groups. Model this using 2D squares to show the leftover cube being split into four equal parts.

KEY LANGUAGE

In lesson: $\frac{1}{4}$, quarters, share/sharing, split, equal groups, how many, same, equally, number sentence

Other language to be used by the teacher: divide/divided, whole, equal part, four parts

RESOURCES

Mandatory: counters, sorting hoops, multilink cubes

Optional: groups of multiple objects, such as pencils, sweets or erasers

 In the eTextbook of this lesson, you will find interactive links to a selection of teaching tools.

Quick recap 

Challenge children to draw a picture that accurately shows a whole split into quarters.

Discover

WAYS OF WORKING Pair work

ASK

- Question **1** a): *If the counters are to be shared fairly, does each child have to have the same number of counters?*
- Question **1** b): *What fraction of the whole will each child receive?*

IN FOCUS The **Discover** image models a scenario where children are required to find a quarter of a given number of objects. Establish how many counters are on the table in the picture. Can children predict how many counters each child will get when they are shared out?

PRACTICAL TIPS You may want to ask four children to come up to the front of the classroom to represent the children in the picture and share out 12 counters to model the question.

ANSWERS

Question **1** a): The counters can be split into 4 equal groups.

Question **1** b): $\frac{1}{4}$ of the 12 counters is 3 counters. Each child will get 3 counters.

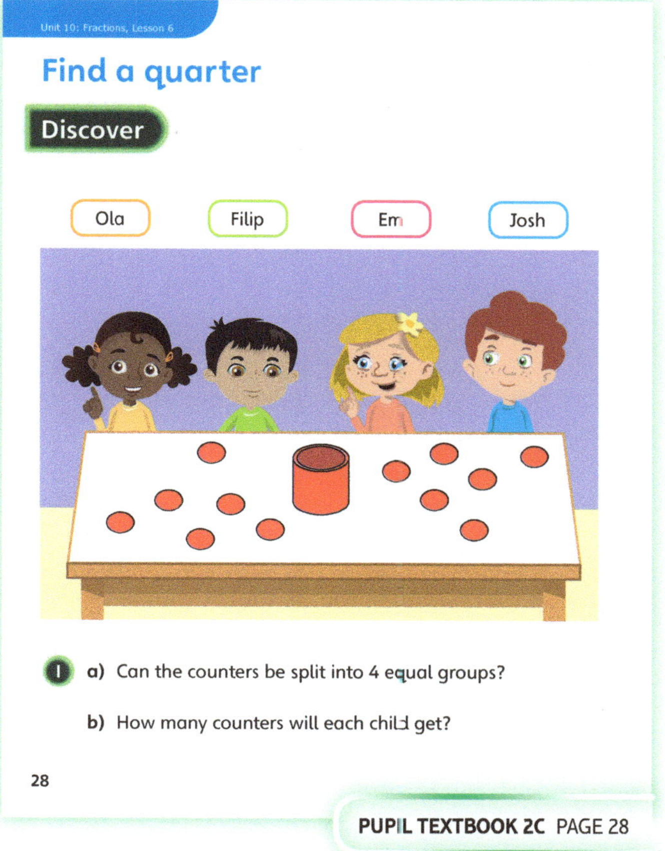

PUPIL TEXTBOOK 2C PAGE 28

Share

WAYS OF WORKING Whole class teacher led

ASK

- Question **1** a): *How many equal groups have the counters been split into?*
- Question **1** a): *Is splitting the counters into four equal groups the same as splitting them into quarters?*
- Question **1** b): *How would you write the fraction of the whole number of counters that each child has been given?*

IN FOCUS Hand out the counters at random and ask children whether they think this is fair or unfair. Discuss whether there is a fair way to hand them out. Then refer to what Astrid says and model handing out the counters to the children one by one, without missing out any of the children, until there are no counters left. Keep referring to the fact that each child has the same number of counters before you give out any more counters.

PUPIL TEXTBOOK 2C PAGE 29

Think together

 Whole class teacher led (I do, We do, You do)

- Question **1**: *How many counters are there in total? Is this number the numerator or the denominator?*
- Question **1**: *How many children are you sharing between? Is this number the numerator or the denominator?*

 Questions **1** and **2** require children to share objects in order to find $\frac{1}{4}$ of different numbers. The groups have been represented by different objects.

In question **3** a), children will find that there is 1 left over when trying to share 13 cubes between 4 groups. Encourage children not to put this leftover cube into one of the groups but to leave it outside the groups.

 For each question, ask children to count out the number of objects that they need using counters or cubes. Ask children to draw 4 circles to represent the number of groups into which they are going to split the cubes. This will model children's working.

 Question **3** b) asks how many more cubes Harry needs to be able to split them into quarters. Ash wonders if there is more than one answer, so refer to this and ask children to find as many different possibilities as they can. They will discover that they need to add 2 cubes, 6 cubes, 10 cubes, and so on to be able to divide by 4. Ask: *Can you see a pattern in those numbers? Why might that be?*

 In question **3** a), can children explain why Lucy is wrong? Do children understand that the groups are not equal? Lucy has got 4 cubes in one group and 3 in the other groups. If she split the first group into a group of 3 and a single cube, do children understand that the cube that is left over cannot be shared equally between the other groups? Do children make that distinction?

Question **1**: There are 20 counters and 4 children. $\frac{1}{4}$ of 20 is 5. Each child will get 5 counters.

Question **2**: There are 16 pencils and 4 pots. $\frac{1}{4}$ of 16 is 4. Each pot will have 4 pencils.

Question **3** a): Lucy is wrong, as 3 groups have 3 counters and 1 group has 4 counters, so they are not shared equally. 13 cannot be split equally into quarters.

Question **3** b): If Harry had 2 more cubes (so 12 cubes in total), he could split them into quarters with 3 in each group, as $\frac{1}{4}$ of 12 = 3. He could also split 16 cubes into quarters with 4 in each group, or 20 cubes into quarters with 5 in each group.

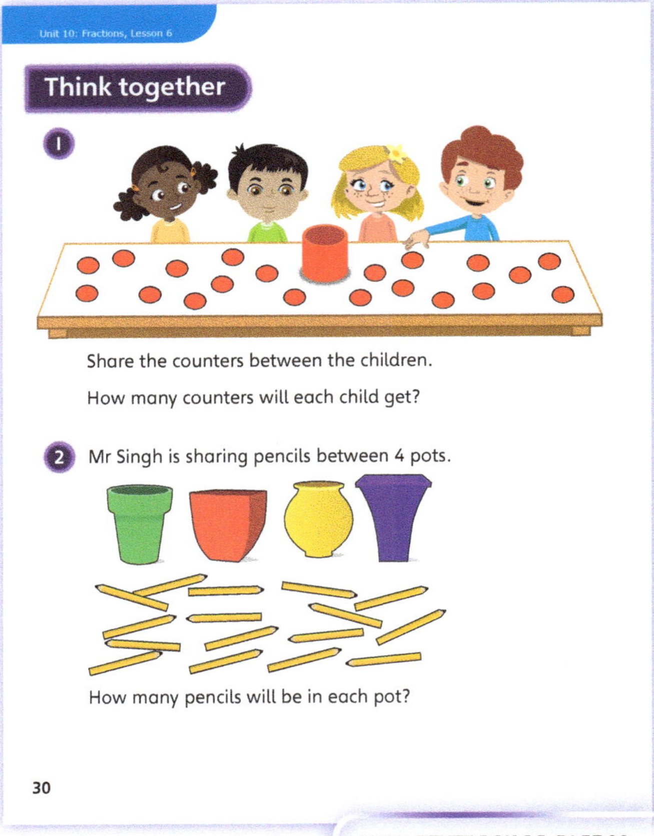

PUPIL TEXTBOOK 2C PAGE 30

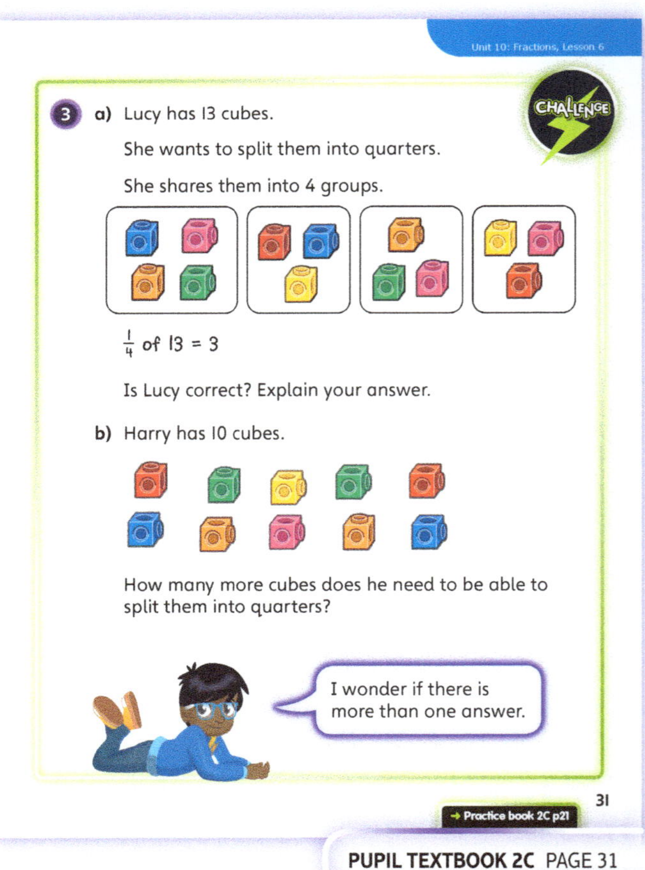

PUPIL TEXTBOOK 2C PAGE 31

Practice

WAYS OF WORKING Independent thinking

IN FOCUS Questions ❶, ❷ and ❸ ask children to share out even numbers. Completing the sentence scaffold to identify $\frac{1}{4}$ of 24 in question ❹, on the other hand, shows objects that have already been split into 4 equal groups. Children are asked to work out the original whole and complete the whole sentence scaffold.

STRENGTHEN Question ❺ asks children to work out $\frac{1}{4}$ of 40 using a picture of 40 stars. Provide children with a 100 square that has been cut off at 40. Children could then fold the cut-off 100 square into 4 equal rows and see that each row ends in a multiple of 10.

DEEPEN Give children different numbers as $\frac{1}{4}$ without providing the 4 parts that represent the whole. Ask children to draw the 4 parts for themselves and put the provided $\frac{1}{4}$ in one of them. This may help children to spot an efficient method of working out the whole, such as doubling and doubling again or multiplying by 4.

ASSESSMENT CHECKPOINT Are children able to work out the whole from $\frac{1}{4}$ in question ❹? Do they count each individual sweet or do they count the number of sweets in one bag and multiply this number by 4? Do children fill in the blank scaffold correctly? In question ❻, do children choose a sensible model to accurately show 28 objects shared into 4 groups of 7 objects.

ANSWERS Answers for the **Practice** part of the lesson can be found in the *Power Maths* online subscription.

Reflect

WAYS OF WORKING Pair work

IN FOCUS Children could complete the **Reflect** part of the lesson individually or with a partner. Children will soon come across numbers that cannot be divided into four groups. Ask children to predict which numbers might and might not be able to be shared equally between four groups.

ASSESSMENT CHECKPOINT Children might predict that all even numbers can be divided by 4. Can children modify this prediction based on attempts to divide 10 or 14 by 4?

ANSWERS Answers for the **Reflect** part of the lesson can be found in the *Power Maths* online subscription.

After the lesson ⏸

- Are children able to see a number both as a whole that can be divided into 4 and as a quarter of a number that needs to be multiplied by 4 to identify the whole?
- Are children secure in understanding that the 4 in $\frac{1}{4}$ means divided by 4?
- Do children call a quarter 'a fourth?'.

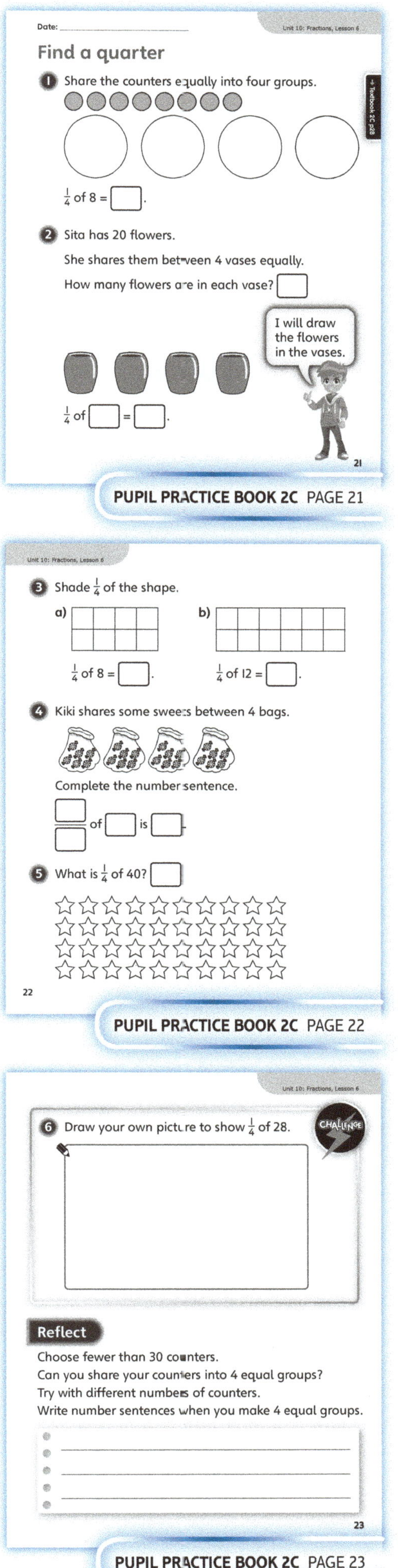

PUPIL PRACTICE BOOK 2C PAGE 21

PUPIL PRACTICE BOOK 2C PAGE 22

PUPIL PRACTICE BOOK 2C PAGE 23

Thirds

Learning focus

In this lesson, children will extend their understanding of fractions to consider thirds as three equal parts of one whole.

Before you teach 

- Can children explain the notation for $\frac{1}{2}$ and $\frac{1}{4}$ in terms of equal parts of one whole?
- Can children draw diagrams that show $\frac{1}{2}$ and $\frac{1}{4}$?

NATIONAL CURRICULUM LINKS

Year 2 Number – fractions

Recognise, find, name and write fractions $\frac{1}{3}$, $\frac{1}{4}$, $\frac{2}{4}$ and $\frac{3}{4}$ of a length, shape, set of objects or quantity.

ASSESSING MASTERY

Children can identify, explain and describe thirds as fractions of a whole that has been split into three equal parts, including drawing their own examples.

COMMON MISCONCEPTIONS

Children may have an understanding of halves and quarters through everyday experience and discussion, but may be less familiar with the concept of thirds. Ask:

• *How many equal parts make halves? How many equal parts make quarters? How many equal parts make thirds?*

STRENGTHENING UNDERSTANDING

Provide iconic visual representations of fractions such as flags and circles split into halves, thirds and quarters in different orientations. Ask: *What is the same? What is different?*

GOING DEEPER

Give children paper squares or rectangles and challenge them to try to show thirds by folding these into three equal parts. Can they identify an effective technique to make three equal parts and can they prove that it is accurate?

KEY LANGUAGE

In lesson: fraction, third, equal, whole, parts

Other language to be used by the teacher: split, half, quarter, share

STRUCTURES AND REPRESENTATIONS

2D shapes, fraction strips

RESOURCES

Optional: visual representations of fractions, paper squares or rectangles, paper strips to fold or cut, flag templates to shade in, counters or cubes

 In the eTextbook of this lesson, you will find interactive links to a selection of teaching tools.

Quick recap 

Provide children with 6 counters and ask them to list all the different ways that they could sort the counters into three equal parts. Ask: *What do you notice?*

Discover

WAYS OF WORKING Pair work

ASK

- Question **1** a): *What is the same and what is different about these flags?*
- Question **1** a): *What do you notice about the parts of each flag?*
- Question **1** b): *How many parts do these flags have?*
- Question **1** b): *Are the parts on these flags equal?*

IN FOCUS Children use the context of flags to revisit the concept of fractions. They identify different fractions of a shape based on the number of equal parts of the whole. Children first consider halves and quarters, which were encountered in previous lessons, before moving on to look at different examples of thirds for the first time. Children should notice that each flag is divided into equal parts.

PRACTICAL TIPS Provide children with simple black and white templates of the flags from the **Discover** scenario. Children copy the picture to shade in the stripes on each flag, counting the number of parts in each flag as they shade.

ANSWERS

Question **1** a): The flag of Monaco is split into 2 equal parts. Each part is one half ($\frac{1}{2}$).
The flag of Mauritius is split into 4 equal parts. Each part is one quarter ($\frac{1}{4}$).

Question **1** b): Both flags are split into 3 equal parts. Each part is one third ($\frac{1}{3}$).

Share

WAYS OF WORKING Whole class teacher led

ASK

- Question **1** a): [Monaco] *Why is there a 2 in this fraction?* [Mauritius] *Why is there a 4 in this fraction?*
- Question **1** b): *Why is there a 3 in this fraction?*
- Questions **1** a) and b): *What does the line mean in the written fraction?*

IN FOCUS The flag images are used to help children to understand and record the fraction notation for thirds. This is first reinforced with halves and quarters that children are already familiar with, before thirds are introduced. Reinforce the link between the number of equal parts that the whole has been split into, and the denominator of the fractional number. Informally, children can think of the line in the fraction meaning 'out of'. For example, $\frac{1}{2}$ could be said as '1 out of 2 equal parts'.

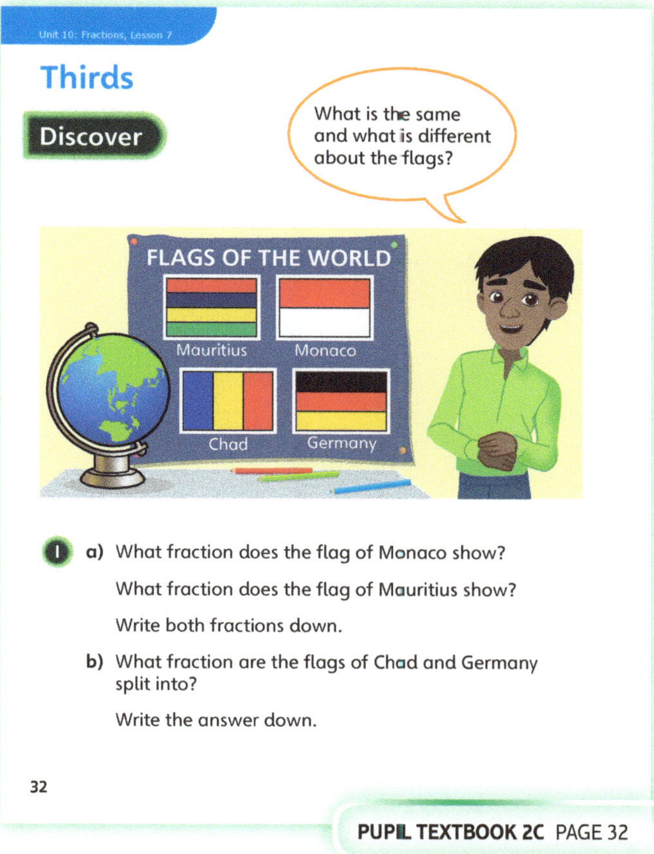

PUPIL TEXTBOOK 2C PAGE 32

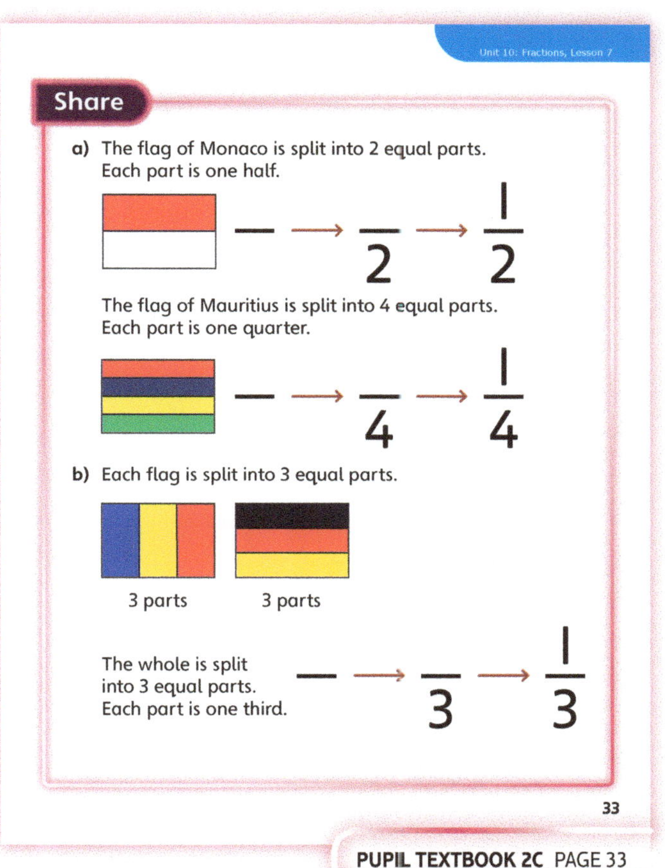

PUPIL TEXTBOOK 2C PAGE 33

Think together

WAYS OF WORKING Whole class teacher led (I do, We do, You do)

ASK

- Question ❶: *How many equal parts is each flag split into?*
- Question ❷: *How many parts does each circle have? Which circles show equal parts?*
- Question ❸: *Can you sort the stars and strawberries into three equal parts?*

IN FOCUS Question ❶ continues to use the context of flags and asks children to identify the number of equal parts in a variety of different flag patterns.

In question ❷, children identify the number of parts that a selection of circles have been split into and judge if the parts of each circle are equal. Agree that a shape only shows thirds if it is split into 3 parts that are all equal.

In question ❸, children move on from finding $\frac{1}{3}$ of a shape to finding $\frac{1}{3}$ of a number. Provide counters or cubes so that children can explore how to arrange the given numbers of objects into 3 equal groups or parts. They then identify how many objects are in 1 part. Children should begin by using the concept of sharing, rather than grouping, because they know the number of equal groups they have to make in each case is 3. They share the objects by allocating one to each group and repeating this until all objects are allocated.

STRENGTHEN Provide several fraction strips, some of which have been cut into thirds and some which have been cut into non-thirds. Ask children to sort the parts in order to check which are equal and to find which of the fraction strips show thirds.

DEEPEN Ask children to practise drawing diagrams of different polygons that are split into three equal parts. Ask: *Do you think any 2D shape can be split into 3 equal parts?*

In question ❸, if children are confident with sharing the stars and strawberries then encourage them to try to make 3 equal parts by grouping. The stars are set out in an array, so they might see that there are 3 equal groups of 2 stars. Can they draw the strawberries in an array to show that there are 3 equal groups of 4 strawberries?

ASSESSMENT CHECKPOINT Question ❷ assesses whether children are able to identify thirds of a shape based on the number of parts that the shape has been split into and whether or not all the parts are equal.

ANSWERS

Question ❶: Gaby and Milo's flags are split into thirds.

Question ❷: Circles A and D have $\frac{1}{3}$ shaded.

Question ❸ a): $\frac{1}{3}$ of 6 is 2.

Question ❸ b): $\frac{1}{3}$ of 12 strawberries is 4 strawberries.

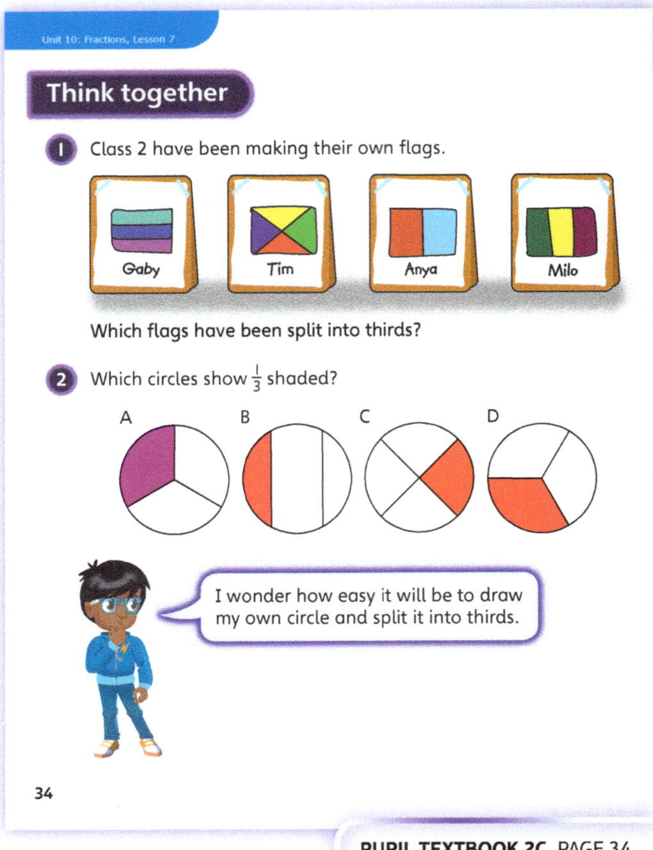

PUPIL TEXTBOOK 2C PAGE 34

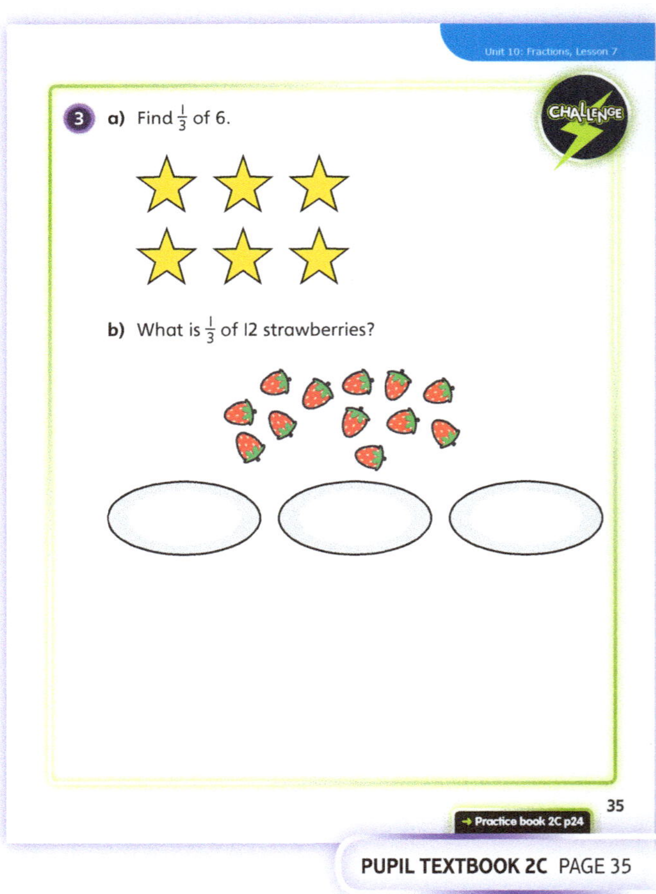

PUPIL TEXTBOOK 2C PAGE 35

Practice

WAYS OF WORKING Independent thinking

IN FOCUS Question **1** gives children the opportunity to practise writing the correct fraction notation for one third. In question **2**, children will need to understand one third as 1 of 3 equal parts in order to correctly shade one third of shapes in a variety of orientations.

Question **3** requires children to use sharing in order to arrange a number of objects in 3 equal groups representing thirds. In question **4**, children draw on lines in order to divide fraction strips into thirds. They should notice that the thirds will be a different size on each strip because each of the strips is a different length.

In question **5**, children work out one third of a given number, and shade in a corresponding grid to represent this.

STRENGTHEN Ask children to practise drawing their own diagrams like those in question **2**, with 2D shapes split into thirds in different orientations. Agree that when showing one thirds, the orientation of the shape does not matter, as long as it is split into exactly 3 equal parts with 1 part shaded.

DEEPEN Challenge children to explore different methods for sharing an amount between 3 equal groups. For example, sharing 2 at a time: 'Two for you, two for you, two for you.'

ASSESSMENT CHECKPOINT Use question **4** to assess whether children can make reasonable judgements about how to split different wholes into thirds.

ANSWERS Answers for the **Practice** part of the lesson can be found in the *Power Maths* online subscription.

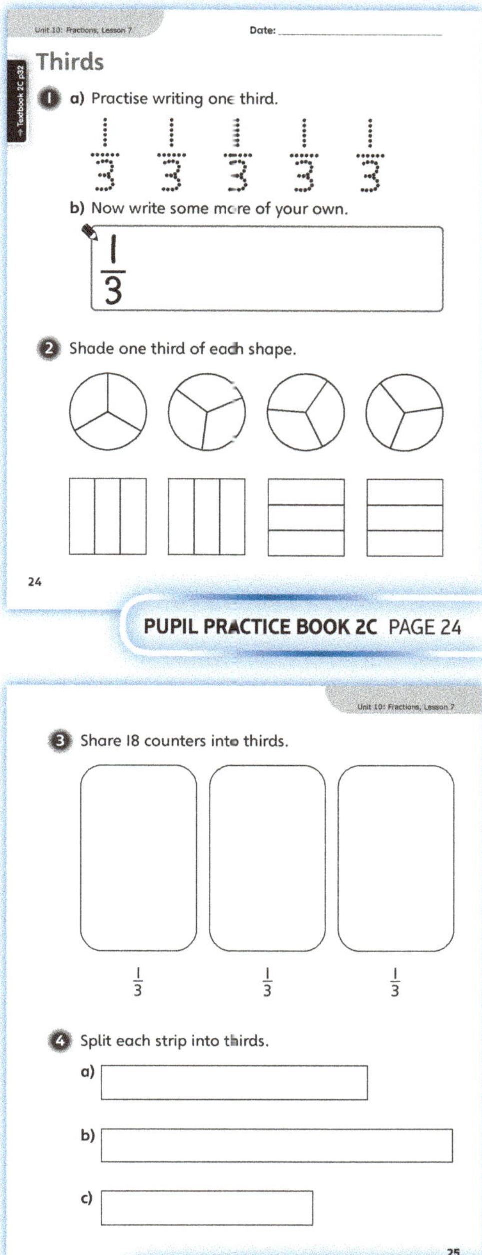

PUPIL PRACTICE BOOK 2C PAGE 24

PUPIL PRACTICE BOOK 2C PAGE 25

Reflect

WAYS OF WORKING Whole class

IN FOCUS The **Reflect** part of the lesson prompts children to apply what they have learnt in order to consider and represent how simple fractions (halves, quarters and thirds) are similar and how they are different.

ASSESSMENT CHECKPOINT Assess whether children can accurately describe and illustrate the difference between halves, quarters and thirds.

ANSWERS Answers for the **Reflect** part of the lesson can be found in the *Power Maths* online subscription.

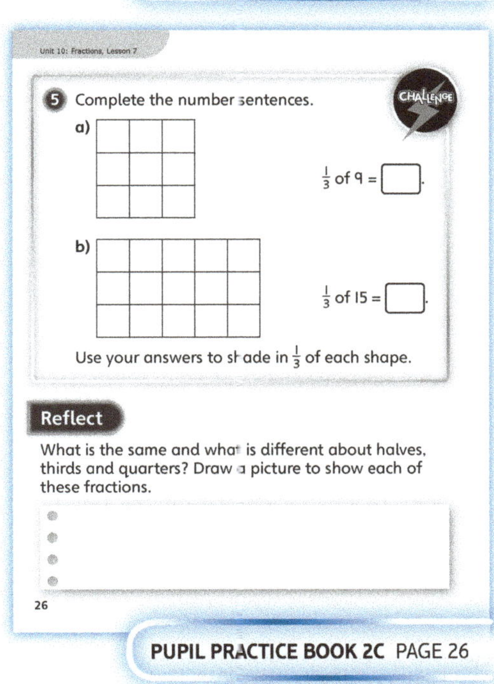

PUPIL PRACTICE BOOK 2C PAGE 26

After the lesson

- Were children able to identify which diagrams were correctly split into thirds and which were not?
- Did children find effective techniques for splitting shapes or numbers into three equal parts or groups?

Find the whole

Learning focus

In this lesson, children will be introduced to non-unit fractions (fractions in which the numerator is not 1).

Before you teach 

- Can children find a fraction of an amount by sharing into equal parts?
- Can children explain how many parts make one whole for halves, quarters and thirds?

NATIONAL CURRICULUM LINKS

Year 2 Number – fractions

Recognise, find, name and write fractions $\frac{1}{3}$, $\frac{1}{4}$, $\frac{2}{4}$ and $\frac{3}{4}$ of a length, shape, set of objects or quantity.

ASSESSING MASTERY

Children can generate the whole number when they are given one of the fraction parts. For example, ask: *If three cubes make one quarter of the whole, how many cubes make the whole?*

COMMON MISCONCEPTIONS

Children may think that you are always told the whole and have to work out the part. Ask:
- *Do you know the whole or the part?*
- *How many parts make the whole for this fraction?*

STRENGTHENING UNDERSTANDING

Provide sorting equipment such as counters or cubes that children can use to create the one known part and then to make each other part, in order to find out how many are required to make the whole.

GOING DEEPER

Challenge children to solve missing number problems that will require them to use a part in order to find the whole.
For example, $\frac{1}{4}$ of ☐ is 10.

KEY LANGUAGE

In lesson: fraction, whole, part

Other language to be used by the teacher: half, quarter, third, sort

STRUCTURES AND REPRESENTATIONS

Fraction representations with 2D shapes, ten frame

RESOURCES

Mandatory: sorting equipment such as cubes or counters

 In the eTextbook of this lesson, you will find interactive links to a selection of teaching tools.

Quick recap 

Ask each child to draw a diagram that shows a whole split into halves, thirds or quarters. Then ask them to compare and discuss their diagrams together.

Discover

WAYS OF WORKING Pair work

ASK

- Question **1** a): *What do you notice about the circle?*
- Question **1** a): *How have the cubes been arranged?*
- Question **1** b): *How many quarters make one whole?*
- Question **1** b): *How many cubes did Meg put in each quarter?*

IN FOCUS Look together at the picture and use this to consider the relationship between the equal parts and the whole. Discuss how the circle has been split to represent quarters and how the cubes have been arranged to show how many in each part and how many in the whole.

PRACTICAL TIPS Play the sorting game from the **Discover** scenario together, asking children to sort a given number of cubes or counters into a circle that has been split into two, three or four equal parts. Ask: *How many in each part? How many make up the whole?*

ANSWERS

Question **1** a): There are 2 cubes in each quarter.
8 cubes make up the whole.

Question **1** b): There are 5 cubes in each quarter.
20 cubes make up the whole.

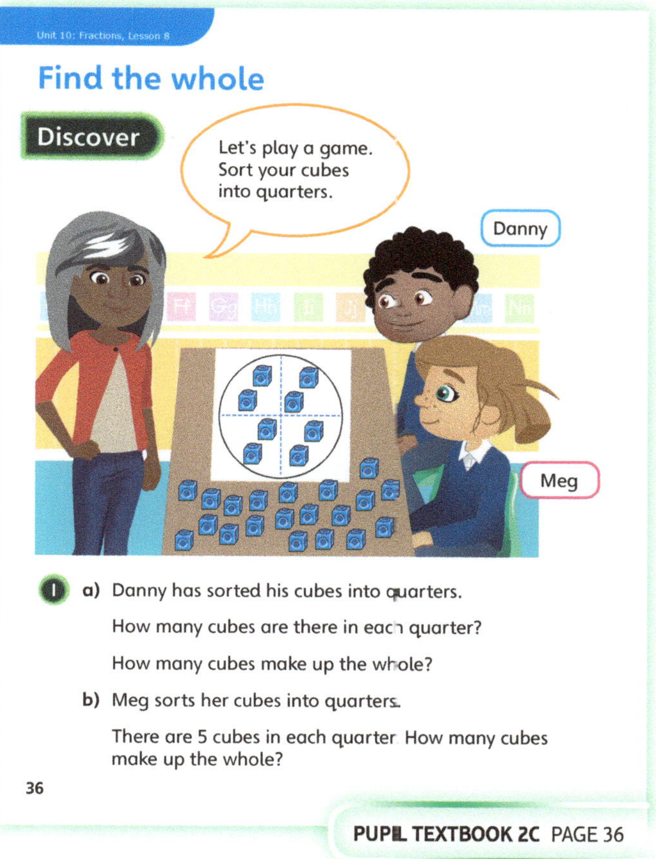

PUPIL TEXTBOOK 2C PAGE 36

Share

WAYS OF WORKING Whole class teacher led

ASK

- Questions **1** a): *How many quarters make a whole?*
- Questions **1** a): *How many cubes are in each quarter?*
- Questions **1** b): *How many cubes are in each quarter now?*
- Questions **1** b): *Why did Flo count in 5s instead of 1s?*

IN FOCUS Look together at the two different circle images in order to explore the part-whole relationship. Compare the number of cubes in one quarter each time and discuss how children will count these to find the total. Count together in 2s up to 8 in question **1** a) and then ask: *Will this work again in part b)?* Agree that it will not because there are more than two cubes in each part this time. Refer children to Flo's comment and count up together in 5s, pointing to a different quarter each time. Stop when you reach 20 and agree that you have counted all four quarters that make up the whole.

PUPIL TEXTBOOK 2C PAGE 37

Think together

WAYS OF WORKING Whole class teacher led (I do, We do, You do)

ASK

- Question ❶: *How many halves make one whole?*
- Question ❷: *Will you count in 1s to find the total? How else could you count?*
- Question ❸: *Can you draw each number sentence using a part-whole model? Is the missing number the whole or the part?*

IN FOCUS In question ❶, children work out how to make a whole when given one half. This is modelled with cubes showing a visual representation of the half. In question ❷, children then make a whole when given one third. In this case, one third is 10, and this is shown with counters on a ten frame. Encourage children to count up in 10s to find the total.

Question ❸ requires children to demonstrate an understanding of fraction number sentences. They can draw their own diagrams to help with this if needed. Children should notice that the missing number changes in each pair of number sentences, so that they are first using the whole to find a part and then using a part to find the whole.

STRENGTHEN Provide cubes or counters for children to sort into equal parts to represent the numbers in each question. In question ❸, children may find it useful to represent each number sentence using a part-whole model. They need to identify whether the given number (10, 6 or 8) is a whole, or a part. If, for example, 10 is a part, they will need another part with 10 in it too. If, instead, 10 is the whole, they will need to determine that 5 and 5 are the 2 equal parts that sum to 10. When they progress to finding $\frac{1}{4}$ of 8, they will need to use a part-whole model with 4 equal parts.

DEEPEN Challenge children to write more pairs of missing number sentences, like those in question ❸, for a partner to solve.

ASSESSMENT CHECKPOINT Question ❷ assesses whether children can count in steps of 10 to find the total and solve the problem efficiently.

ANSWERS

Question ❶: One half is 5 cubes, so the whole is 10 cubes.

Question ❷: One third is 10 counters, so the whole is 30 counters.

Question ❸ a): $\frac{1}{2}$ of 10 = 5; $\frac{1}{2}$ of 20 = 10
These number sentences both find a half, but one finds a half of 10 and the other finds a half with 10 as the answer, so it is double 10 (20).

Question ❸ b): $\frac{1}{2}$ of 6 = 3; $\frac{1}{2}$ of 12 = 6
$\frac{1}{4}$ of 8 = 2; $\frac{1}{4}$ of 32 = 8

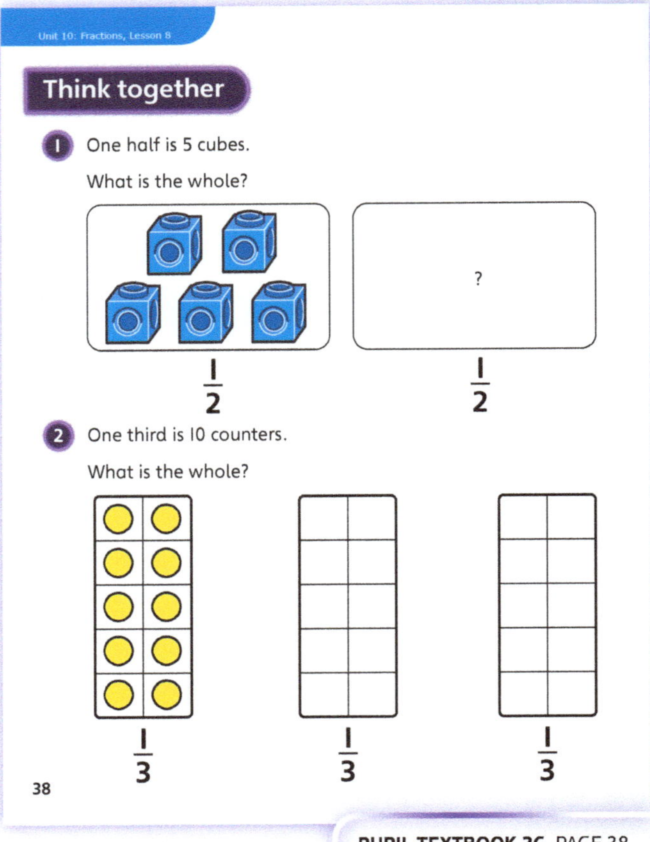

PUPIL TEXTBOOK 2C PAGE 38

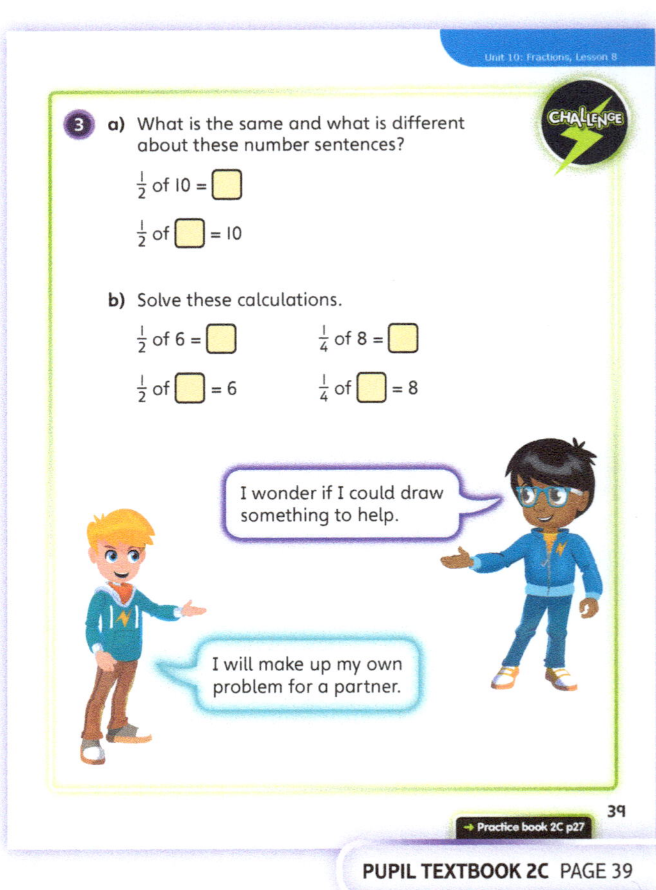

PUPIL TEXTBOOK 2C PAGE 39

Practice

WAYS OF WORKING Independent thinking

IN FOCUS Questions ① to ④ require children to work out the whole when given one half, one quarter or one third. Visual representations of one part are given. Children draw the other parts and then count up in order to find the whole. Sentence scaffolds are provided to support children with this.

Question ⑤ is presented as a word problem. Children should identify that they are given one half and need to use this to work out the whole. They can draw any diagram that they will find helpful for this.

Question ⑥ is more abstract and requires children to complete fraction number sentences with missing numbers. They first use the whole to find a part and then use the part to find the whole.

STRENGTHEN Provide sorting items such as cubes and counters that children can manipulate for each question to support their engagement with the concept of parts and whole.

DEEPEN Deepen children's understanding of the relationship between the parts and the whole by providing them with number sentence questions like those in question ⑥ where they use the known whole to find a given part. Challenge children to then write the corresponding number sentence where they will be using a part to find the whole.

ASSESSMENT CHECKPOINT Use question ① to assess whether children can understand the concept of generating one whole from a given part and can accurately complete a diagram and a number sentence to show this.

ANSWERS Answers for the **Practice** part of the lesson can be found in the *Power Maths* online subscription.

PUPIL PRACTICE BOOK 2C PAGE 27

PUPIL PRACTICE BOOK 2C PAGE 28

Reflect

WAYS OF WORKING Pair work

IN FOCUS The **Reflect** part of the lesson prompts children to think of their own part-whole problem. They represent it with a picture and then challenge a partner to find the whole from a given fractional part.

ASSESSMENT CHECKPOINT Assess whether children can understand the structure of a part-whole problem and can use this to generate their own fraction question.

ANSWERS Answers for the **Reflect** part of the lesson can be found in the *Power Maths* online subscription.

After the lesson ⏸

- Were children able to work from the known part to find the whole and vice versa?
- Did children draw accurate diagrams to represent the parts and the whole?

PUPIL PRACTICE BOOK 2C PAGE 29

Unit and non-unit fractions

Learning focus

In this lesson, children will be introduced to non-unit fractions (fractions in which the numerator is not 1).

Before you teach 

- Could children explain a numerator and a denominator when studying unit fractions?
- Were children confident at recognising and finding unit fractions $\frac{1}{2}$, $\frac{1}{4}$ and $\frac{1}{3}$ from previous lessons?

NATIONAL CURRICULUM LINKS

Year 2 Number – fractions

Write simple fractions, for example $\frac{1}{2}$ of 6 = 3, and recognise the equivalence of $\frac{2}{4}$ and $\frac{1}{2}$.

ASSESSING MASTERY

Children can recognise non-unit fractions and distinguish between these and unit fractions, and can explain the denominator and numerator of each fraction. Children can write the non-unit fractions they see in a picture and can draw their own non-unit fraction when they see it written numerically.

COMMON MISCONCEPTIONS

Children may mix up the meaning of the numerator and the denominator. For example, children may think that $\frac{3}{4}$ means having three parts and then get confused when they try to shade four parts. When seeing a number of parts greater than one shaded for the first time, children may also take that to mean that the number of equal parts has increased. Ask:
- *How many equal parts are there? Does that change if you shade more parts?*

STRENGTHENING UNDERSTANDING

Use physical resources to represent a whole divided into four equal parts and count up to a whole in quarters by moving each of the equal parts in turn while counting $\frac{1}{4}$, $\frac{2}{4}$, $\frac{3}{4}$ and so on. Ask children questions about what happens to the denominator and numerator each time you move a part.

GOING DEEPER

Once children have practised counting up in unit fractions, do they recognise that $\frac{2}{2}$ or $\frac{3}{3}$ or $\frac{4}{4}$ are each the same as one whole? Give children a scenario, such as a cake split into quarters that were all eaten, and ask them to write the fraction that has been eaten.

KEY LANGUAGE

In lesson: equal parts, numerator, denominator, equal parts, **unit fraction**

Other language to be used by the teacher: non-unit fraction

STRUCTURES AND REPRESENTATIONS

2D shapes, fraction strips

RESOURCES

Mandatory: 2D shapes made out of paper or card

Optional: yellow and blue card, counters, multilink cubes

In the eTextbook of this lesson, you will find interactive links to a selection of teaching tools.

Quick recap 

Discuss simple fractions as a class. Ask: *How many halves make one whole? How many thirds make one whole? How many quarters make one whole?*

Discover

WAYS OF WORKING Pair work

ASK

- Question **1** b): *How many equal parts is each kite split into? What is the denominator of each kite?*
- Question **1** b): *Why does the numerator of the fraction change for each kite?*

IN FOCUS This picture portrays non-unit fractions for the first time. In question **1** a), children know the number of equal parts on each kite, so they will write 4 as the denominator. Children may be able to count the number of yellow parts on each kite, but not know how to express this as a non-unit fraction. For example, they may not know how to represent the 3 in $\frac{3}{4}$. Try to avoid giving too much guidance and see what children do for a kite showing $\frac{3}{4}$ shaded yellow. Do they link the number of yellow parts to the numerator of the fraction?

PRACTICAL TIPS Arrange triangles of blue and yellow card to represent the three kites so that children can manipulate the parts as they count them.

ANSWERS

Question **1** a): The kites are split into 4 equal parts, or quarters.

Question **1** b): Molly's kite is $\frac{2}{4}$ plain yellow. Josh's kite is $\frac{3}{4}$ plain yellow. Ola's kite is $\frac{4}{4}$ plain yellow.

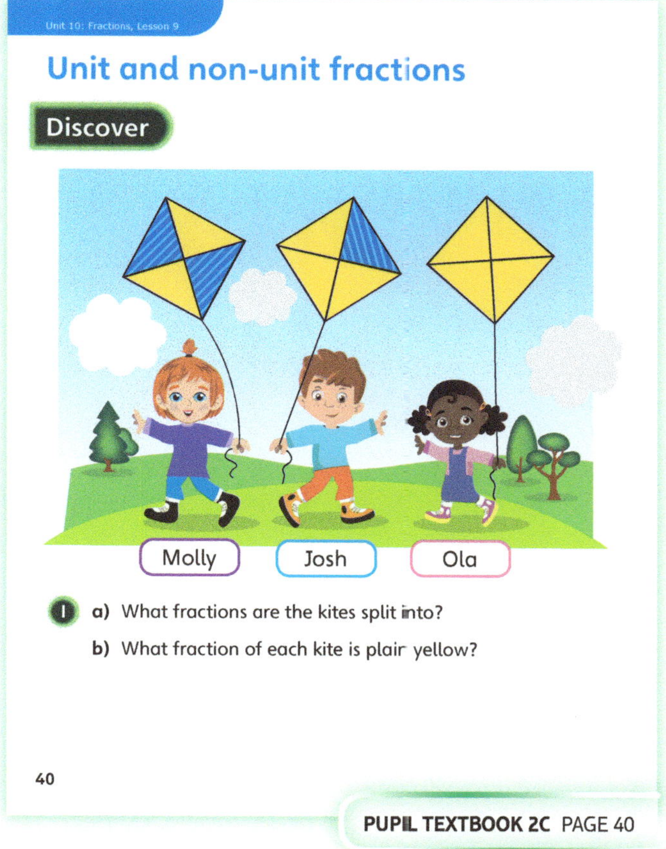

PUPIL TEXTBOOK 2C PAGE 40

Share

WAYS OF WORKING Whole class teacher led

ASK

- Question **1** a): *How many equal parts are there in each kite? What fraction of the whole is each part?*
- Question **1** b): *Why does the numerator increase by 1 each time? Is $\frac{4}{4}$ the same as one whole? How do you know?*

IN FOCUS Questions **1** a) and b) give a step-by-step explanation for the unit fraction and then each non-unit fraction and also recap denominators. This is the first time children have met non-unit fractions, so they will need help with the reading. Explain that $\frac{2}{4}$ is read as two quarters and $\frac{3}{4}$ is read as three quarters. Each section of the kite is labelled with its unit fraction name. Explain that the total number of parts has not changed but that the number of yellow parts has changed. Once you have explained and discussed each example, ask: *What do you notice about the numerator in non-unit fractions?*

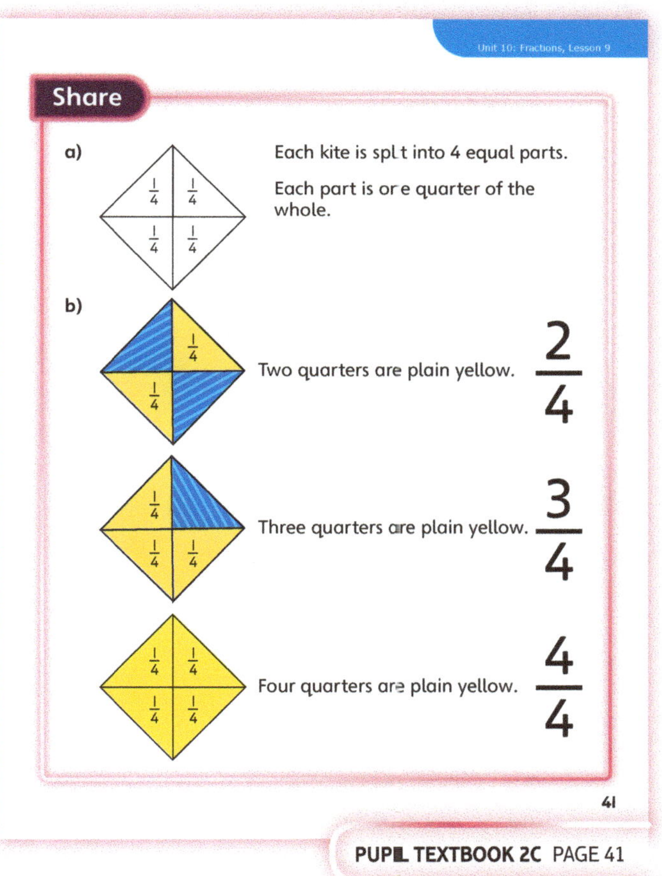

PUPIL TEXTBOOK 2C PAGE 41

Think together

WAYS OF WORKING Whole class teacher led (I do, We do, You do)

ASK

- Question **1**: *How many equal parts has the circle been split into? How many parts are shaded?*
- Question **2**: *What do you notice about the numerator and the denominator in your fraction?*

IN FOCUS Question **1** asks children to describe the structure of a non-unit fraction with thirds, requiring them to make the link between the visual representation and the fraction.

Question **2** requires children to distinguish between a unit fraction and a non-unit fraction. They should also notice that a non-unit fraction where the numerator and the denominator are the same represents one whole.

Question **3** gives children the opportunity to count up in unit fractions in order to generate non-unit fractions. They do this by shading unit fraction parts of fraction strips until they have shaded one whole.

STRENGTHEN Provide children with sentence scaffolds to structure their discussions in question **1**. For example:

- The circle is made up of ☐ equal parts.
- The denominator is ☐.
- ☐ of the parts are shaded.
- The numerator is ☐.
- The non-unit fraction is ☐.

DEEPEN Refer to what Ash says about choosing a different shape. Challenge children to show and shade two halves with as many different shapes as they can.

ASSESSMENT CHECKPOINT Do children use their knowledge about numerators from this and previous lessons to distinguish between unit and non-unit fractions? Are children able to apply previous knowledge to identify or show unit and non-unit fractions?

ANSWERS

Question **1**: The denominator is the number of equal parts the whole is split into and goes below the line, so 3. The numerator is the number of those parts that are shaded, and goes above the line, so 2.

Question **2**: Children should draw a picture, split it into halves and shade both parts. Children should write the fraction $\frac{2}{2}$. Children should notice that, in $\frac{2}{2}$, the numerator and denominator are the same, so the whole shape is split into 2 parts and both are shaded. This is the same as saying the whole is shaded, so $\frac{2}{2}$ = one whole.

Question **3** a): Children should first shade $\frac{1}{2}$ of the strip. They should then shade the other half, giving the shaded fraction of $\frac{2}{2}$.

Question **3** b): Children will first shade $\frac{1}{4}$, then $\frac{2}{4}$ (or one half), then $\frac{3}{4}$ and finally $\frac{4}{4}$, or the whole.

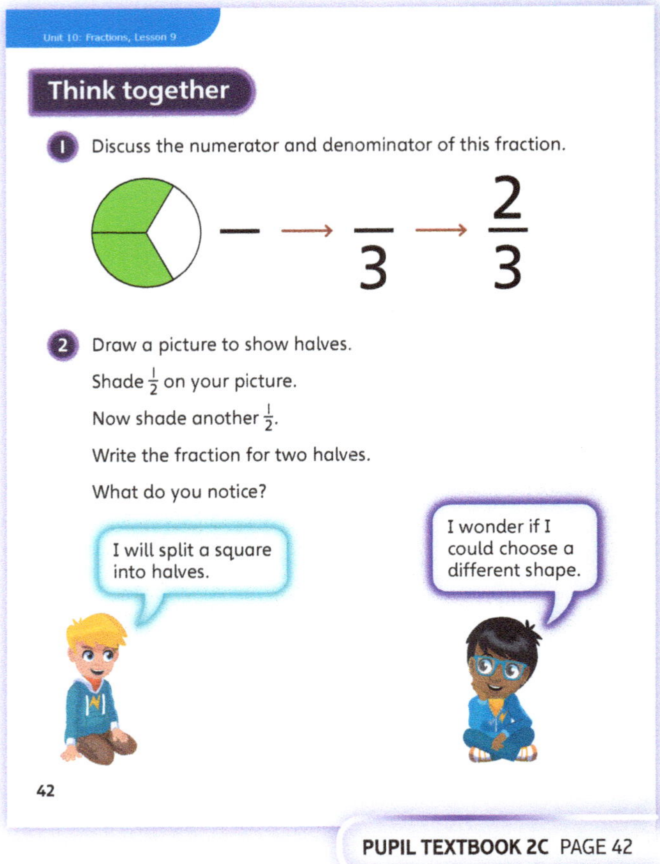

PUPIL TEXTBOOK 2C PAGE 42

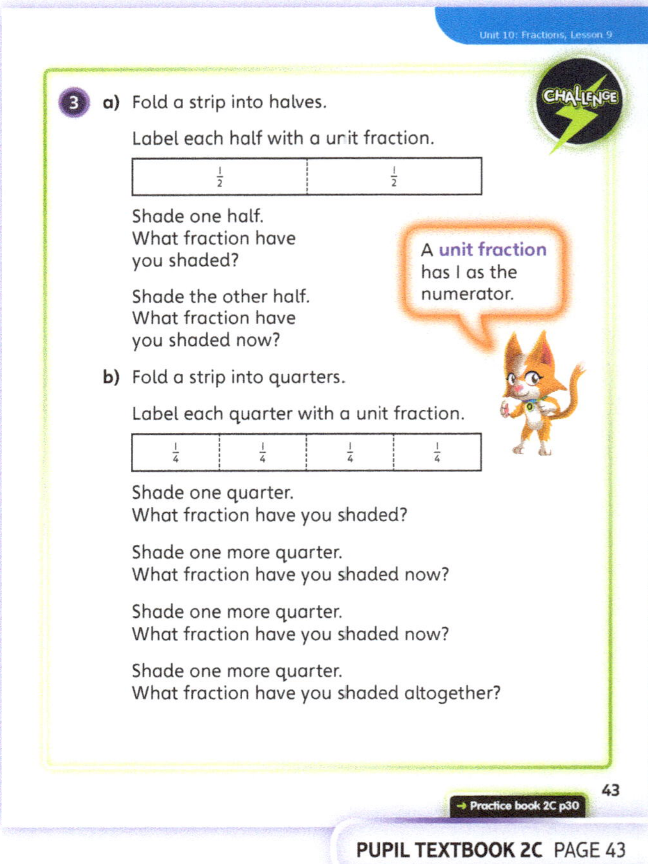

PUPIL TEXTBOOK 2C PAGE 43

Practice

WAYS OF WORKING Independent thinking

IN FOCUS Questions **1** and **2** ask children to identify and write the unit and the non-unit fractions that they see. Put this into real terms by referring to shaded and unshaded parts of each shape.

Questions **3** and **4** allow children to show an understanding of what non-unit fractions are by shading given unit and non-unit fractions.

Question **5** asks children to match up fractions, which requires them to write down what fraction each shape represents and to visualise what each written fraction looks like as a shape.

STRENGTHEN To help children draw pictures for question **4**, ask them to describe the fraction before they start to sketch it. Ask: *What shape will you draw? How many equal parts will you split your shape into? How many of these parts will you shade? Will there be any parts unshaded? Why?*

DEEPEN If children have successfully shaded the unit and non-unit fractions in question **3**, ask them whether they can work out what fraction is not shaded each time.

ASSESSMENT CHECKPOINT Use question **1** to assess whether children can interpret visual representations of unit and non-unit fractions and can use these to identify the numerator and the denominator each time.

ANSWERS Answers for the **Practice** part of the lesson can be found in the *Power Maths* online subscription.

Reflect

WAYS OF WORKING Independent thinking or pair work

IN FOCUS Children are asked to identify and draw unit fractions. No pictorial representation is given to help children find the unit fractions, requiring children to look closely at the numbers and apply what they have learnt to the written fraction. Encourage children to explain why they have circled the fractions that they have circled, using the words 'numerator', 'denominator' and 'equal parts'.

ASSESSMENT CHECKPOINT Can children instantly recognise unit fractions by looking at the numerator or do they still look at the denominator? Which fraction do children pick to draw? Do children draw a model or an image that they have seen previously or are they able to create one for themselves?

ANSWERS Answers for the **Reflect** part of the lesson can be found in the *Power Maths* online subscription.

After the lesson 

- Were children confident in the distinction between unit and non-unit fractions and why there is a difference in the numerators?
- Were children easily able to include thirds in their thinking in this lesson as well as halves and quarters.

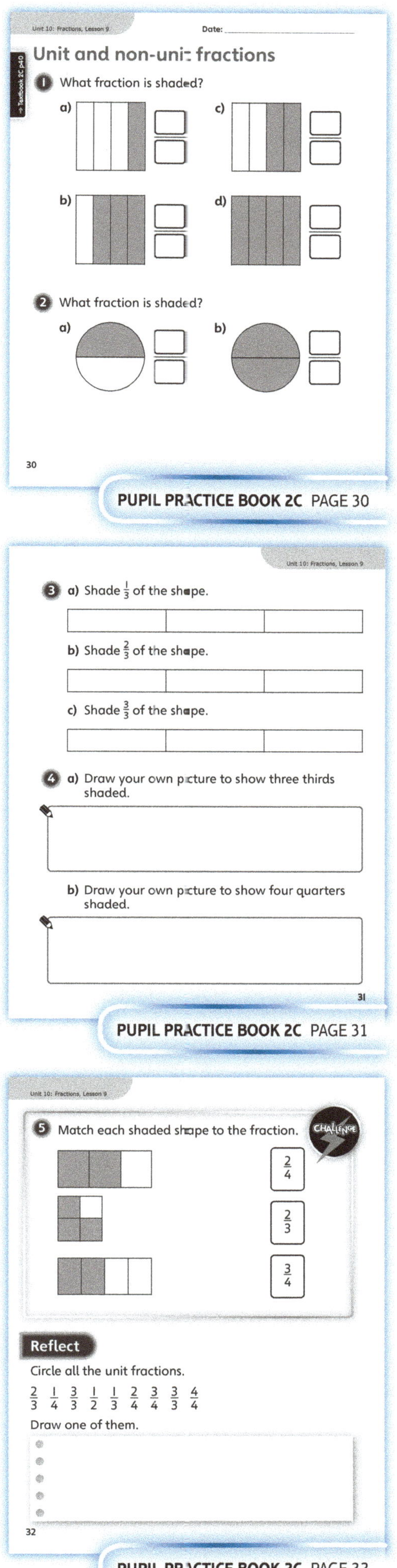

PUPIL PRACTICE BOOK 2C PAGE 30

PUPIL PRACTICE BOOK 2C PAGE 31

PUPIL PRACTICE BOOK 2C PAGE 32

Recognise the equivalence of a half and two quarters

Learning focus

In this lesson, children will learn that $\frac{1}{2}$ and $\frac{2}{4}$ are equivalent fractions. Children will prove this using physical resources and different numbers of objects.

Before you teach 

- Have children shown any awareness of equivalence in fractions before this lesson?
- Are children able to find half of a number? Do they understand that this means dividing the number by 2?

NATIONAL CURRICULUM LINKS

Year 2 Number – fractions

Write simple fractions, for example $\frac{1}{2}$ of 6 = 3, and recognise the equivalence of $\frac{2}{4}$ and $\frac{1}{2}$.

ASSESSING MASTERY

Children can recognise that $\frac{1}{2}$ and $\frac{2}{4}$ are equivalent in a range of contexts, such as shapes, pieces of paper and counters. Children can draw $\frac{1}{2}$ and $\frac{2}{4}$ using the same whole, can explain the similarities and differences between them and understand why the numerators and denominators are different while having the same value within each context.

COMMON MISCONCEPTIONS

Children may initially think that $\frac{2}{4}$ is more than $\frac{1}{2}$ because both numbers in $\frac{2}{4}$ are bigger. Break down each fraction and model them by folding one piece of paper in half and then folding one of the halves in two, showing that one half is the same as two quarters. This will show children that the only visual difference is that $\frac{2}{4}$ has an additional fold in the middle. Ask:
- *Does each side of paper take up more space or less space than the other? Do bigger numbers mean a bigger fraction of the whole?*

STRENGTHENING UNDERSTANDING

There are lots of practical contexts to help children see the equivalence of $\frac{1}{2}$ and $\frac{2}{4}$. For example, you could measure the height of a child in the class. Then, divide this height into 2 to find $\frac{1}{2}$ of the child's height. You could also divide the child's height into 4 and multiply by 2 to find $\frac{2}{4}$ of their height. Children will see that $\frac{1}{2}$ and $\frac{2}{4}$ of the child's height are the same.

GOING DEEPER

 This picture demonstrates the equivalence between $\frac{1}{2}$ and $\frac{2}{4}$. The shape has been divided into half and then each half split into a different half to show very different-looking quarters. Ask: *If you cut up two of the quarters and placed them on the other half, would they take up the same amount of space as the other two quarters?*

KEY LANGUAGE

In lesson: equivalent, fraction, half, halves, same, quarter, different, bar model

Other language to be used by the teacher: numerator, denominator, split, into

STRUCTURES AND REPRESENTATIONS

Bar model, arrays, 2D shapes

RESOURCES

Mandatory: multilink cubes, counters, pieces of paper for folding, colouring pencils

Optional: metre sticks, any objects in the classroom that can be used to demonstrate $\frac{1}{2}$ and $\frac{2}{4}$ such as pencil pots, resource trays, reading book baskets or file dividers

 In the eTextbook of this lesson, you will find interactive links to a selection of teaching tools.

Quick recap 

Fold a sheet of paper in half and then in half again. Ask: *What do you expect to see when I open the paper back up?*

Discover

WAYS OF WORKING Pair work

ASK

- Question **1** a): *When you fold the paper in half and then in half again, how many parts have you folded it into?*
- Question **1** b): *Do you get the same fraction when folding a different-shaped piece of paper?*

IN FOCUS As children follow the instructions in the picture, ask them to explain what fraction the piece of paper is being split into each time they fold it. Are they aware that you make a quarter by folding a half in half? Ask children to compare the halves and quarters that they have created. Do they spot that the half they have shaded in is the same amount of space as $\frac{2}{4}$?

PRACTICAL TIPS Give children pieces of paper cut into the shapes shown in question **1** b) so that children can investigate this using the same technique.

ANSWERS

Question **1** a): The paper is now divided into quarters. $\frac{1}{2}$ of the paper is shaded. $\frac{1}{2}$ and $\frac{2}{4}$ are the same, or equivalent.

Question **1** b): It is the same with other shapes: $\frac{1}{2}$ and $\frac{2}{4}$ are the same, or equivalent.

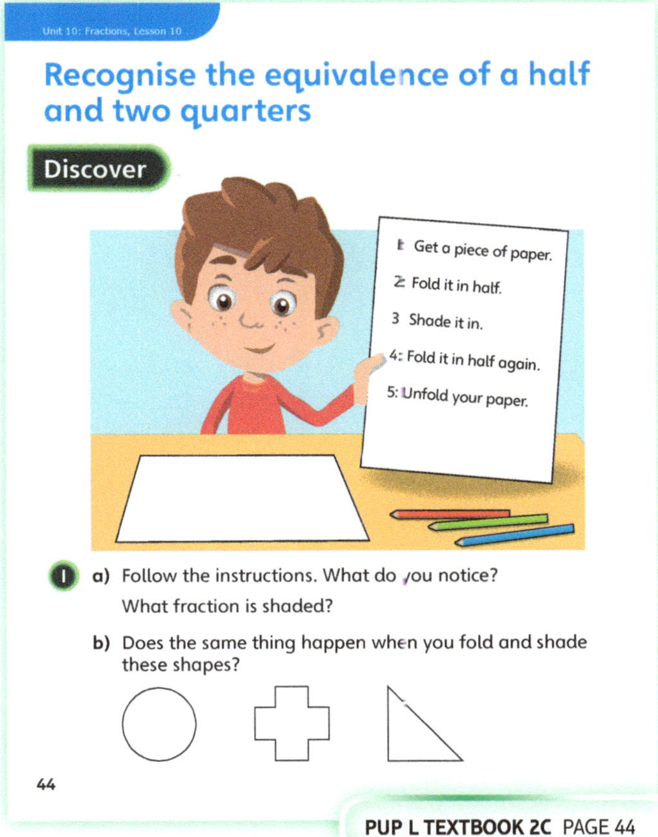

PUPIL TEXTBOOK 2C PAGE 44

Share

WAYS OF WORKING Whole class teacher led

ASK

- Question **1** a): *Do one half and two quarters take up the same amount of space when laid on top of each other?*
- Question **1** b): *There is more than one way to fold the cross. Is there more than one way to fold the piece of paper from question **1** a) into halves and quarters?*
- Question **1** b): *Would $\frac{1}{2}$ of the circle be equivalent to $\frac{2}{4}$ of the cross?*

IN FOCUS Using the pieces of paper that children folded and shaded in question **1** a), introduce the word 'equivalent'. Ask: *Is one half equivalent to one quarter?* Encourage children to check this by cutting out their two quarters and laying them on top of the other half, as Astrid suggests. Ask children to demonstrate that this works for the other shapes. Do children understand that they need to use a circular piece of paper to find a fraction of a whole circle?

Children may not realise that the triangle can be halved twice to make quarters: you have to fold it vertex-to-vertex twice. Can children draw how the triangle looks after two vertex-to-vertex folds?

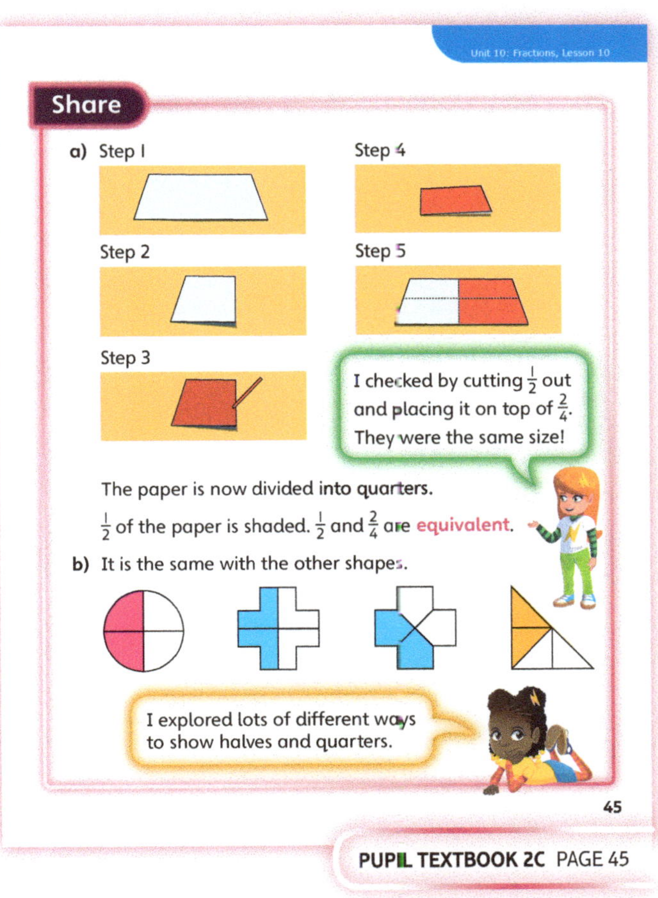

PUPIL TEXTBOOK 2C PAGE 45

Think together

WAYS OF WORKING Whole class teacher led (I do, We do, You do)

ASK

- Question **1**: *Does each whole have to be the same size to compare?*
- Question **1**: *How many different ways can you show this with the same piece of paper?*

IN FOCUS In question **1**, children should realise that to show $\frac{1}{2}$ and $\frac{2}{4}$ are equal, they must first fold both pieces of paper in half *in the same way*. If they fold them in different orientations, they will not be able to demonstrate that $\frac{1}{2} = \frac{2}{4}$. Continue on from this by giving children a different shaped piece of paper and ask them to do the same thing.

Question **2** asks children to model the same concept of equivalence but with counters. Eight counters have been circled into halves and quarters.

STRENGTHEN In questions **2** and **3**, encourage children to recreate what is shown on the page using multilink cubes, showing the different parts by using different coloured cubes or by stacking cubes. Comparing one half of the number of cubes to two quarters of the number of cubes in different colours will show that the amounts are the same but are made of different parts.

DEEPEN Ask children to use the numbers and pictorial representations in questions **2** and **3** to draw their own bar model with labelled parts, showing each half being split in half to make quarters. Demonstrate this first on a part-whole model to structure children's understanding of where each number comes from.

ASSESSMENT CHECKPOINT In question **1**, do children understand that the two wholes have to be the same size to start with? When comparing a group of counters that have been arranged into halves and quarters in question **2**, do children remember that they are comparing two of the quarters and put another circle around them? Do children have to use a sharing strategy to work out $\frac{1}{2}$ and $\frac{1}{4}$ of 8? Do they use their answer for $\frac{1}{2}$ to fill in question **2** b) or do they work it out afresh?

ANSWERS

Question **1**: The paper needs to be folded into equal $\frac{1}{2}$ and then $\frac{1}{4}$ pieces. For example:

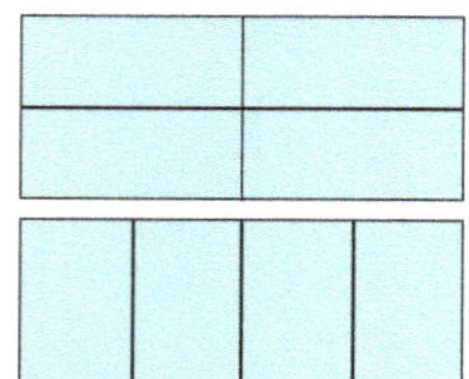

Question **2**: a) $\frac{1}{2}$ of 8 is 4.

Question **2**: b) $\frac{1}{4}$ of 8 is 2.

$\frac{2}{4}$ of 8 is 4.

Question **3**: It can help, as the bar model splits 12 into halves and $\frac{1}{2}$ is equivalent to $\frac{2}{4}$. $\frac{1}{2}$ of 12 is 6, so $\frac{2}{4}$ of 12 is also 6.

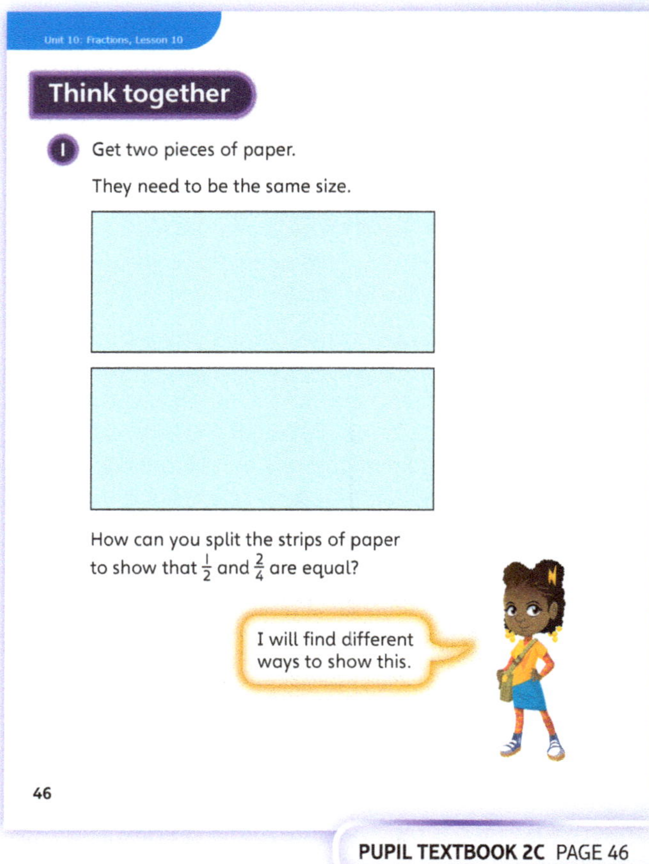

PUPIL TEXTBOOK 2C PAGE 46

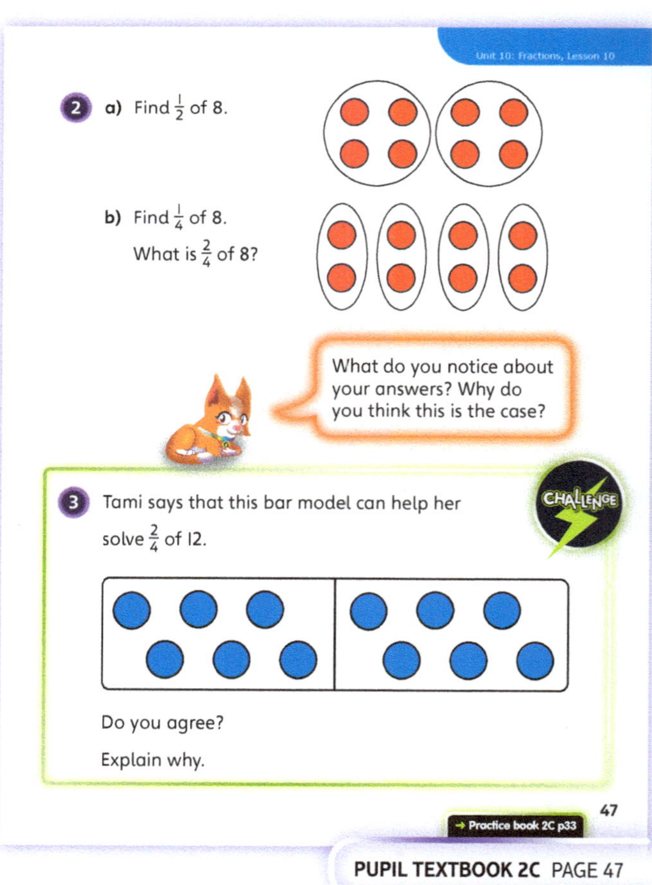

PUPIL TEXTBOOK 2C PAGE 47

Practice

WAYS OF WORKING Independent thinking

IN FOCUS There are two approaches that children may take when answering question **5**. Some children may first find $\frac{2}{4}$ of 20 (=10) and $\frac{1}{2}$ of 20 (=10) and conclude that there are no cherries left. Other children may conclude that since $\frac{2}{4} = \frac{1}{2}$, Harry and Ali each have $\frac{1}{2}$ of the strawberries. Between them, they have a whole (or all the strawberries), so there are none left over. In question **6**, children investigate which numbers can be divided into halves and quarters. Encourage children to model both problems using cubes. It may help children to put cubes into a large part-whole model.

STRENGTHEN Encourage children to circle one half of the cherries in question **5**. They have done this with a shape and this question asks them to practise with a quantity. Do children conclude that, as long as 10 cherries are circled, this is one half of 20?

DEEPEN For children who are confident that $\frac{1}{2}$ is equal to $\frac{2}{4}$ give them a $\frac{1}{4}$ and see if they can work out a half. For example, ask: *If $\frac{1}{4}$ is 5, what is $\frac{1}{2}$?* Some children may find the whole first (20), then halve it to get a half (10). Children may then see that they can simply double $\frac{1}{4}$ to get a $\frac{1}{2}$.

THINK DIFFERENTLY Question **3** requires children to reason about the size of different fractions and to work out which ones are equivalent. Encourage children to use the pictures to help them decide.

ASSESSMENT CHECKPOINT In question **2**, do children shade two cubes for $\frac{1}{2}$ or do they shade one cube because the numerator is 1? Do they shade the same cubes as they shaded for $\frac{2}{4}$ or do they use a count one, leave one strategy?

In question **5**, how do children work out $\frac{1}{2}$ and $\frac{2}{4}$ of 20? Do they share out cubes; do they use a 'circle one, leave one' strategy; or do they simply know it as a number fact?

ANSWERS Answers for the **Practice** part of the lesson can be found in the *Power Maths* online subscription.

Reflect

WAYS OF WORKING Whole class

IN FOCUS Children choose their own method to demonstrate equivalence between $\frac{1}{2}$ and $\frac{2}{4}$. They can fold paper, cut paper, stack cubes, and so on. Ensure that, whichever method they choose, children label each part and clearly show how one half has been halved again.

ASSESSMENT CHECKPOINT Listen for the words that children use in their explanation, such as the words 'equivalent' or 'equivalence' or see if they make references to the numerator and denominator. Do children understand that the whole has to be the same each time in order to compare the fractions?

ANSWERS Answers for the **Reflect** part of the lesson can be found in the *Power Maths* online subscription.

After the lesson ⏸

- Did children mention the relationship between the numerator and denominator in these equivalent fractions?
- Did children show that they understood one quarter as half of a half? Did children apply this to working out one quarter of a number?
- Did children understand that wholes have to be the same for the fractions to be equivalent?

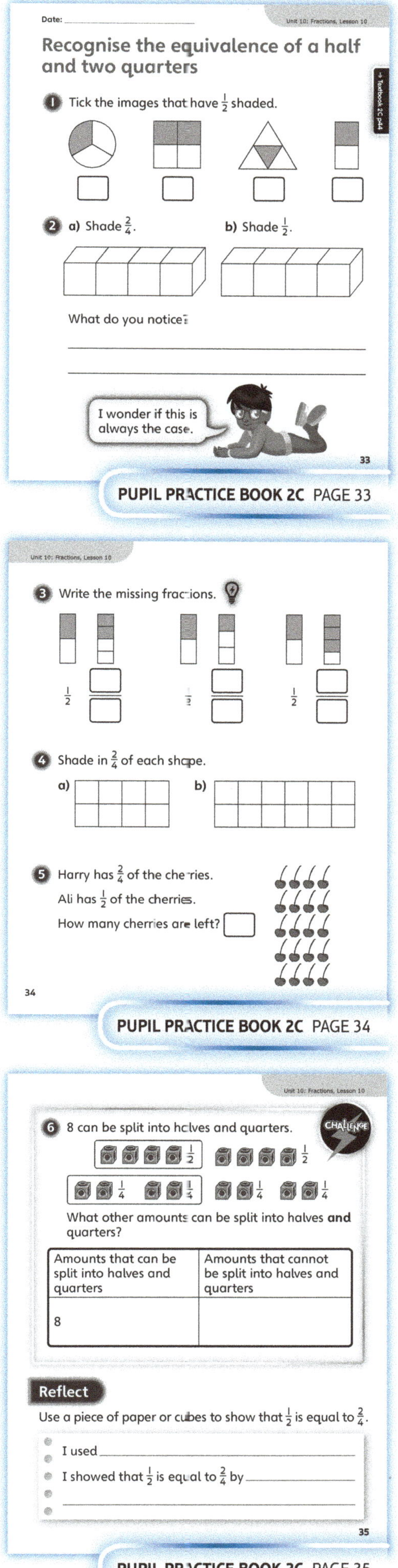

PUPIL PRACTICE BOOK 2C PAGE 33

PUPIL PRACTICE BOOK 2C PAGE 34

PUPIL PRACTICE BOOK 2C PAGE 35

Recognise three quarters

Learning focus

In this lesson, children will apply their knowledge of unit and non-unit fractions to work out one quarter and three quarters of numbers up to 20.

Before you teach 

- Did children understand fractions where the numerator was greater than 1 in previous lessons?
- Do children have a strategy for working out a fraction of a number, such as sharing physical resources?

NATIONAL CURRICULUM LINKS

Year 2 Number – fractions

Recognise, find, name and write fractions $\frac{1}{3}$, $\frac{1}{4}$, $\frac{2}{4}$ and $\frac{3}{4}$ of a length, shape, set of objects or quantity.

ASSESSING MASTERY

Children can work out $\frac{1}{4}$ of a number and use this information to work out $\frac{3}{4}$. Children can understand that the number they are dividing by is the denominator of the fraction they are trying to find and can work out the whole from a non-unit fraction such as $\frac{3}{4}$.

COMMON MISCONCEPTIONS

The scaffolds in this lesson guide children through the process of working out a non-unit fraction step-by-step, but they may still become confused as to what each number in the fraction represents and what to do with it. Children might think that to get from $\frac{1}{4}$ to $\frac{3}{4}$ they multiply both the numerator and the denominator by 3 and get $\frac{3}{12}$. Ask:

- *These are all quarters. Does that change when you have more quarters?*

STRENGTHENING UNDERSTANDING

Provide physical resources such as paintbrushes and pots so that children can model the questions. Share out the resources in quarters and ask children to count how many there are in three quarters.

GOING DEEPER

Working out a whole from a non-unit fraction requires strong conceptual understanding of what the denominator and the numerator represent. Give children a number that represents a non-unit fraction – for example, $\frac{3}{4}$. With the help of pictures or physical resources, children will see that they divide that fraction by 3 in order to find one equal part because the non-unit fraction only represents three parts. Show children that they just have to add one of those parts to find the whole.

KEY LANGUAGE

In lesson: fraction, quarter, $\frac{3}{4}$, **three quarters**, divided, equal parts, share, how many, half, whole, unit fraction, **non-unit fraction**

Other language to be used by the teacher: numerator, denominator

RESOURCES

Mandatory: multilink cubes, counters

Optional: paper plates, classroom objects such as pots and paintbrushes or pencils and pencil cases

 In the eTextbook of this lesson, you will find interactive links to a selection of teaching tools.

Quick recap 

Challenge children to draw as many different diagrams as they can that show quarters.

Discover

 Pair work

- Question **1** a): *How many equal parts has the pizza been split into? How many equal parts have mushrooms on?*
- Questions **1** a) and b): *What is the same and what is different about the answers to part a) and part b)?*

 This picture is a reminder of unit and non-unit fractions. Question **1** a) requires children to work out that the pizza has been divided into 4 equal parts (quarters). Show children that 3 of those equal parts have mushrooms on and that this is represented as $\frac{3}{4}$.

 Give each child a paper plate. Ask them to draw the pizza onto their plate and then to cut out the parts. Agree that there are four equal parts representing quarters and that three of these parts have mushrooms.

Question **1** a): 3 out of the 4 equal slices of pizza have mushrooms. $\frac{3}{4}$ of the pizza has mushrooms.

Question **1** b): 1 out of the 4 equal slices of pizza does not have mushrooms. $\frac{1}{4}$ of the pizza does not have mushrooms.

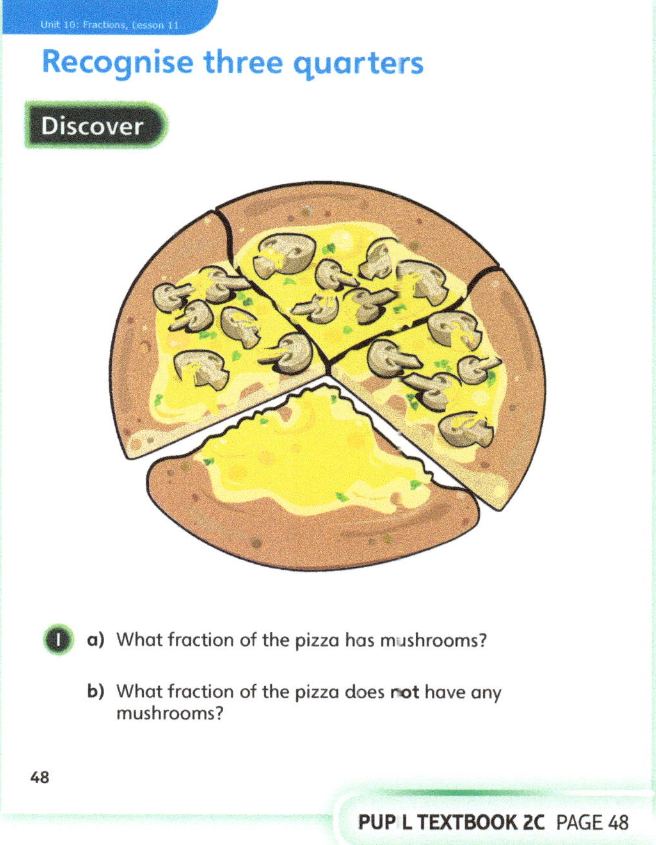

PUPIL TEXTBOOK 2C PAGE 48

Share

 Whole class teacher led

- Questions **1** a) and b): *What is the same and what is different about the fractions in questions **1** a) and b)?*
- Questions **1** a) and b): *Why does the numerator change?*

 When both fractions have been written up, work together with the whole class to compare them. Refer to what Sparks says about $\frac{3}{4}$ being a non-unit fraction. Ask children to explain what a non-unit fraction is and where they have seen one before.

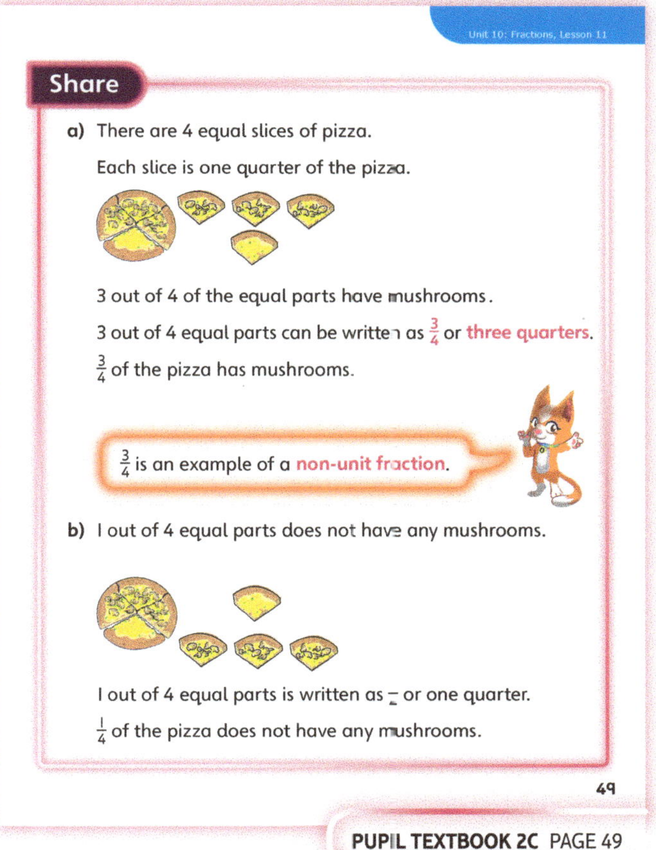

PUPIL TEXTBOOK 2C PAGE 49

Think together

WAYS OF WORKING Whole class teacher led (I do, We do, You do)

ASK

- Question ② : *Does each whole have to be the same size to be compared?*

IN FOCUS Question ① is scaffolded to help children work out $\frac{3}{4}$ of a quantity. Question ② requires children to recognise $\frac{3}{4}$ from visual representations. Question ③ shows that $\frac{1}{2}$ and a further $\frac{1}{4}$ shaded makes $\frac{3}{4}$ shaded. Children should recognise this and may split the $\frac{1}{2}$ into quarters.

STRENGTHEN Encourage children to use multilink cubes or counters to represent the slices of pizza shared in question ① . Ask children to draw four sorting hoops, label each circle $\frac{1}{4}$ and share the resources out one by one into the circles until they have none left.

Encourage children to recreate the square in question ③ by folding a piece of paper. When they fold the paper in half and then in half again, the folds will show them how to split the shaded half into quarters.

DEEPEN Look at the answers to question ① . Do children spot a similarity between or pattern within the numbers?

ASSESSMENT CHECKPOINT Do children use the answer to question ① a), which they know is also $\frac{1}{4}$, to help them to work out $\frac{3}{4}$ in question ① b)? Do children work out $\frac{3}{4}$ another way, such as by sharing physical resources into four equal groups and counting the number of resources in three of the groups?

ANSWERS

Question ① a): Gino will give 2 slices of pizza to 1 person.

Question ① b): Gino will give 6 slices of pizza to 3 people.

Question ② : Shapes a), b) and d) have $\frac{3}{4}$ shaded.

Question ③ : If we change the $\frac{1}{2}$ to $\frac{2}{4}$, then $\frac{3}{4}$ of the shape is shaded.

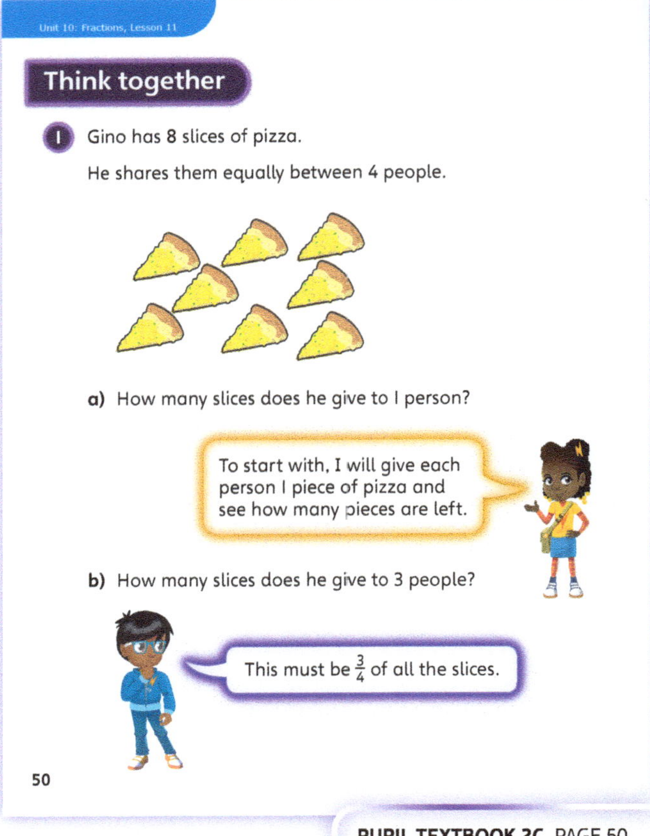

PUPIL TEXTBOOK 2C PAGE 50

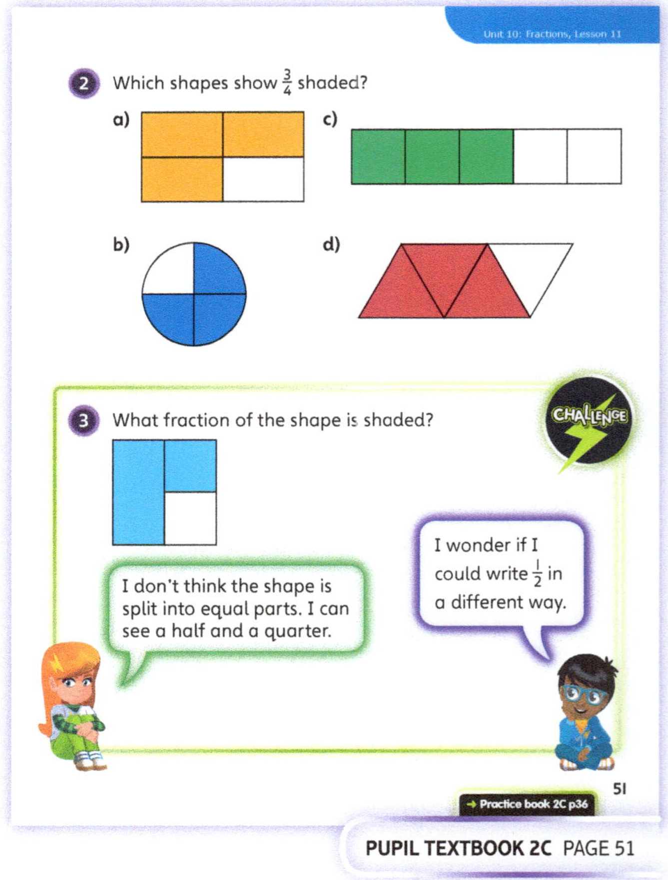

PUPIL TEXTBOOK 2C PAGE 51

Practice

WAYS OF WORKING Independent thinking

IN FOCUS The shapes in questions **1** and **2** have been split into different numbers of equal parts for children to shade $\frac{3}{4}$ of. The rectangle in question **2** a) has been split into eight parts, requiring children to see that a column of two squares is one of four equal parts. To shade $\frac{3}{4}$ of the rectangle in **2** c), children will need to shade 9 squares. This is slightly more difficult as they will not be able to shade whole rows or columns, but will have to share the 12 parts into 4 equal groups first, to determine that there are 3 squares in each quarter. Multilink cubes or counters could be used for support. Questions **3** and **4** ask children to work out $\frac{1}{4}$ and $\frac{3}{4}$ of different numbers. You may need to guide children to explain what each number in the fraction represents.

STRENGTHEN Question **6** tells children that $\frac{3}{4}$ is 9 and shows a picture demonstrating this, and then asks them to identify $\frac{1}{4}$ and the whole. One of the quarters is empty, so ask children how they know the number of counters that should fill that quarter. Can they look at the number of counters in each of the other quarters and use their knowledge of equal parts to conclude that the final quarter must also have 3 counters? Alternatively, can they do this by simply dividing 9 by 3 to conclude that there are 3 counters in each part?

DEEPEN Give children a number that cannot be divided into 4 and ask them to try to split that number of cubes into 4 equal groups. Do children try to split the leftover cubes into halves and share those out equally?

THINK DIFFERENTLY Question **5** is an abstract question asking children to find $\frac{3}{4}$ of a number without the context of shapes or equal groups of objects.

ASSESSMENT CHECKPOINT Use question **5** to assess whether children can find $\frac{3}{4}$ of a number.

ANSWERS Answers for the **Practice** part of the lesson can be found in the *Power Maths* online subscription.

Reflect

WAYS OF WORKING Pair work

IN FOCUS This explores the misconception that a shape split into any 4 parts with 3 parts shaded represents $\frac{3}{4}$. Children might suggest instead dividing the circle into sectors (slices) to ensure that each part is equal.

ASSESSMENT CHECKPOINT Do children recognise that although three of the four parts are shaded, the four parts are not equal and so they do not represent quarters.

ANSWERS Answers for the **Reflect** part of the lesson can be found in the *Power Maths* online subscription.

After the lesson ⏸

- Were children able to find $\frac{3}{4}$ of an amount that has more than 4 equal parts?
- Did children require a pictorial representation to help them work out a fraction? Did they draw one for themselves if no pictorial representations were provided?
- Did children remember that $\frac{1}{2}$ is equivalent to $\frac{2}{4}$ from the previous lesson and use this to help them in this lesson?

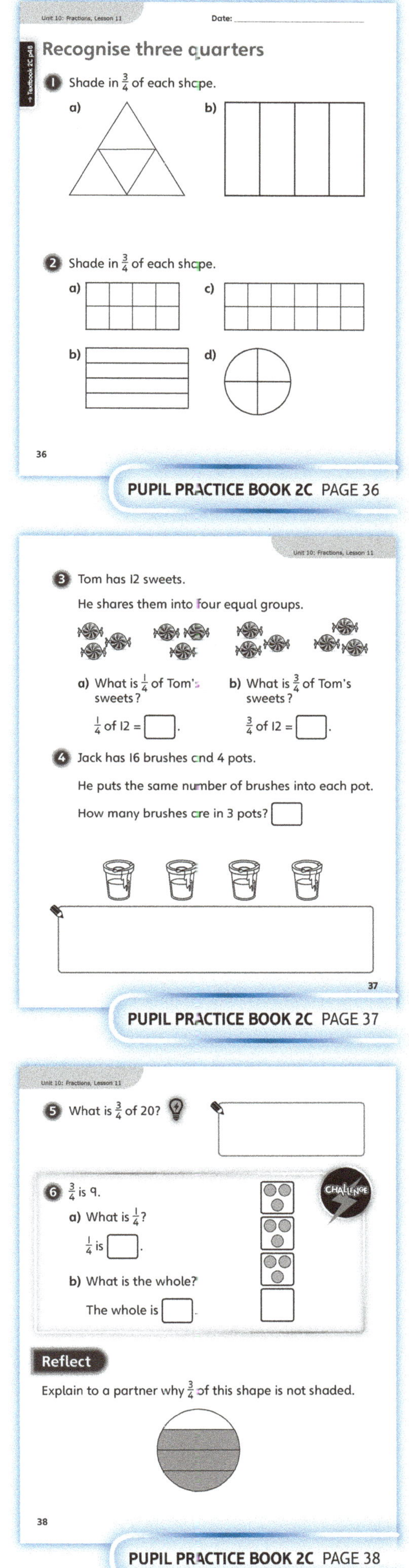

PUPIL PRACTICE BOOK 2C PAGE 36

PUPIL PRACTICE BOOK 2C PAGE 37

PUPIL PRACTICE BOOK 2C PAGE 38

Count in fractions up to a whole

Learning focus

In this lesson, children will learn about non-unit fractions becoming one whole.

Before you teach 

- Were children confident with equivalence when comparing $\frac{1}{2}$ and $\frac{2}{4}$ in Lesson 10? Could they apply this understanding of equivalence to $\frac{3}{3}$ and $\frac{4}{4}$?
- Were children confident when answering missing number questions in other units?

NATIONAL CURRICULUM LINKS

Year 2 Non-statutory guidance

Pupils should count in fractions up to 10, starting from any number and using the $\frac{1}{2}$ and $\frac{2}{4}$ equivalence on the number line (for example, $1\frac{1}{4}$, $1\frac{2}{4}$ (or $1\frac{1}{2}$), $1\frac{3}{4}$, 2).

ASSESSING MASTERY

Children can recognise a whole by seeing whether the numerator and denominator are the same and can understand that a whole as a fraction with the same numerator and denominator can represent the same whole as another fraction made up of a different number of equal parts. Children can fill in missing number sentence scaffolds in order to make one whole.

COMMON MISCONCEPTIONS

When comparing fractions that make one whole, such as $\frac{3}{3}$ and $\frac{4}{4}$, children may think that the whole has to look different. They may think that the whole that has been split into quarters has to be bigger because there are more parts. Show children two different wholes that are the same size but have been split into different parts. Ask:
- *Are the wholes the same size? What is different about them? Does one have to be bigger than the other?*

STRENGTHENING UNDERSTANDING

Part-whole models split into fractions are a useful model to demonstrate one whole being split into multiple parts. They can be shown as a whole with multiple equal parts. Alternatively, they can show the whole partitioned into a unit fraction and non-unit fraction, with the non-unit fraction further partitioned into its unit fractions.

GOING DEEPER

There are multiple ways of making and drawing one whole. Ask children to complete the sentence scaffold 1 = ☐ as many times as they can, with accompanying drawings. The whole can change each time as well as using a different number of equal parts. Children could also write the same number sentences using real-world contexts, such as a cake that is divided into quarters, all four of which are eaten, meaning that the whole cake is gone.

KEY LANGUAGE

In lesson: fraction, different, same, half, halves, whole, equals, quarters, numerator, denominator, diagram, number sentence, thirds

Other language to be used by the teacher: non-unit fraction, split, equal parts, unit fraction, equivalent

STRUCTURES AND REPRESENTATIONS

2D shapes

RESOURCES

Mandatory: multilink cubes, counters

Optional: objects to cut up into different parts such as sandwiches, cakes, oranges or card shapes

 In the eTextbook of this lesson, you will find interactive links to a selection of teaching tools.

Quick recap ↺

Ask children to each draw a number line that starts at 0 and goes up in 1s.

Discover

WAYS OF WORKING Pair work

ASK

- Question **1** b): *Are the sandwiches different sizes? What is different about the sandwiches?*
- Question **1** b): *Does Jack have more than Maya? Does Maya have more than Jack?*

IN FOCUS Encourage children to look carefully at the sandwiches in the picture and talk about what they look like. Ask children to discuss how many equal pieces each sandwich has been split into. The sandwiches are the same whole but have been split into halves and quarters. One sandwich is made up of more equal parts but both sandwiches are the same overall whole.

PRACTICAL TIPS Show children a square sandwich or piece of card. Make one cut to show halves, like Jack's sandwich. Make a second cut to show quarters, like Maya's sandwich. Agree that the whole is still the same but that there is a different number of equal parts.

ANSWERS

Question **1** a): Jack's sandwich is in halves.
2 halves = 1 whole, so $\frac{2}{2} = 1$.

Question **1** b): Maya's sandwich is in quarters.
4 quarters = 1 whole, so $\frac{4}{4} = 1$.

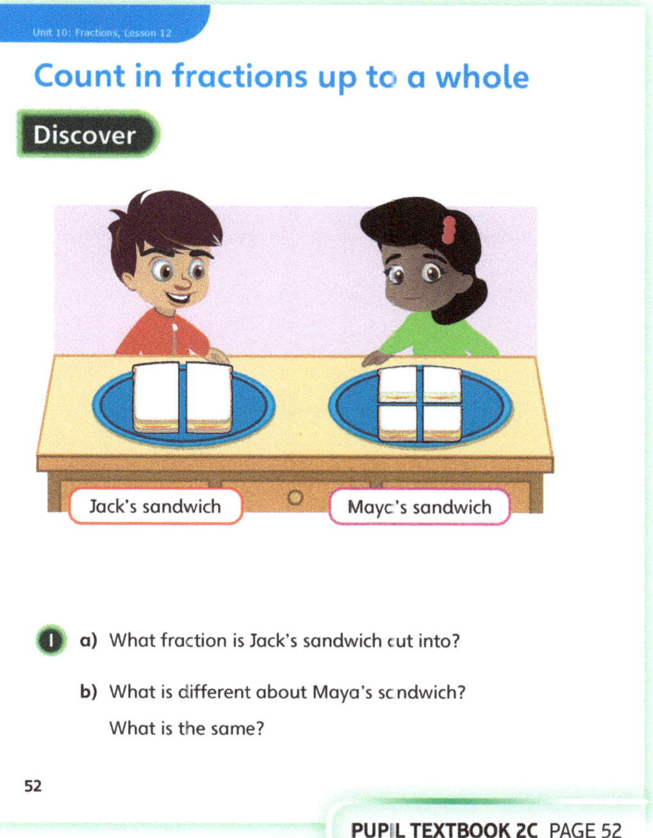

PUPIL TEXTBOOK 2C PAGE 52

Share

WAYS OF WORKING Whole class teacher led

ASK

- Questions **1** a) and b): *Does each plate show one whole sandwich?*
- Questions **1** a) and b): *What is the same about Jack's sandwich and Maya's sandwich? What is different?*

IN FOCUS This part of the lesson explains how many different parts each sandwich has been cut into and introduces non-unit fractions where the numerator is the same as the denominator. Ash wonders if the fraction is always one whole when the numerator and denominator are the same. Ask children to look at some other examples of fractions with the same numerator and denominator to decide whether this is correct.

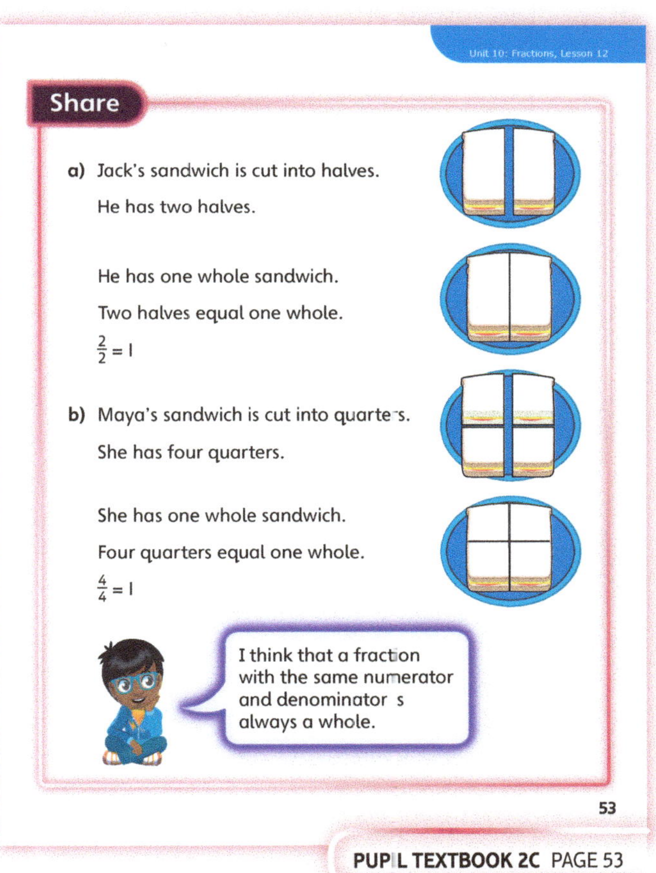

PUPIL TEXTBOOK 2C PAGE 53

Think together

WAYS OF WORKING Whole class teacher led (I do, We do, You do)

ASK

- Question **1**: *What do you notice about the denominator in each of the fractions?*
- Question **2**: *How do you know which fractions show one whole without drawing or making them?*

IN FOCUS Question **1** requires children to work through the pictures and write the shaded fraction of each shape. The question guides children to point out that the last image with all the parts shaded is one whole.

In question **3**, children complete a missing number sentence scaffold by writing the fraction that would make a whole. Children can use the accompanying pictures to help them, by assigning the parts of each shape to the different parts of the missing number sentence scaffold.

STRENGTHEN There are multiple ways to fill in the missing number sentence scaffolds in question **3**. For example, question **3** b) can make two different number sentences: $\frac{1}{3} + \frac{2}{3}$ and $\frac{2}{3} + \frac{1}{3}$. If children have physical resources to model the fractions they are working with, can they group the resources differently each time to model the parts that they are adding?

DEEPEN Building on work that children have done in previous lessons, can they add fractions to make one whole where the denominators are different? For example, $\frac{1}{2} + \frac{2}{4} = 1$.

ASSESSMENT CHECKPOINT In question **2**, do children draw or make each fraction before deciding whether it is one whole or do they recognise a whole from the fact that the numerator and denominator are the same? In question **3**, do children know that the denominators have to be the same in order to add them to make the same whole?

ANSWERS

Question **1** a): A is $\frac{1}{4}$, B is $\frac{2}{4}$, C is $\frac{3}{4}$, D is $\frac{4}{4}$

Question **1** b): The whole is shaded in D, $\frac{4}{4} = 1$.

Question **2**: $\frac{3}{3}$, $\frac{2}{2}$ and $\frac{4}{4}$ are equal to one whole.

Question **3** a): $\frac{1}{2} + \frac{1}{2} = 1$

Question **3** b): $\frac{1}{3} + \frac{2}{3} = 1$

Question **3** c): $\frac{3}{4} + \frac{1}{4} = 1$

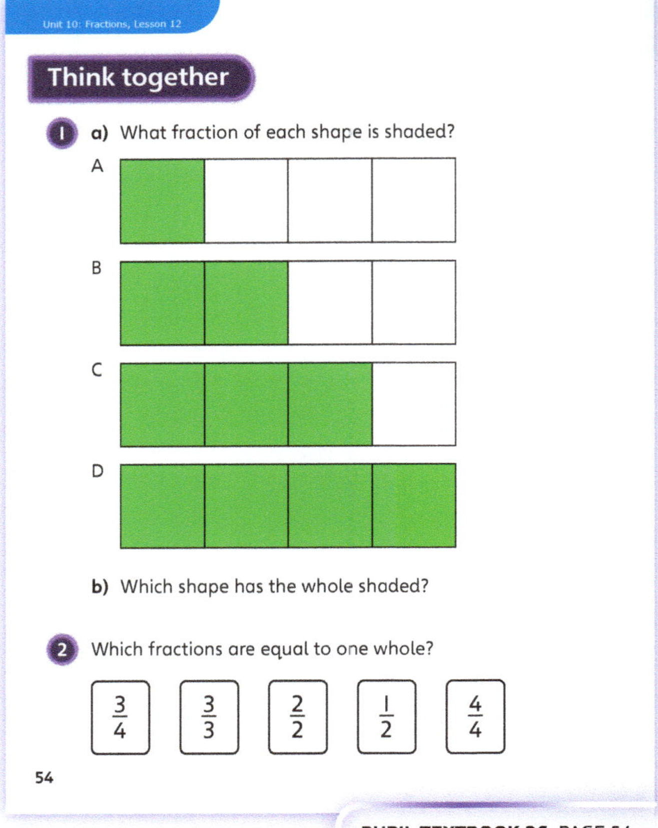

PUPIL TEXTBOOK 2C PAGE 54

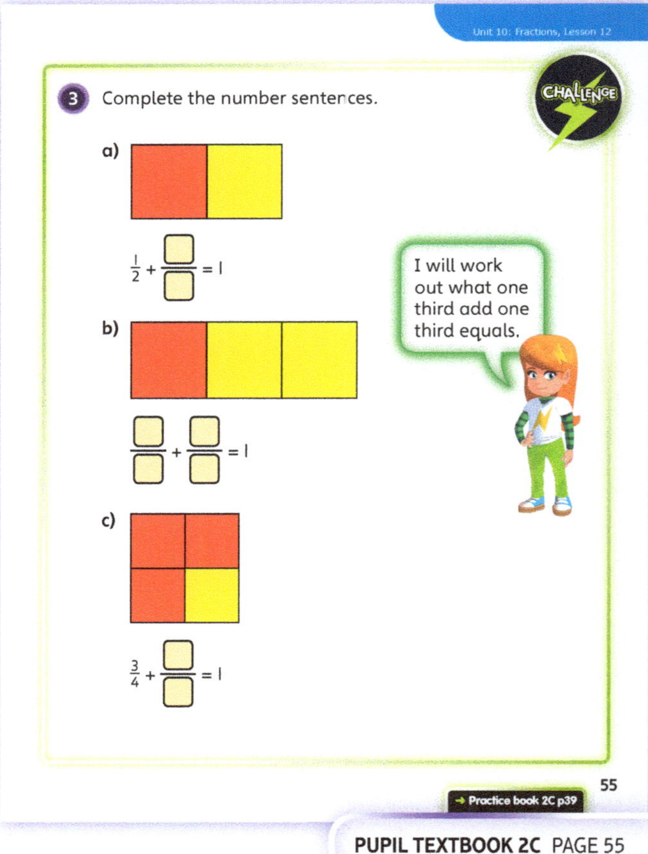

PUPIL TEXTBOOK 2C PAGE 55

Practice

WAYS OF WORKING Independent thinking

IN FOCUS In question ③, children look for the correct number of shaded parts to make a whole. Question ⑤ asks children to make one whole by writing a non-unit fraction and a unit fraction based on the pictures. Children write the answer with the same numerator and denominator and see that this is equal to 1.

STRENGTHEN Use different physical resources to model the different wholes in questions ② and ④. Ask: *What objects in the classroom can be split into two halves, three thirds or four quarters? Does the whole need to be different each time or can it be the same?*

DEEPEN Question ⑦ requires children to understand that they have to start with the same whole but split it into different numbers of equal parts so that each child can have the same amount of cake. Ask children to draw a picture to represent this question and check that they have drawn two wholes that are the same size rather than five parts that are the same size. Ask the same question with the addition of a child eating four equal parts while eating the same amount of cake as Jemima and Sam.

THINK DIFFERENTLY In question ⑥, children find the missing fraction needed to make one whole. Refer children to what Ash asks and guide them to conclude that the missing fraction in question ⑥ c) can be written as $\frac{1}{2}$ or $\frac{2}{4}$.

ASSESSMENT CHECKPOINT In question ①, do children label each equal part on the image to help them fill in the box? In question ⑤, do children rely on the pictures to work out what fractions to add together or are they aware of the pattern of numerators making the same total as the denominator?

ANSWERS Answers for the **Practice** part of the lesson can be found in the *Power Maths* online subscription.

Reflect

WAYS OF WORKING Independent thinking

IN FOCUS Encourage children to think about this question and explanation for themselves. Ask children to write down fractions with the same numerator and denominator, so that they can use them either to prove their opinion or to help inform their mathematical opinion.

ASSESSMENT CHECKPOINT Do children automatically know the answer or do they have to draw fractions to check? Are they able to explain why the numerators and denominators have to be the same for the fraction to equal one whole? Do they attempt to write other fractions that make one whole based on this knowledge, such as $\frac{5}{5}$ or $\frac{6}{6}$, even if they do not know the names of these fractions?

ANSWERS Answers for the **Reflect** part of the lesson can be found in the *Power Maths* online subscription.

After the lesson ⏸

- Did children recognise a whole as a written fraction?
- Did children represent wholes that were the same but that were split into different parts?
- Did children understand the relationship between the numerator and the denominator in a whole and explain why they are the same?

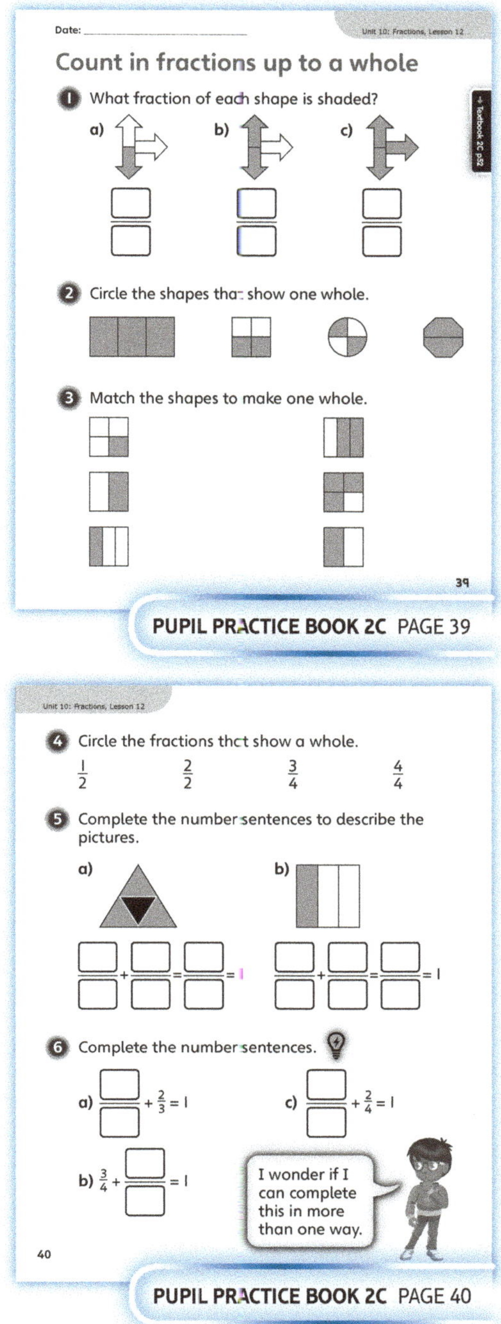

PUPIL PRACTICE BOOK 2C PAGE 39

PUPIL PRACTICE BOOK 2C PAGE 40

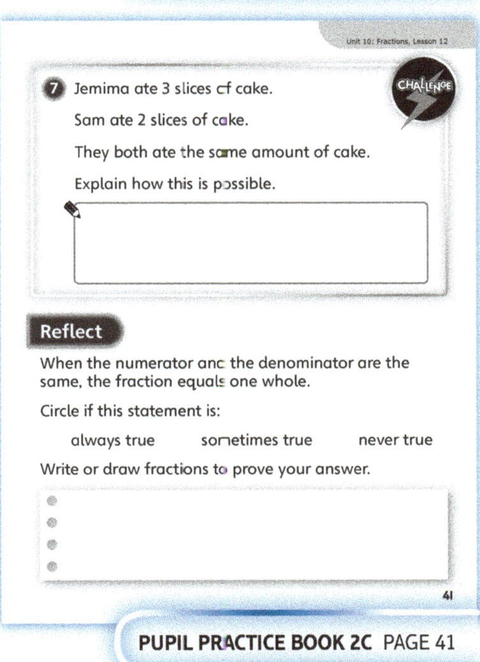

PUPIL PRACTICE BOOK 2C PAGE 41

End of unit check

Don't forget the unit assessment grid in your *Power Maths* online subscription.

WAYS OF WORKING Group work adult led

IN FOCUS Question 2 focuses on calculating half of a number, reinforcing that children need to share 8 into two equal parts.

Question 3 requires children to work out the relationship between the numbers 12 and 4 and to recognise that these numbers represent the whole and one equal part of the whole. They can use grouping here as they are given the total (12) and the number in each group (4). By arranging 12 objects in groups of 4, they should find that there are 3 equal groups.

Question 4 requires children to find an equivalent non-unit fraction to the unit fraction $\frac{1}{2}$.

Question 5 tests children's knowledge of non-unit fractions. The fact that the denominator in all of the fractions is 4 requires children to correctly link the number of pieces that are eaten to the numerator.

Think!

WAYS OF WORKING Pair work

IN FOCUS This task shows unit and non-unit fractions. Ensure that children can read them out loud and understand what each fraction represents.

Children sort fractions creatively using words from the word bank, revising the key concepts covered in this unit: denominators as equal parts, equivalent fractions, fractions making one whole, unit and non-unit fractions, and so on. Get children to check other pairs' categories.

Encourage children to come up with one sentence to describe each of their categories before writing their answer in **My journal**. For example, 'On this side, all of the fractions have the denominator 4 because they have all been divided into four equal parts'.

ANSWERS AND COMMENTARY To show mastery, children will be able to see a number as both being a whole and a part. For example, if the number 8 is the whole, 4 would be $\frac{1}{2}$; if 8 is $\frac{1}{2}$, 16 would be the whole. Assess if children can go further and say what the whole would be if 8 were $\frac{1}{3}$ or $\frac{1}{4}$. Show this on a set of bar models, each split into equal parts representing 8.

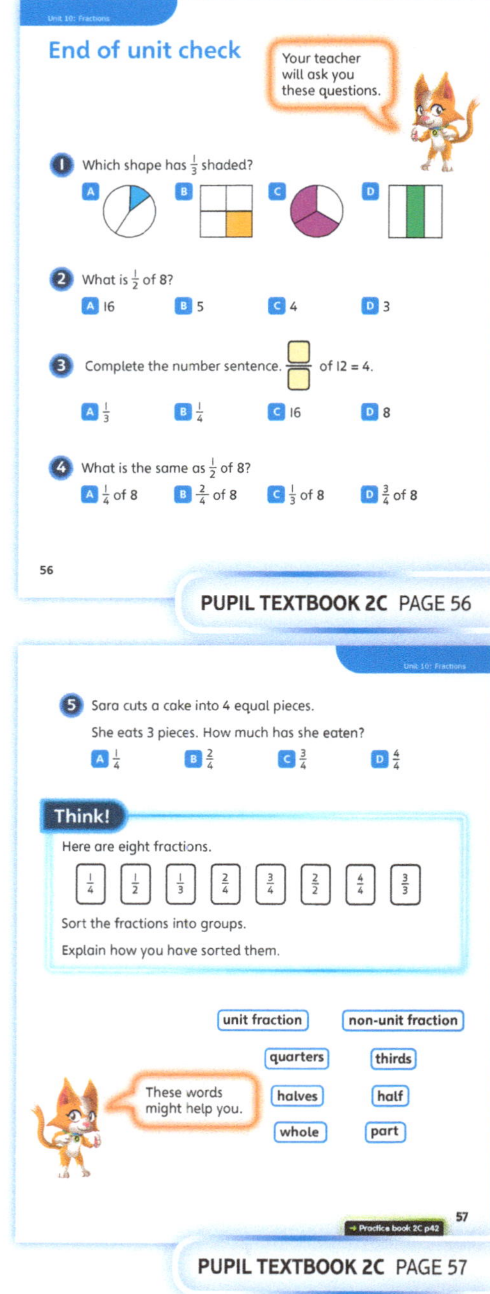

PUPIL TEXTBOOK 2C PAGE 56

PUPIL TEXTBOOK 2C PAGE 57

Q	A	WRONG ANSWERS AND MISCONCEPTIONS	STRENGTHENING UNDERSTANDING
1	D	A has been split into three parts but the parts are not equal. Children who choose A do not understand the concept of equal parts.	For questions 2 to 5, cubes or counters can be used to physically represent each question. These resources can be arranged on paper so that children can draw around them to represent the whole and the equal parts. Each part can be labelled with the correct fraction name, such as $\frac{1}{4}$.
2	C	Children may mistakenly think that 8 is $\frac{1}{2}$ and choose A.	
3	A	Children may think that B is correct because $\frac{1}{4}$ has a 4 in it and the equal part is 4. Encourage children to see that there are 12 objects and 4 objects in each group. How many groups are there?	
4	B	Children may choose A or C because they recognise that the numerator (1) is the same as in $\frac{1}{2}$.	
5	C	Children may choose $\frac{1}{4}$ because they relate the fact that the cake is split into four equal pieces to each piece being one quarter.	

My journal

WAYS OF WORKING Independent thinking

ANSWERS AND COMMENTARY

This task allows children to write about one of the chosen categories by which they have sorted fractions. Children will have rehearsed this with their partner in **Think!** and now are asked to write this in their own words. Encourage children to strengthen their explanation by drawing bar models or pictures of what each category represents, such as drawing one shaded whole to illustrate the category 'fractions that make one whole'.

If children are struggling to articulate a pattern, give them sentence scaffolds such as 'All of the denominators are the same/different' or 'These fractions all make the same/a different amount'.

Power check

WAYS OF WORKING Independent thinking

ASK

- *Do you think you could use physical resources to show that something has been divided into equal parts?*
- *Are you confident that you could draw a fraction?*
- *Could you draw the same fraction in different ways?*

Power play

WAYS OF WORKING Pair work

IN FOCUS This game gives children the opportunity to practise recognising, writing and drawing a fraction. It asks them to recognise fractions from shapes as well as from the numerator and denominator. Assess whether children can work together in pairs to decide what whole to start with when drawing a fraction.

ANSWERS AND COMMENTARY Children may say fractions incorrectly and read out the numbers they see, such as reading out $\frac{3}{4}$ as 'three fours'.

When writing down equivalent fractions, children may write $\frac{1}{2}$ when identifying two quarters. Does their partner realise what they have done and can they explain why they have done it?

With fractions such as $\frac{4}{4}$, children may draw one whole and shade it in without indicating that it is split into four equal parts. Can children see why they must split it into four equal parts in order to show that it does not represent another fraction, such as $\frac{3}{3}$ or $\frac{2}{2}$?

After the unit ⏸

- Could children recognise fractions in a range of contexts?
- Did children confidently spot equivalent fractions when written down or did they require a visual image to see the equivalence?
- Were children able to make any generalisations about patterns that they spotted between the numerator and denominator in equivalent fractions or in fractions that made one whole?

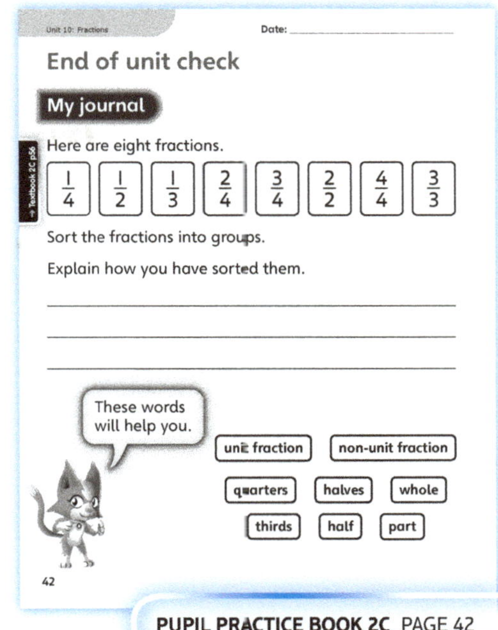

PUPIL PRACTICE BOOK 2C PAGE 42

PUPIL PRACTICE BOOK 2C PAGE 43

Strengthen and **Deepen** activities for this unit can be found in the *Power Maths* online subscription.

Unit 11
Time

Don't forget to watch the Unit 11 video!

WHY THIS UNIT IS IMPORTANT

This unit will develop children's ability to tell and write the time to five minutes, including quarter past and quarter to the hour. Children will know the number of minutes in an hour and hours in a day.

Children will solve problems using these new concepts and previous learning, including word problems, and comparing and sequencing questions.

WHERE THIS UNIT FITS

→ Unit 10: Fractions

→ **Unit 11: Time**

→ Unit 12: Problem solving and efficient methods

This unit builds on the concepts of time learnt in Year 1 and will draw on comparing and ordering skills, whilst linking to knowledge of the part-whole model.

Before they start this unit, it is expected that children:
- can find o'clock and half-past times on an analogue clock
- can count on and count back reliably in 5s up to 60
- recognise and understand the word 'quarter'.

ASSESSING MASTERY

Children who have mastered this unit will be able to read, write and show the time on a clock to five minutes. They will confidently use the vocabulary 'past', 'to', 'o'clock', 'half past', 'quarter past' and 'quarter to'. They will be able to use these elements to confidently solve mathematical problems. Children will recognise that there are 60 minutes in an hour and 24 hours in a day and be able to explain how these are represented and shown on an analogue clock.

COMMON MISCONCEPTIONS	STRENGTHENING UNDERSTANDING	GOING DEEPER
Children may confuse the two clock hands, either reading them incorrectly or drawing them incorrectly. Children may find difficult the concept of recognising a quarter of the analogue clock or which quarter is 'past' and which is 'to'. For children who are struggling with this concept, you could provide a colour-coded clock face with the quarters labelled clearly on it.	It may be beneficial to provide children with a colour-coded clock to manipulate (to help further, the hands could be labelled). If the colour-coding matched that in the Textbook then this would help secure the link between the concrete and pictorial representation. The clock could also be labelled and colour-coded, where appropriate, with key vocabulary such as 'quarter', 'half', 'past', 'to'.	Children could be challenged regularly to improve their fluency and recognition of key vocabulary by quizzing them on the words and coloured segments labelled on the clock. To do this you could take away one or two of the labels. Ask: *What is missing from the clock's labels?* *How do you know?* *What does that label mean?*

Unit II: Time

UNIT STARTER PAGES

Use these pages to introduce the unit focus to children. You can use the characters to explore different ways of working too.

STRUCTURES AND REPRESENTATIONS

Clock tool: Pictures of clock faces are used regularly to represent times. They are used for demonstration purposes and also as the basis of problems to solve. When presented with a clock face with no hands, children will be encouraged to complete these representations to demonstrate their understanding.

 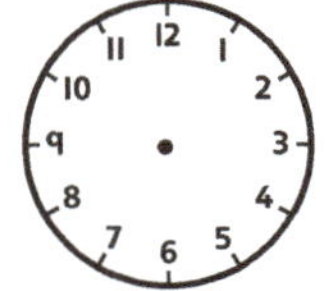

Part-whole model: This model helps children partition a number of minutes greater than 60 into hours and minutes.

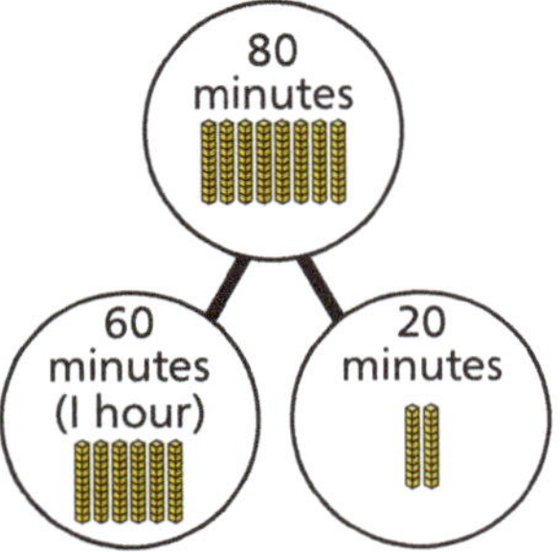

KEY LANGUAGE

There is some key language that children will need to know as part of the learning in this unit:

→ hands, face, hours, minutes, analogue

→ minute hand, hour hand

→ o'clock, past, to, half past, quarter past, quarter to, quarter of an hour

→ five, ten, fifteen, twenty, twenty-five, thirty, thirty-five, forty, forty-five, fifty, fifty-five, sixty

→ 5, 10, 15, 20, 25, 30, 35, 40, 45, 50, 55, 60

→ 24 hours, day, daytime, night time, around the clock, am, pm

→ midday, midnight, morning, afternoon

PUPIL TEXTBOOK 2C PAGE 58

PUPIL TEXTBOOK 2C PAGE 59

O'clock and half past

Learning focus

In this lesson, children will recap their learning about measuring time. They will read and describe times to the hour and the half hour.

Before you teach ⏸

- Is your classroom environment set up to allow children to read the time easily and regularly?
- Could you provide children with experience of different types of analogue clock to develop their fluency?

NATIONAL CURRICULUM LINKS

Year 1 Measurement – time

Tell the time to the hour and half past the hour and draw the hands on a clock face to show these times.

ASSESSING MASTERY

Children can confidently read analogue clocks showing times to the hour and the half hour. Children can identify and explain the meanings of the hour and minute hands, and can draw hands showing o'clock and half-past times on blank clock faces.

COMMON MISCONCEPTIONS

Children may confuse the two clock hands. Provide a colour-coded clock to manipulate showing, for example, 2 o'clock. Ask:
- *What tells you that it is 2 o'clock? Which hand is showing that it is exactly to the hour and no minutes?*

Children may incorrectly assume that the hour hand 'jumps' from one hour to the next, without travelling gradually through the hour. This may lead children to draw half-past times with the hour hand pointing directly at the hour. Show children a real clock, pointing at an o'clock time. Ask:
- *What happens to the hour hand as we move to half past 2? Does the hour hand jump from one hour directly to the next?*

STRENGTHENING UNDERSTANDING

To strengthen understanding, give children pictures of different activities linked to a time of day, such as brushing teeth or going to bed, and get them to order these events on a storyboard and match them to the clock times shown. Alternatively, ask children to draw their own pictures to create their own storyboard.

GOING DEEPER

Give children an investigation to find out what is happening around the school at different times of the day. Provide a timetable with o'clock and half-past times listed against a blank space for them to record each activity. Children should be given the responsibility of checking the clock to know when to complete their investigations. While this will develop their ability to recognise the times, it will also have the added benefit of developing their awareness of how long an hour or half an hour 'feels'.

KEY LANGUAGE

In lesson: minute hand, **o'clock**, half past, hour hand

Other language to be used by the teacher: time, hour, minute, analogue

STRUCTURES AND REPRESENTATIONS

Sketches of analogue clocks on whiteboard or in children's workbooks

RESOURCES

Mandatory: analogue clock tool

Optional: vocabulary flash cards, analogue clock and written time flash cards, sorting circles

 In the eTextbook of this lesson, you will find interactive links to a selection of teaching tools.

Quick recap 

Discuss time together as a class. Ask: *What times of the day do you know? What might be happening at these times?*

Discover

 Pair work

ASK

- Question **1** a): *What times do you recognise in this picture?*
- Question **1** a): *What do you notice about where the hour hands are pointing in the half-past times?*
- Question **1** b): *Can you recognise and describe any patterns you can see in the clock times? How can this help you find the time that is hidden?*

IN FOCUS Use the picture to recap o'clock and half-past times and briefly assess children's current understanding. Begin by reinforcing the idea that the hour hand does not jump from hour to hour, but travels between each number gradually. Children can see this when they look at a time showing half past the hour. The hour hand is half-way between two numbers on the clock face. For question **1** b), make sure children look at the clock faces in the correct sequence – along the top row and then along the bottom row.

At this stage, children may find it difficult to determine that there is a half-hour gap between the times of departure, but you can help them determine the hidden time by counting in sequence as a class, for example: *Let's say the times together: 'half past 11', '12 o'clock', 'half past 12', '1 o'clock', 'half past 1', '2 o'clock'. What comes next?*

PRACTICAL TIPS Slowly move the minute hand through one hour on an analogue clock. Draw children's attention to the hour hand moving gradually from one hour to the next. Pause at the half-past time and agree that the hour hand is half-way between one number and the next.

ANSWERS

Question **1** a): half past 11, 12 o'clock, half past 12, 1 o'clock, half past 1, 2 o'clock

Question **1** b): The last clock will show half past 2.

Share

WAYS OF WORKING Whole class teacher led

ASK

- Question **1** a): *What would the clock look like if the boat left at x o'clock? Or half past x?*
- Question **1** a): *Can you draw or make x o'clock. Or half past x?*
- Question **1** a): *Why does the minute hand point to the 6 when it is a half-past time?*
- Question **1** b): *What is the next half-past time after 2 o'clock?*

IN FOCUS Use this opportunity to recap the two potential misconceptions listed on page 98:
- Children may confuse the two clock hands
- Children may incorrectly assume that the hour hand 'jumps' from one hour to the next.

Allow children the opportunity to show their understanding of times by showing times on geared clocks, drawing times onto blank clock faces or calling out times when shown different clocks.

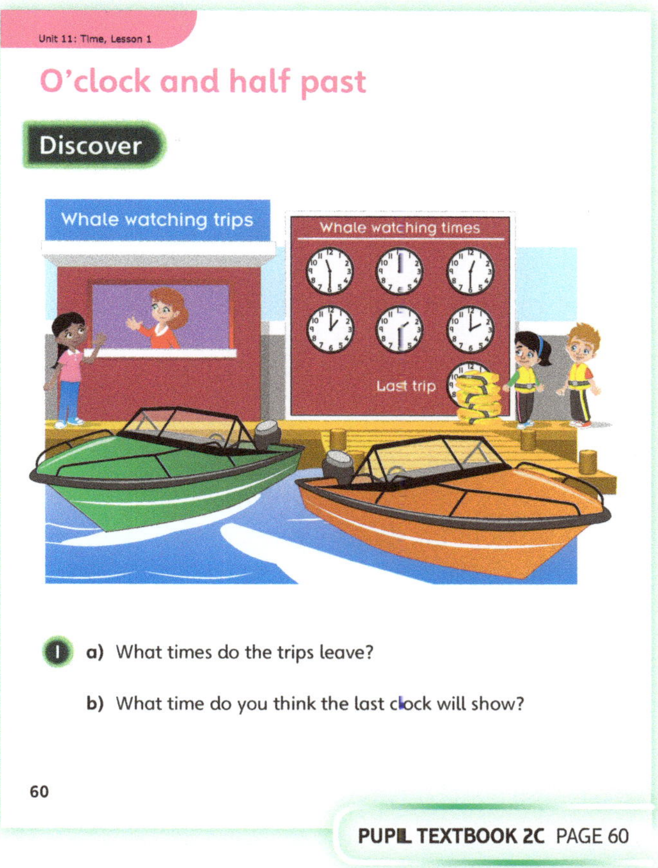

1 a) What times do the trips leave?

b) What time do you think the last clock will show?

60

PUPIL TEXTBOOK 2C PAGE 60

PUPIL TEXTBOOK 2C PAGE 61

Think together

WAYS OF WORKING Whole class teacher led (I do, We do, You do)

ASK

- Question ❶: *How can you tell which hour it is? What hand should you look at?*
- Question ❶: *What might you be doing at that time of day?*
- Question ❷: *What is different about these clocks compared to the clocks in question ❶?*
- Question ❷: *What would you be doing at this time in the afternoon or evening? What about in the morning?*
- Question ❷: *What do you recognise about the hour hand on both these clocks?*

IN FOCUS Questions ❶ and ❷ give children the opportunity to practise their recognition of o'clock and half-past times. Discuss with children the clues they can use to easily identify an o'clock time and a half-past time.

STRENGTHEN For all questions in this section of the lesson, encourage children to make the times shown. Additionally, have flash cards available with o'clock and half-past times on them, shown both on an analogue clock and as a time written in words. Ask: *Can you find the matching time and explain how you know it is the same? Is it an o'clock or half-past time?*

DEEPEN If children are successful in sorting all the clock faces shown in question ❸, ask if they are all the possible times that could be sorted into those categories. Ask: *Can you draw 4 more missing times and put them in the correct circles? Can you explain how you know you have organised them correctly?*

ASSESSMENT CHECKPOINT At this point in the lesson, children should be able to confidently identify both the hour hand and minute hand on an analogue clock. They should be able to read o'clock times and half-past times. Children should also be able to explain how the hour hand changes as the time moves from being on the hour to half past the hour.

ANSWERS

Question ❶ a): 3 o'clock

Question ❶ b): 7 o'clock

Question ❷ a): Half past 3

Question ❷ b): Half past 9

Question ❸ a): O'clock times: A, D, F; half-past times: B, C, E, G.

Question ❸ b): The minute hand points to 12 for o'clock times.

Question ❸ c): The minute hand points to 6 for half-past times.

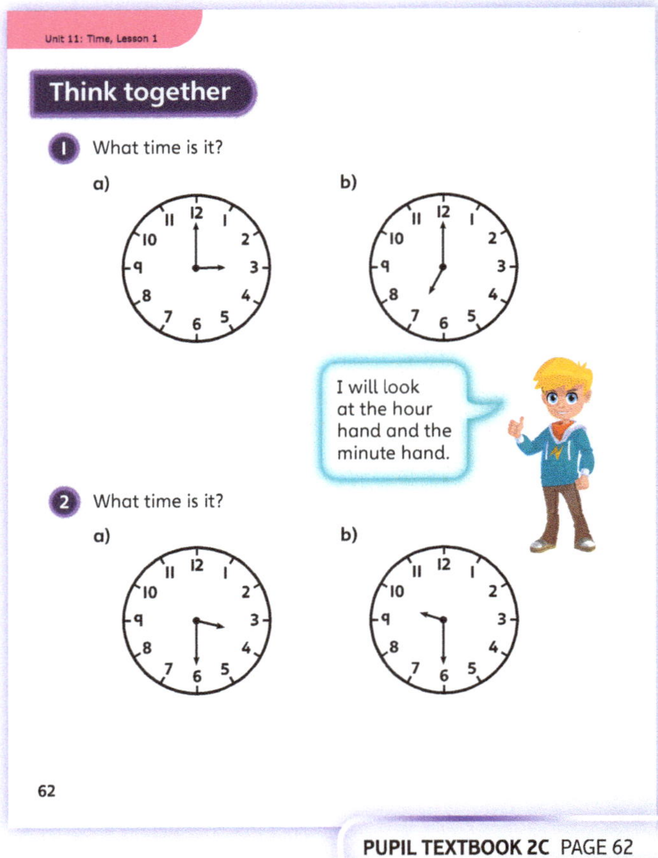

PUPIL TEXTBOOK 2C PAGE 62

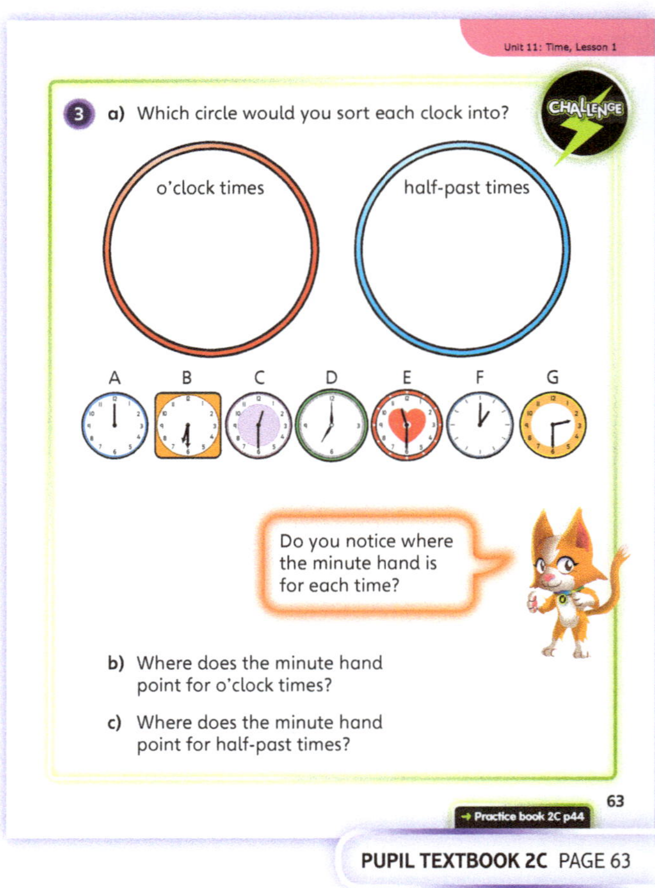

PUPIL TEXTBOOK 2C PAGE 63

Practice

WAYS OF WORKING Independent thinking

IN FOCUS Question **1** offers the opportunity for children to recognise the differences between o'clock and half-past times. Ask: *What is different between o'clock and half-past times? How are the hands different?*

STRENGTHEN Question **2** requires children to practise recording the vocabulary of the times they have been studying in the lesson. Provide flash cards with the vocabulary written on them to assist children with this. For children struggling with question **3**, offer them either plastic clock manipulatives to make the time shown or flash cards with different times shown as a picture of an analogue clock and in words.

DEEPEN Question **5** asks children to identify an o'clock time based on written clues with no visual representation. This could be deepened by requiring children to prove their ideas using evidence. Ask children to prove their ideas are correct using pictures or resources.

THINK DIFFERENTLY Question **4** addresses the misconception of confusing the hour hand and the minute hand. For children who have recognised the mistake that Sam has made, ask if they can write some advice for Sam. What will they say to help her understand her mistake? Encourage the use of pictures or resources to show her where she has gone wrong.

ASSESSMENT CHECKPOINT Children should be showing confident understanding of the minute and hour hands on an analogue clock. They should be able to clearly explain how the two hands look during an o'clock time and a half-past time and should be able to describe the similarities and differences they can see. Children should be able to confidently and accurately draw these times on to blank clock faces or make them using manipulatives.

ANSWERS Answers for the **Practice** part of the lesson can be found in the *Power Maths* online subscription.

Reflect

WAYS OF WORKING Pair work

IN FOCUS Give children time to come up with their own ideas about what is the same about the two types of times. Once they have done so and recorded their thoughts, provide them with the opportunity to share their ideas with a partner. If they have different ideas, can they convince their partner that they are correct?

ASSESSMENT CHECKPOINT At this point in the lesson, children should be able to recognise the similarities between the two types of time. Listen to children's discussions and reasoning as, through discussing the similarities, they are also likely to discuss the differences.

ANSWERS Answers for the **Reflect** part of the lesson can be found in the *Power Maths* online subscription.

After the lesson ⏸

- Are children able to confidently explain the differences between how the minute and hour hands work?
- How will you build in more opportunities to practise these skills throughout the school day?

PUPIL PRACTICE BOOK 2C PAGE 44

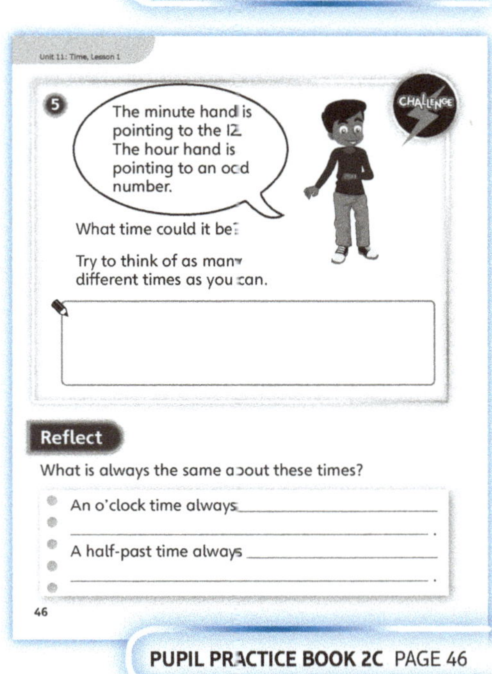

PUPIL PRACTICE BOOK 2C PAGE 45

PUPIL PRACTICE BOOK 2C PAGE 46

Quarter past and quarter to

Learning focus

In this lesson, children will describe times using the vocabulary of 'quarter past' and 'quarter to'. They will confidently read and record times on an analogue clock.

Before you teach 

- Have the children made the link between their learning in this unit and the learning in the unit on fractions?
- How could you make this link explicit in your teaching?

NATIONAL CURRICULUM LINKS

Year 2 Measurement – time

Tell and write the time to five minutes, including quarter past/to the hour, and draw the hands on a clock face to show these times.

ASSESSING MASTERY

Children can identify 'quarter past' and 'quarter to' using the hour and minute hands, and can match this vocabulary against pictures of appropriate clock faces. Children can recognise that the word 'past' refers to the previous hour and 'to' refers to the following hour, as well as being able to use their understanding to accurately record the hands on a blank analogue clock.

COMMON MISCONCEPTIONS

Children may find it difficult to recognise a quarter of the analogue clock or which quarter is 'past' and which is 'to'. Provide a colour-coded clock face with the quarters labelled clearly on it. Ask:
- *What hour has this quarter just past? What hour is this quarter leading to?*

STRENGTHENING UNDERSTANDING

To help children develop their fluency with half past, o'clock, quarter past and quarter to, draw a large chalk circle on the playground with those points of the clock labelled. Children should run to the correct point on the clock face when the vocabulary is called out by the teacher.

GOING DEEPER

Deepen understanding by asking children questions such as:
- *How many quarters of an hour are there in one hour?*
- *If the time was half past 2, how many 'quarters of an hour' would there be until half past 3?*
- *How about from 3 o'clock to quarter past 4?*

KEY LANGUAGE

In lesson: quarter past, quarter to

Other language to be used by the teacher: hour, minute, hand, time, o'clock, half past, almost

STRUCTURES AND REPRESENTATIONS

Sketches of analogue clocks on whiteboard or in children's workbooks

RESOURCES

Mandatory: analogue clock tool

Optional: vocabulary flash cards, analogue clock and written time flash cards, rope

 In the eTextbook of this lesson, you will find interactive links to a selection of teaching tools.

Quick recap 

Look together at the class clock. Ask: *Which is the hour hand? Which is the minute hand?*

Discover

 Pair work

ASK

- Question **1** a): *What times do you recognise?*
- Question **1** a): *What do the two hands mean on the clock?*
- Question **1** b): *What time will the monkeys be fed?*
- Question **1** b): *What does the hour hand look like on each clock face? Does it always point directly at an hour?*
- Question **1** b): *Are there any times you cannot read? What is the same and different about those ones and the ones you do know?*

IN FOCUS Use this picture as an opportunity to begin discussing the similarities and differences between the times the children have already learnt and those that are new. Recap the potential misconception of not recognising how the hour hand moves during the hour to ensure children avoid the misconception later in the lesson.

In Unit 10, children learnt to divide shapes into quarters. Ask: *How many equal parts are there if I divide a circle into quarters?* Relate this to the clock face. Show them that when the time is quarter to or quarter past, the minute hand divides the circle of the clock face into two portions: one-quarter and three-quarters.

PRACTICAL TIPS Slowly move the minute hand on an analogue clock face from 5 o'clock to 6 o'clock. Ask children to stop you as you reach each of the times shown on the sign in the **Discover** image, and to name any times that they recognise.

ANSWERS

Question **1** a): The tiger will be fed at quarter past 5.

Question **1** b): The penguin will be fed at quarter to 6.

Share

 Whole class teacher led

ASK

- Question **1** a): *How is the minute hand changing for each clock time?*
- Question **1** a): *What do you notice about where the hour hand is at quarter-past and quarter-to times?*
- Question **1** a): *Once the tiger has been fed, what time will the next quarter of an hour be? What would the clock face show?*
- Question **1** b): *What does 'quarter past' mean? Past what?*
- Question **1** b): *What does 'quarter to' mean? To what?*

IN FOCUS During this stage of the lesson, give children different times of when zoo exhibits open and close. When given the times, children could be asked to show the time using either a picture or an analogue clock they can manipulate. Be sure to reinforce the fact that the hour hand travels through the hours and does not jump from one to the next. To make this section more practical, 12 children could be arranged in a circle to mark the divisions on an analogue clock. Using two pieces of rope, one longer than the other, four other children could create the hands.

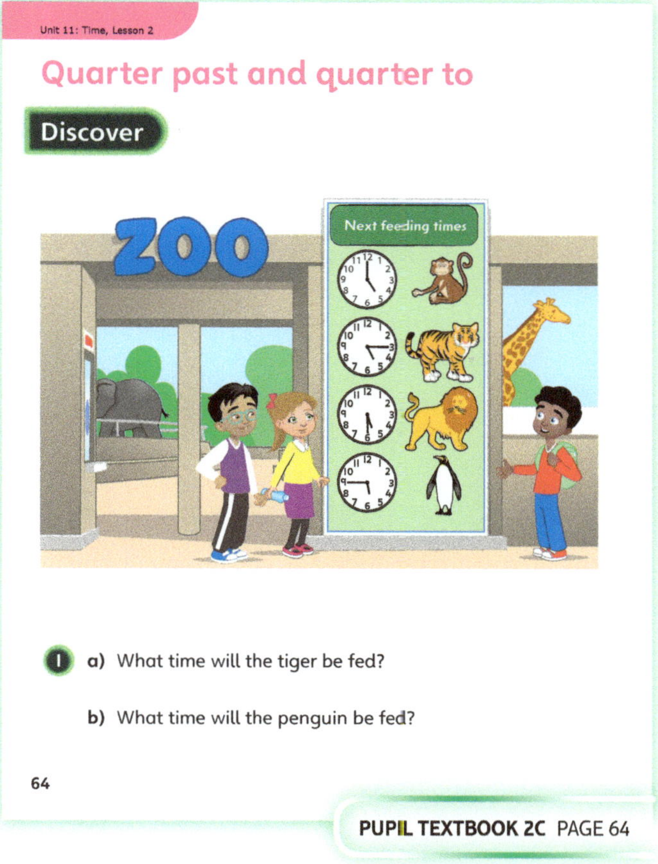

PUPIL TEXTBOOK 2C PAGE 64

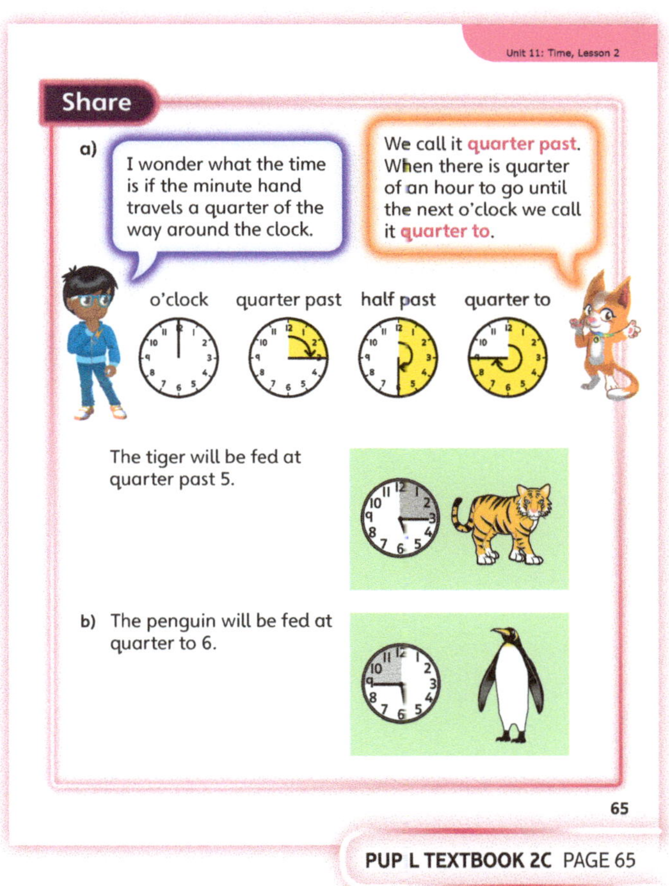

PUP L TEXTBOOK 2C PAGE 65

Think together

 Whole class teacher led (I do, We do, You do)

- Question ❶: *How will you read the analogue clock?*
- Question ❶: *What clues can you look for to know if the time reads quarter to or quarter past?*
- Question ❶: *Where is the minute hand pointing in the first clock? What about in the second clock?*

 The image in question ❶ reinforces the idea of quarter to and quarter past. Use the shaded part of the picture to reinforce the concept.

 In question ❷, if children are struggling to read the times, encourage them to use an analogue clock to help show the times. Ask them to start at 11 o'clock and move the minute hand through to quarter past. Does their clock match one of the clocks in the picture?

 In question ❸, ask children to investigate how many different quarter-past and quarter-to times they can find. Can they organise and explain their answer?

 At this point children should be more confident when reading quarter-past and quarter-to times. They should be able to explain what is meant by this vocabulary and be able to record these times on both analogue clocks and pictures of blank analogue clocks. Look for their confident and fluent understanding and use of the minute and hour hands.

Question ❶ a): The reptile house opens at quarter past 10.

Question ❶ b): The reptile house closes at quarter to 4.

Question ❷: You can meet the macaws at quarter past 11 and quarter to 3.

Question ❸: Check children can accurately display quarter-past and quarter-to times on a clock or drawing. Can they read the times shown on a partner's clock correctly?

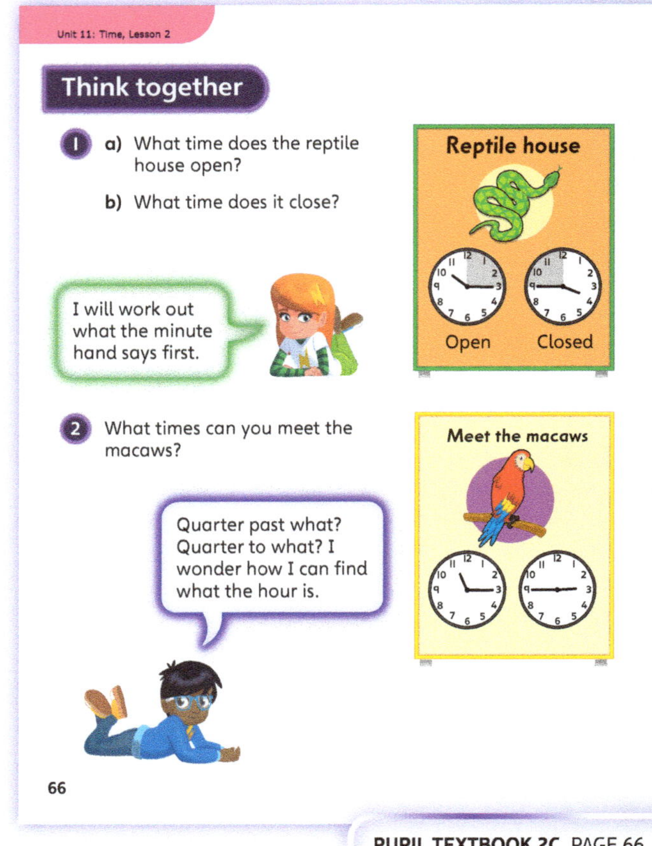

PUPIL TEXTBOOK 2C PAGE 66

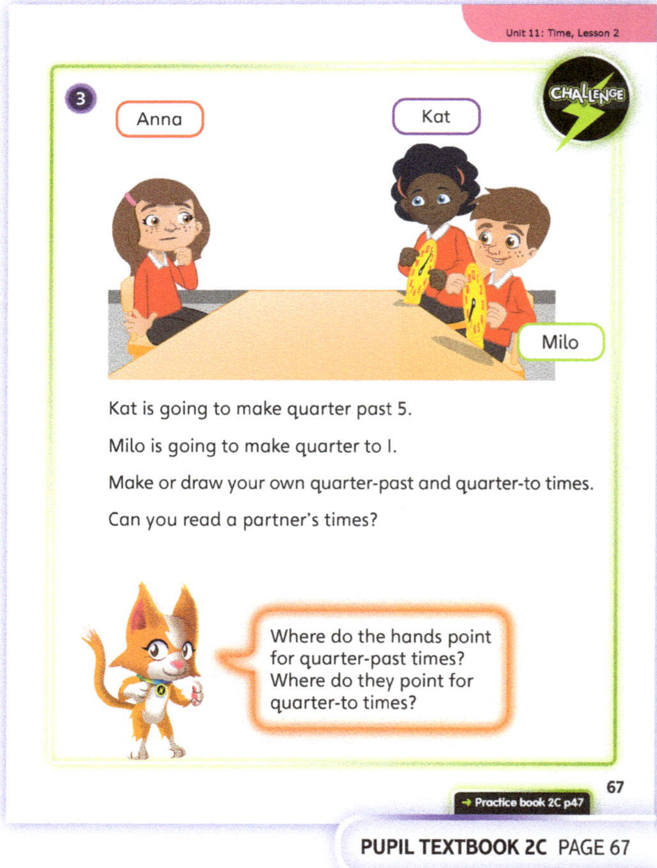

PUPIL TEXTBOOK 2C PAGE 67

Practice

WAYS OF WORKING Independent thinking

IN FOCUS Questions **1** and **2** scaffold children's understanding of 'quarter to' and 'quarter past' by providing them with shaded representations of the times. They also allow children the opportunity to work with clocks of different shapes and sizes to help develop their fluency.

STRENGTHEN For children struggling to match the clocks with the right times in questions **1**, **2** and **3**, it may be beneficial, as in the previous lesson, to offer children flash cards with the pictures and written names of times on them. Can children find the matching times? Ask children what clues they will look for in the pictures and in the written names. How can they use the clues to help them find the matching time more quickly?

DEEPEN Use question **6** to deepen children's reasoning and explanation skills. Ask children to give examples of times that support their explanations and to draw accompanying images on analogue clock faces, including shading the quarter past or quarter to.

THINK DIFFERENTLY Question **5** explores the misconception of confusing quarter past and quarter to. Ask children what part of the lesson they think Malik did not understand. Ask: *What would you say to Malik to help him understand?* Can children use a resource or picture to help Malik understand his mistake?

ASSESSMENT CHECKPOINT At this point in the lesson, children should be able to confidently recognise and record quarter-to and quarter-past times on an analogue clock. They should be able to describe what is meant by 'to' and 'past' and how these relate to the preceding and following hours.

ANSWERS Answers for the **Practice** part of the lesson can be found in the *Power Maths* online subscription.

Reflect

WAYS OF WORKING Pair work

IN FOCUS This question offers a good opportunity to observe children's reasoning. Pay particular attention to children's reasoning when it comes to quarter to 6 and quarter past 6. Ask children to explain why a clock showing quarter past 6 is not showing quarter to 6. Can they explain how they are different?

ASSESSMENT CHECKPOINT Look for children able to explain the similarities and differences between the clock faces shown in the question. Children should be able to clearly explain why only one of the clocks shows the correct time.

ANSWERS Answers for the **Reflect** part of the lesson can be found in the *Power Maths* online subscription.

After the lesson ⏸

- How confident are children with the concept of 'to' and 'past'?
- How many opportunities were there to problem solve in this lesson?

PUPIL PRACTICE BOOK 2C PAGE 47

PUPIL PRACTICE BOOK 2C PAGE 48

PUPIL PRACTICE BOOK 2C PAGE 49

Tell the time to 5 minutes

Learning focus

In this lesson, children continue to develop their ability to read an analogue clock by learning to read the five-minute intervals. They will link this to their prior learning about half past, quarter to and quarter past.

Before you teach 

- Are children confident counting in 5s?
- How will you make this lesson as hands-on as possible? What resources could you provide beyond the clocks?

NATIONAL CURRICULUM LINKS

Year 2 Measurement – time

Tell and write the time to five minutes, including quarter past/to the hour, and draw the hands on a clock face to show these times.

ASSESSING MASTERY

Children can recognise any time where the minute hand is pointing directly to a number on the clock face. Children can read and record times on an analogue clock and link their understanding to the vocabulary they have already learnt in the unit so far.

COMMON MISCONCEPTIONS

Children may confuse the numbers around the edge of the clock, showing the hours, with a representation of the minutes. For example, they may read 'five minutes past 10' as 'one minute past 10'. Ensure children are able to access a clock that has each individual minute marked around the clock face. Ask:

- *What do these small marks around the edge show? How many minutes have you counted when you get to the first number? How about the second number? How many minutes are in between each number?*

STRENGTHENING UNDERSTANDING

Children could be given opportunities to count in 5s through activities like counting games, counting songs and specially prepared dot-to-dots.

GOING DEEPER

Encourage children to look around them for times that they recognise. This could include investigating a specially prepared television scheduling magazine that shows the times either as written names or on analogue clocks. Can children create a timetable of programmes they would watch?

KEY LANGUAGE

In lesson: past, to, five

Other language to be used by the teacher: half, quarter, minutes, hour, o'clock, ten, twenty, twenty-five, thirty-five, fifty-five

STRUCTURES AND REPRESENTATIONS

Sketches of analogue clocks on whiteboard or in children's workbooks

RESOURCES

Mandatory: analogue clock tool

Optional: vocabulary flash cards, analogue clock and written time flash cards

 In the eTextbook of this lesson, you will find interactive links to a selection of teaching tools.

Quick recap 

As a class, count up together in 5s from 0 to 60.

Discover

WAYS OF WORKING Pair work

ASK

- Question **1** a): *What time does the watch say?*
- Question **1** a): *What time is five minutes from the time shown?*
- Question **1** b): *If the bus is due to arrive in five minutes, but it takes another fifteen minutes after that, how many minutes will it be, in total, from now? What is 5 + 15?*
- Question **1** b): *What time would the bus have arrived if it had got there five minutes early?*

IN FOCUS When looking at this picture, present children with an analogue clock that has minute divisions around the outside. Ensure that children recognise how the numbers around the outside do not match the minutes that the minute hand shows. In question **1** b), help children understand that they should not simply add 15 minutes to the current time. Instead, if the bus is 15 minutes late, it will arrive 15 minutes after the scheduled time. That is 5 + 15 = 20 minutes from now.

PRACTICAL TIPS Show the time from the watch on an analogue clock face and ask a child to tell you what time it says. Then model moving the minute hand on by 5 minutes to reveal the time that the bus will arrive. Ask: *What time does it say now?*

ANSWERS

Question **1** a): The bus will arrive at 25 minutes past 1.

Question **1** b): If the bus is 15 minutes late, it will arrive at 20 minutes to 2.

Share

WAYS OF WORKING Whole class teacher led

ASK

- Question **1** a): *What are the numbers on the clock for? Do they show the number of minutes? How can you use them to help?*
- Question **1** a): *What does x minutes past y look like on a clock?*
- Question **1** a): *What does it mean when the minute hand points to the number x?*
- Questions **1** a) and b): *Why is it useful to know how to skip count in 5s when you are telling the time?*
- Questions **1** a) and b): *What time is five minutes more than x past y?*
- Questions **1** a) and b): *How many minutes past the hour is half past x? Or quarter past x? Or quarter to x?*

IN FOCUS Be sure to make this part of the lesson as hands-on as possible. Ask children to match flash cards to the correct part of a clock. For example, a flash card saying 'twenty to' should be matched to the number 8 on the analogue clock. Children could also move around a large clock drawn out on the playground, counting the five-minute intervals as they move.

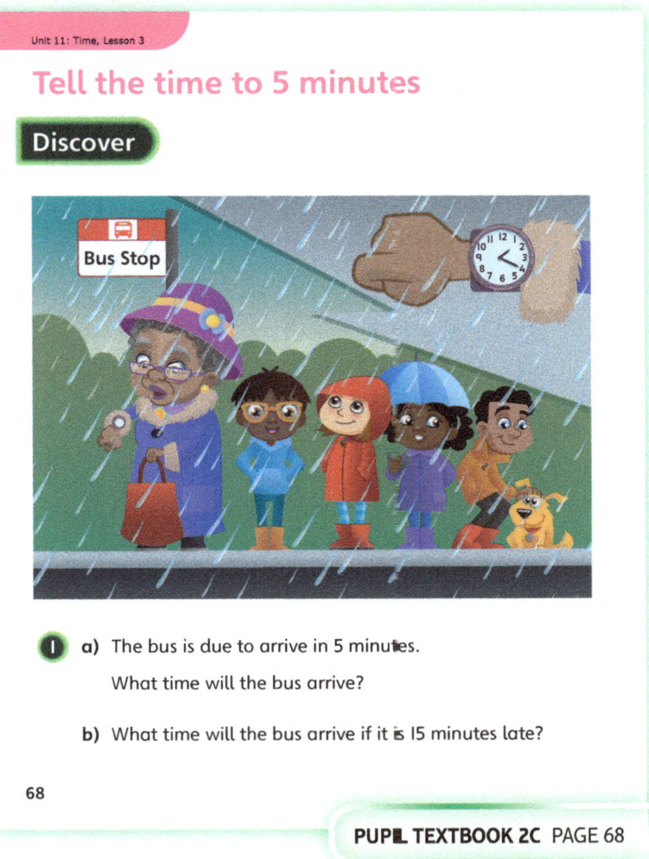

PUPIL TEXTBOOK 2C PAGE 68

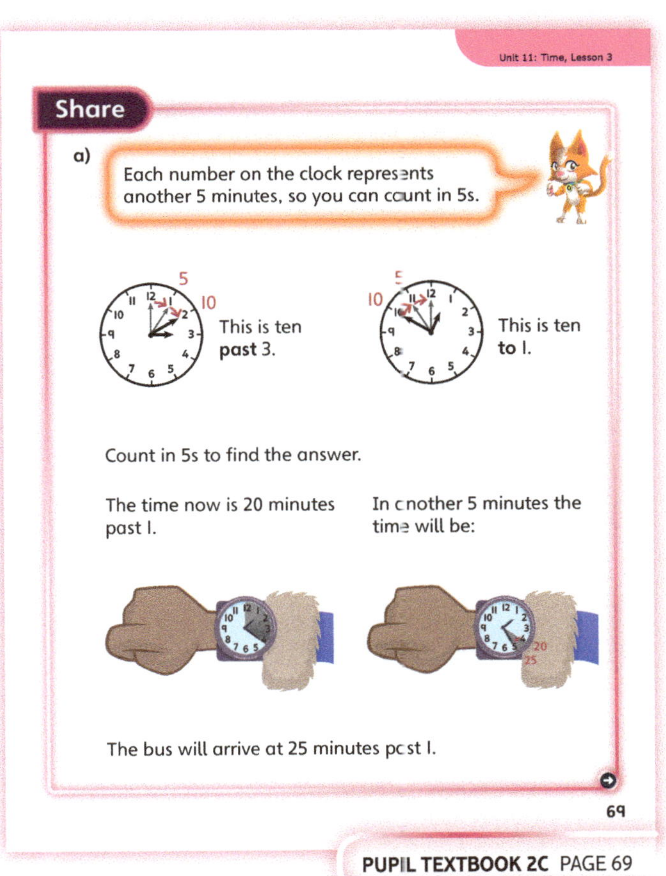

PUPIL TEXTBOOK 2C PAGE 69

Think together

WAYS OF WORKING Whole class teacher led (I do, We do, You do)

ASK

- Questions ① a) and c): *How can you tell how many minutes past 6 it is? Can you count in 5s, following the arrows?*
- Questions ① b) and d): *How can you tell how many minutes to 7 it is? Can you count back from 7 o'clock (five to 7, ten to 7, etc.)?*
- Question ②: *What could you count in to make it easier to read the minutes?*

IN FOCUS Question ① will help scaffold the children's counting in five-minute intervals. To secure their understanding that each segment represents 5 minutes, ask children to match the picture to an analogue clock showing all the minute divisions.

STRENGTHEN If children are struggling to tell the time in questions ① and ②, offer them an analogue clock to make the time with. As children move the minute hand round through the correct hour, count the five-minute intervals with them. Make sure to reinforce the 'past' and 'to' elements of the clock face. Additionally, offer children flash cards showing different minutes past and to. Can they match the picture they are looking at with one of the flash cards? Using the flash card, can they name the time in full?

DEEPEN If children have found all of the times between the two times given in question ③, deepen their understanding by asking if there are any times that could be labelled in more than one way. Additionally, ask children if all hours have the same number of solutions. Can they explain?

ASSESSMENT CHECKPOINT At this point in the lesson, children should be more confident when counting in five-minute intervals. They should be increasingly fluent in their understanding and use of 'past' and 'to', recognising and explaining when to use each. They should be able to link their understanding of the vocabulary from the previous lesson with that of this lesson.

ANSWERS

Question ① a): 5 minutes past 6

Question ① b): 25 minutes to 7

Question ① c): 10 minutes past 6

Question ① d): 20 minutes to 7

Question ② a): 20 minutes past 3

Question ② b): 10 minutes to 4

Question ② c): 25 minutes past 3

Question ② d): 5 minutes to 4

Question ③: Check children can draw 'past' and 'to' times on a clock between 8 o'clock and 9 o'clock. Can children write the correct times they have drawn on their clocks?

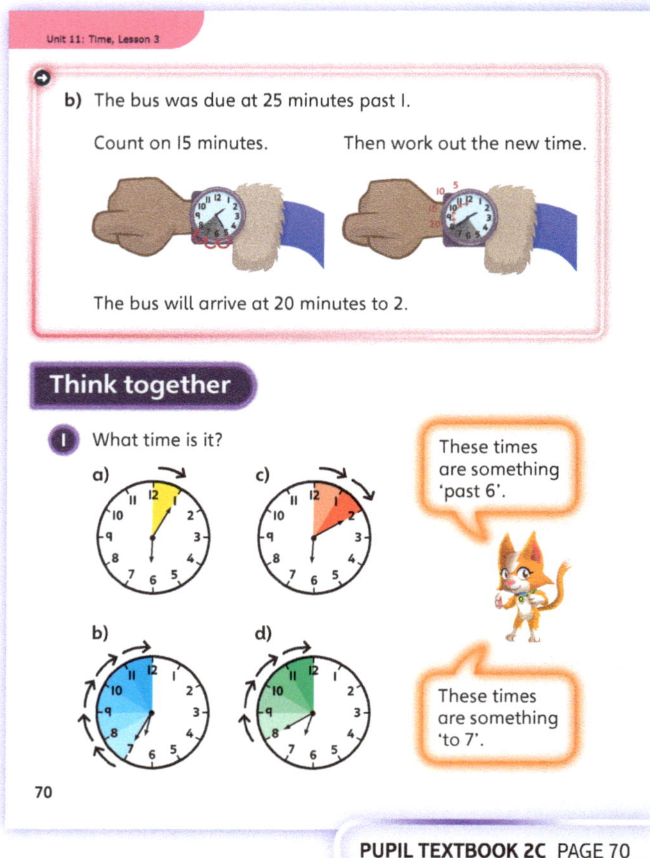

PUPIL TEXTBOOK 2C PAGE 70

PUPIL TEXTBOOK 2C PAGE 71

Practice

WAYS OF WORKING Independent thinking

IN FOCUS Question **2** links the concept to children's real-life experience. It may be beneficial to have the pictures and vocabulary available as cut-outs so children can manipulate and order them more easily. For questions **1** to **4**, encourage children to make the times on an analogue clock so they can manipulate them.

STRENGTHEN In question **4**, ask children to think about the position of each hand on the clock face separately. For example, if the time is 20 minutes 'to' 3, what hour will they be counting within? Will the hour hand be nearer to the 2 or the 3? What number will the minute hand be pointing to?

DEEPEN Use Ash's question in question **5** to deepen children's thinking. How many times can children say in two different ways? Are there any times children can say in more than two ways? Is this the same for every hour?

ASSESSMENT CHECKPOINT At this point in the lesson, children should be able to confidently count around the clock in five-minute intervals. They should be able to use 'past' and 'to' fluently to describe the times they read and they should be able to link the vocabulary they have learnt in previous lessons with the vocabulary learnt in this lesson.

ANSWERS Answers for the **Practice** part of the lesson can be found in the *Power Maths* online subscription.

Reflect

WAYS OF WORKING Pair work

IN FOCUS Give children an opportunity to develop their own line of thinking. Ask them what representations of time they have worked with and how they could use them now to show the time

ASSESSMENT CHECKPOINT Look for children recognising that at twenty past the minute hand should be pointing at the 4. Children's reasoning should make mention of the concepts they have covered so far. For example, they may have used a concrete representation or picture, counted in 5s on their fingers and/or looked back at their prior work.

ANSWERS Answers for the **Reflect** part of the lesson can be found in the *Power Maths* online subscription.

After the lesson ⏸

- Have children recognised how the concepts they have learnt about in the past three lessons link?
- How will you reinforce this link?

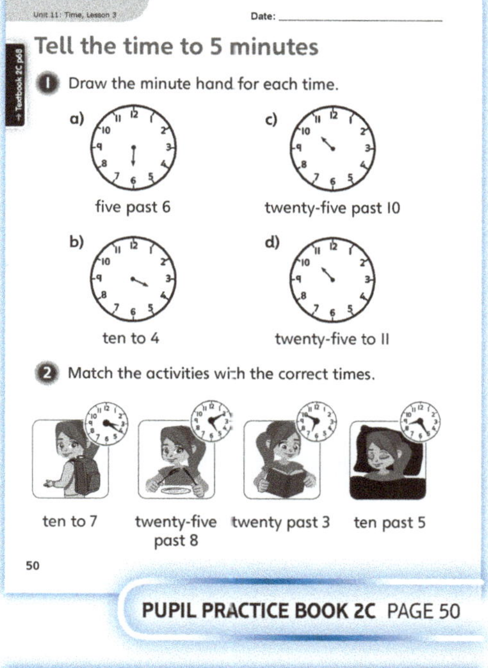

PUPIL PRACTICE BOOK 2C PAGE 50

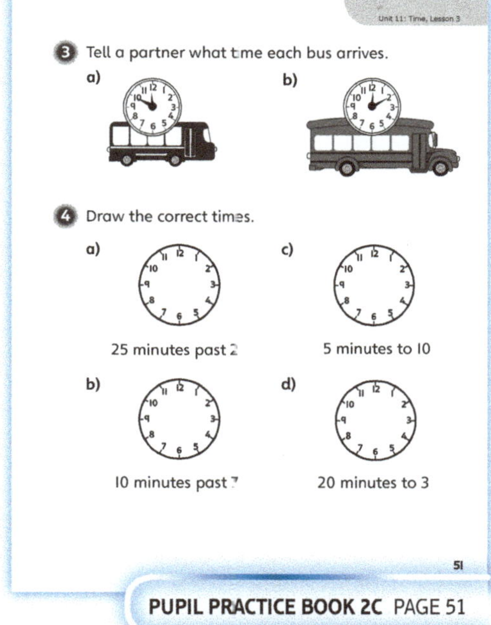

PUPIL PRACTICE BOOK 2C PAGE 51

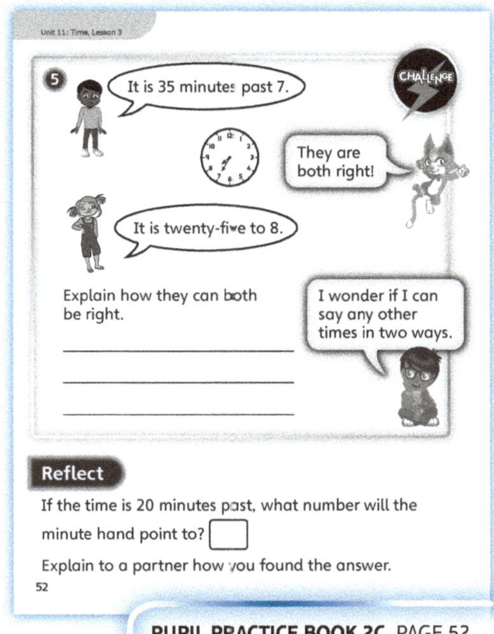

109

Minutes in an hour

Learning focus

In this lesson, children will develop their understanding of how many minutes there are in an hour. They will use this understanding, and the representations of part-whole and bar models, to help them solve mathematical problems.

Before you teach ⏸

- Have children recognised that there are 60 minutes in an hour?
- How will you challenge those children in the opening part of the lesson?

NATIONAL CURRICULUM LINKS

Year 2 Measurement – time

Know the number of minutes in an hour and the number of hours in a day.

ASSESSING MASTERY

Children can recognise that there are 60 minutes in an hour and can show this understanding on an analogue clock. Children can use their understanding to calculate and convert durations expressed in minutes, or hours and minutes, up to 100 minutes.

COMMON MISCONCEPTIONS

Children may not recognise that they can continue counting in minutes once they reach the full 60 minutes in an hour. To help children understand this concept, link the representation of a clock face to the representation of a number line. Ask:
- *How are these two representations the same? How are they different? What would happen if you counted on from 60 on a number line? Could you do this on a clock face? How is the process different on a clock face to a number line? How is it the same?*

STRENGTHENING UNDERSTANDING

Offer children opportunities to count up in 5s and 10s around a clock face with 12 interval markings but no numbers. This will help to reinforce the idea of counting around the circle, without needing to stop at the 12. When counting, ask: *Does this representation looks like anything you have met before? How is it the same to what you have seen? How is it different?*

GOING DEEPER

Challenge children to find how many minutes are in a set number of hours. Ask: *How many minutes are in one hour? Two hours? Three? Four?* Ask children to describe any patterns they find.

KEY LANGUAGE

In lesson: hour, minutes, longer, same, units

Other language to be used by the teacher: last, clock, convert

STRUCTURES AND REPRESENTATIONS

Sketches of analogue clocks, part-whole model, bar model

RESOURCES

Mandatory: analogue clock tool

Optional: base 10 equipment

 In the eTextbook of this lesson, you will find interactive links to a selection of teaching tools.

Quick recap 

As a class, count up together in 5s around a clock face from 0 to 60.

Discover

WAYS OF WORKING Pair work

ASK

- Question **1** a): *How many hours and minutes are in 80 minutes?*
- Question **1** b): *Where would the minute hand point to if it was 55 minutes past the hour?*
- Question **1** b): *Is a race result of 1 hour and 10 minutes faster or slower than 55 minutes?*

IN FOCUS Once children know that there are 60 minutes in an hour, question **1** a) helps them learn to partition a number of minutes greater than 60 into hours and minutes. It is important that children use a structure like a bar model or a part-whole model to support their learning here.

PRACTICAL TIPS To help children use the part-whole models, they could also use base 10 equipment (counters or cubes) to partition 80 into 60 minutes and 20 minutes. They can draw on their knowledge of number bonds to 10 to do this. To help children see that they can continue counting around a clock face for as long as they like, use a clock face like the one in the picture to count the minutes in Eve's time. Children should follow the minutes with their fingers to help cement their understanding.

ANSWERS

Question **1** a): Joe took 1 hour and 20 minutes.

Question **1** b): Anya took 55 minutes.
Eve took 1 hour 10 minutes, which is 70 minutes. Eve took longer.

Share

WAYS OF WORKING Whole class teacher led

ASK

- Question **1** a): *How can you prove that there are 60 minutes in one hour? Can you count in 5s to show me?*
- Question **1** a): *If a whole hour is 60 minutes, can you use this to show how many minutes half of an hour is?*
- Question **1** b): *If something lasts for 1 hour and 5 minutes, how many minutes is that? What about 1 hour and 10 minutes?*
- Question **1** b): *If a whole hour is 60 minutes, can you use this to show how many minutes half of an hour is?*
- Question **1** b): *Do you think it is easier to think of a time as 75 minutes or 1 hour and 15 minutes?*
- Question **1** b): *How is 80 minutes less than 1 hour and 30 minutes when 80 is a bigger number than both 1 and 30?*

IN FOCUS At this point in the lesson, it is important to make sure that children are confident with the concept that minutes can continue to be counted after reaching the 60th minute. At all times during this section, make sure the link between the measurement of minutes and of hours and minutes is made explicit. Ensure children are given sufficient opportunity to practise making this link.

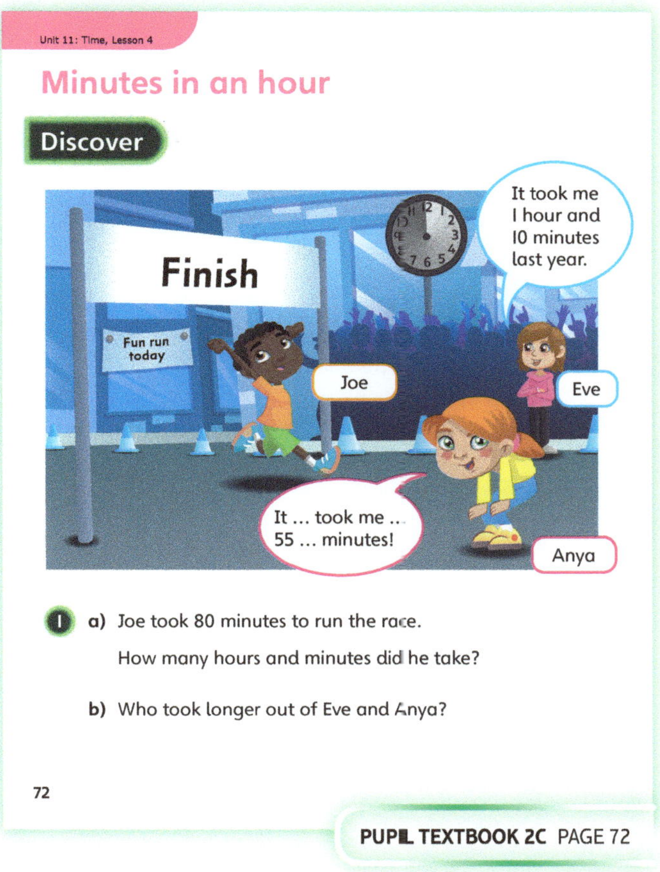

PUPIL TEXTBOOK 2C PAGE 72

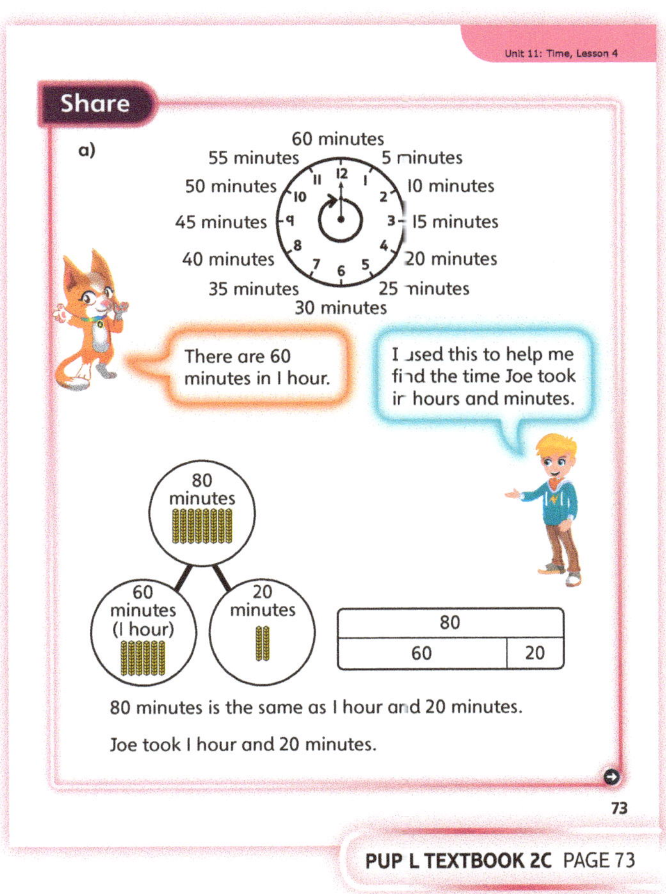

PUPIL TEXTBOOK 2C PAGE 73

Think together

WAYS OF WORKING Whole class teacher led (I do, We do, You do)

ASK

- Question **1** a): *How can you find out how many minutes are in one hour?*
- Question **1** a): *Why is the full clock face coloured?*
- Question **1** b): *What do you need to do if you are counting the minutes in more than one hour?*
- Question **1** b): *Why is there more than one clock face? Is that what it would look like if you were counting this in real life?*
- Question **1** b): *How else could you represent the times?*
- Question **1** b): *How does base 10 equipment help you represent the times?*

IN FOCUS In question **1** b), make sure children understand the image of the two clock faces as a *representation of time*, not to mean that there would be more than one clock face in real life. Support this idea by encouraging children to move the minute hand around a clock to match the question. Identify that this is possible on the same clock face.

STRENGTHEN If children are struggling with question **2**, offer them the concrete and pictorial representations of times they have worked with so far to support their thinking. Ask children to count the minutes around a clock face. How many minutes did they count? Could they represent the minutes using another resource or picture?

DEEPEN Deepen conceptual understanding of question **3** by challenging children to create a mini lesson that explains how the part-whole model shows how many hours and minutes are in *x* minutes. Ask children to imagine they are going to introduce this to someone who has never seen it before. Can they prepare a short lesson that would help someone understand how the part-whole model shows how many hours and minutes are in *x* minutes? What concrete resources or pictures will they use? Children could be given a video recording device to record the mini lesson.

ASSESSMENT CHECKPOINT At this point in the lesson, look for children recognising they can count the minutes around a clock face, crossing the hour mark as many times as necessary. Children should be more confident at converting between hours and minutes and should be able to support their reasoning with concrete and pictorial representations, including base 10 equipment, part-whole models and bar models.

ANSWERS

Question **1** a): 1 hour is the same as 60 minutes.

Question **1** b): 1 hour and 5 minutes is the same as 65 minutes.

Question **2**: 1 hour and 15 minutes is the same as 75 minutes.

Question **3**: 95 minutes is 1 hour and 35 minutes.

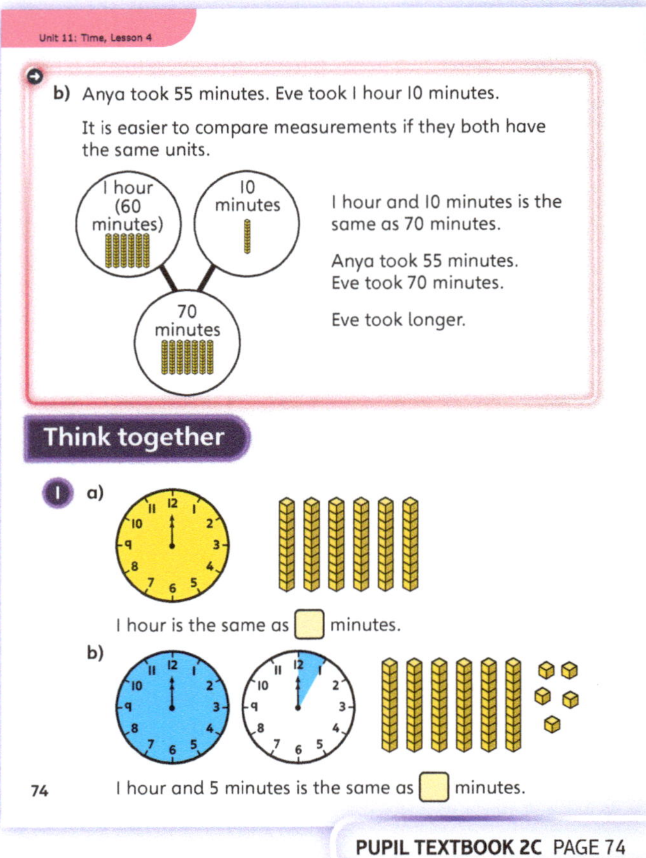

PUPIL TEXTBOOK 2C PAGE 74

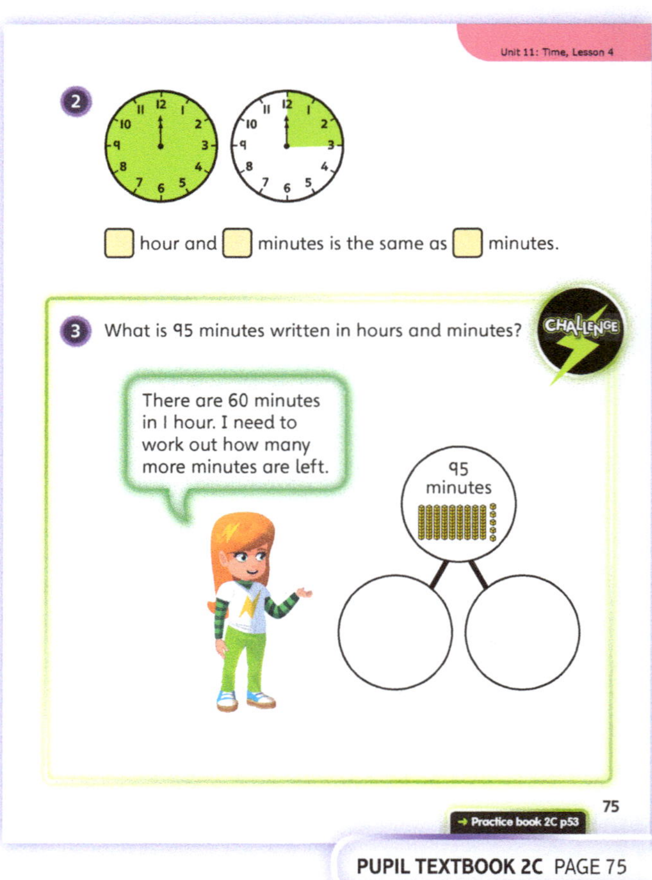

PUPIL TEXTBOOK 2C PAGE 75

Practice

WAYS OF WORKING Independent thinking, pair work

IN FOCUS Questions **1**, **2** and **3** scaffold children's ability to find a number of minutes on the clock. To help children make the links between the part-whole model and the clock face, provide clock tools that children can manipulate in order to count the minutes. As children move the minute hand, encourage them to colour in each five-minute segment.

STRENGTHEN If children are struggling with question **4**'s abstract representations of time, provide children with the concrete and pictorial representations they have been using thus far to create the times listed. Ask: *How many minutes are there in 1 hour? How many minutes are there from 1 hour to 77 minutes? Can you represent this on a clock face? Or a bar model? Which representation helped them compare times most clearly?*

DEEPEN Challenge children to devise a word problem similar to question **4** and then swap problems with a partner. Can they use resources to show how they found the solution?

ASSESSMENT CHECKPOINT At this point in the lesson, children should be confident when converting hours to minutes and vice versa. Look for children to be using the concrete and pictorial representations they have worked with to support their reasoning. Children should also use these representations to help them compare and order times. Children should count the minutes around the clock confidently and fluently, and recognise when a given number of minutes will be greater or less than an hour.

ANSWERS Answers for the **Practice** part of the lesson can be found in the *Power Maths* online subscription.

Reflect

WAYS OF WORKING Pair work

IN FOCUS Introduce children to the question and give them time to consider what their answer would be. Once they have done so, ask them to *prove* how many minutes are in an hour. Once children have fed back, give them time to discuss with a partner how they would help someone overcome the misconception that there are 12 minutes in one hour.

ASSESSMENT CHECKPOINT Children should recognise that the numbers around the clock face do not directly link to the number of minutes they represent. They should be able to explain, and show, how they know there are 60 minutes in one hour.

ANSWERS Answers for the **Reflect** part of the lesson can be found in the *Power Maths* online subscription.

After the lesson ⏸

- Are children confident with how many minutes are in an hour?
- Are children secure converting from hours to minutes and from minutes to hours?

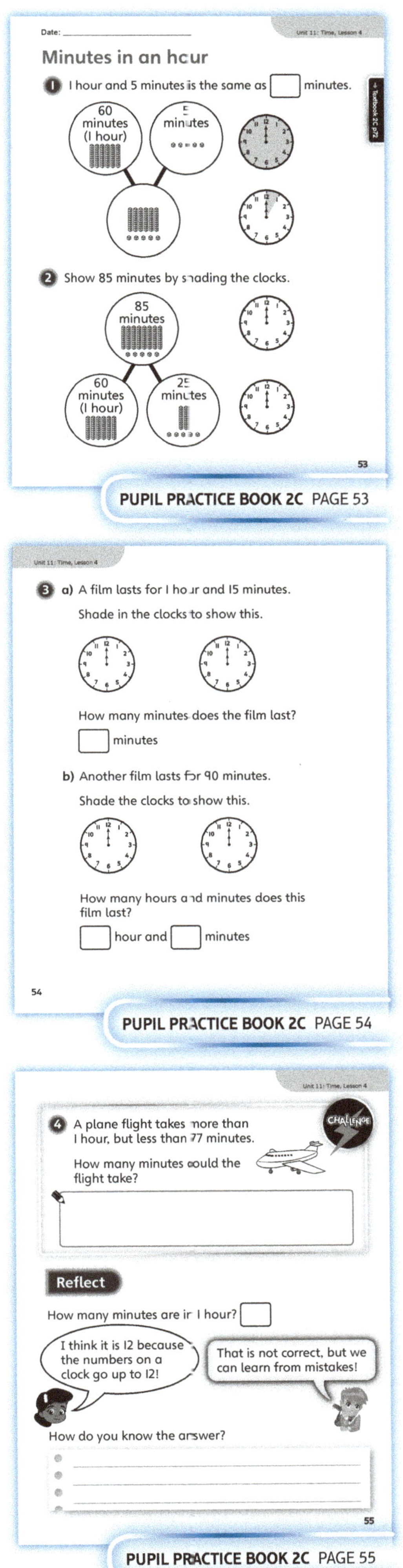

PUPIL PRACTICE BOOK 2C PAGE 53

PUPIL PRACTICE BOOK 2C PAGE 54

PUPIL PRACTICE BOOK 2C PAGE 55

Hours in a day

Learning focus

In this lesson, children will learn that there are 24 hours in a day. They will learn how there are morning and afternoon times, and use this to solve simple problems.

Before you teach 

- What other experiences could you offer children to help them engage with and understand times of day that they may not have real-life experience of?

NATIONAL CURRICULUM LINKS

Year 2 Measurement – time

Know the number of minutes in an hour and the number of hours in a day.

ASSESSING MASTERY

Children can demonstrate their understanding that there are 24 hours in a day. Children can recognise that 2 o'clock in the morning is different from 2 o'clock in the afternoon, and can use this to solve mathematical problems.

COMMON MISCONCEPTIONS

Children may think that one day is shown by the hour hand moving once around the clock to the next instance of the same time (for example, 11 o'clock to 11 o'clock). Ask:

- *How many hours are there in a day? How many hours have you counted? Do those two amounts match? Which is greater? How many more hours do you need to count until you have counted a full day?*

Children may have a lack of understanding of the parts of the day they have little experience of (for example, very late at night) and may associate the start of the day with dawn and the end of the day with sunset. Ask:

- *If the sun comes up at 5 o'clock in the morning and goes down at 9 o'clock at night, for how many hours has it been daylight?*
- *Is that a full 24 hours? What happens with the other hours? How many hours pass while you are asleep?*

STRENGTHENING UNDERSTANDING

Prepare children for this lesson by giving them opportunities to count around the clock face in hours. Children could role-play what they would be doing at different times during the day as they count around a clock. Alternatively, they could use a teddy bear to act out the situations, to allow them to concentrate on counting hours.

Children could also be given 24 pictures of activities done during the day. Ask them to order these around a clock. Use this as an opportunity to discuss why there are 24 pictures and not 12.

GOING DEEPER

Children could be given the challenge of investigating how many hours there are in three days. Ask children if they can they prove their responses. At this level, children should try to do this by repeated addition of 24 hours rather than using multiplication.

KEY LANGUAGE

In lesson: hours, day, o'clock, midday, daytime, midnight, around the clock

Other language to be used by the teacher: morning, afternoon, twice, nighttime, am, pm, duration

STRUCTURES AND REPRESENTATIONS

Sketches of analogue clocks, bar model, part-whole model

RESOURCES

Mandatory: analogue clock tool

Optional: laminated pictures of clock faces, weekday timeline (noting hours and weekdays), 24 pictures of activities linked to each hour on the clock, soft toys for role play

 In the eTextbook of this lesson, you will find interactive links to a selection of teaching tools.

Quick recap

As a class, discuss what children know about hours and minutes. Ask: *How long is one hour? What can you think of that takes about an hour?*

Discover

WAYS OF WORKING Pair work

ASK

- Question **1** a): *How many times does the minute hand travel around the clock in one day?*
- Question **1** b): *What times do the two clocks show?*
- Question **1** b): *How long does Sunil have to drink his smoothie?*

IN FOCUS The picture in this part of the lesson offers an interesting opportunity to discuss the duration of time between the two pictures. Emphasise that it is daylight outside in both pictures and that the second picture says 'Next day' to encourage children to realise that at least 24 hours have passed.

PRACTICAL TIPS Show the time 1 o'clock on an analogue clock. Move the minute hand round, saying each o'clock time until you reach 4 o'clock. Pause and ask if this is what the scenario shows. Agree that the time in the image is the next day so it must be another 24 hours later.

ANSWERS

Question **1** a): The hour hand goes around the clock twice in one day. There are 24 hours in one day.

Question **1** b): 4 o'clock is more than 24 hours later. It should have been drunk by 1 o'clock the next day, so it is not safe to drink.

PUPIL TEXTBOOK 2C PAGE 76

Share

WAYS OF WORKING Whole class teacher led

ASK

- Question **1** a): *How can you show how many hours there are in a day? Can you say each o'clock time as I turn the hand and tell me to stop when you think a day has gone past? Let's do the same activity again – this time also tell me what you would be doing at each o'clock time.*
- Question **1** a): *How many 8 o'clock times are in a whole day?*
- Question **1** a): *When will 24 hours from now be?*
- Question **1** b): *Give me a time that is more or less than 24 hours from now.*
- Question **1** b): *Where have you seen the words '24 hours' written in real life?* [For example: '24 hour cash', 'open 24 hours'.] *What does it mean?*

IN FOCUS To help children relate to the times they will have little experience of, likely to be 9 pm to 5 am, it may be worthwhile discussing what might be happening during these times. You could provide pictures of nocturnal animals, motorway workers, street cleaners, and so on, and link them to the times as you count through the clock.

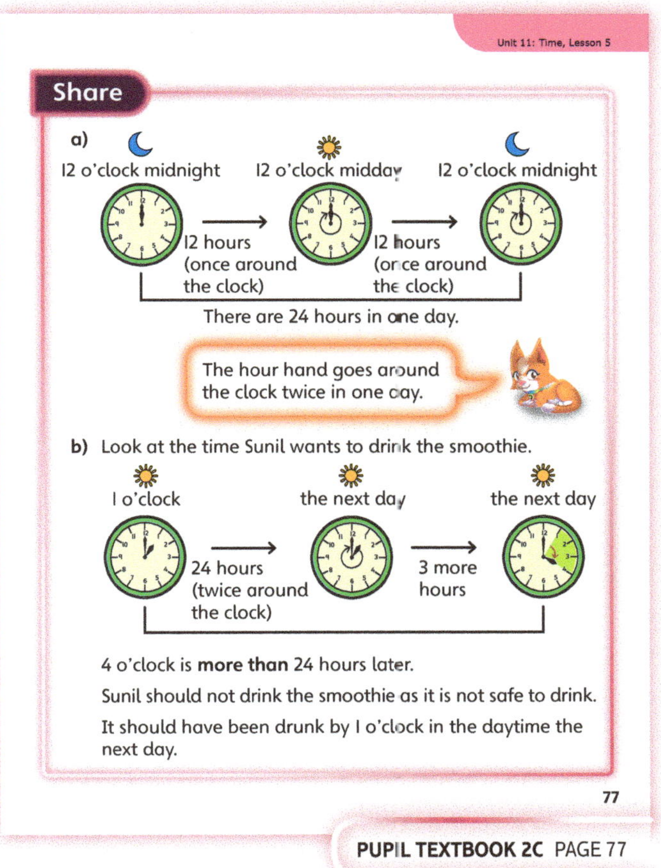

PUPIL TEXTBOOK 2C PAGE 77

Think together

 Whole class teacher led (I do, We do, You do)

ASK

- Question ❶: *How do you know when to start counting in hours?*
- Question ❶: *What time will it be after 1 hour? 2 hours? Do you notice a pattern?*
- Question ❷: *To move through 24 hours, how many times will the hour hand travel around the clock?*
- Question ❸: *What do you need to know to begin working out 24 hours from now?*

IN FOCUS Question ❶ offers a good opportunity to reinforce the understanding that a day is composed of 24 hours, which is twice around the clock. Use the images of the clocks, coupled with the day and night symbols, to reinforce and cement this understanding. Encourage children to use analogue clock manipulatives to move through 24 hours, counting the hours as they do so.

STRENGTHEN In question ❷, some children may find it difficult to count the hours between 1:25 on Monday and 1:15 on Tuesday. To help children, ask: *Can you count 24 hours from twenty-five past 1 on Monday? Is this before or after quarter past 1 on Tuesday?* If children solve this, extend their learning by asking how many more minutes the patient needs to wait.

DEEPEN Once children have explained what information is missing in question ❸, deepen the investigation by asking whether there are any potential solutions they could find with the information they have. Can they explain how they came to their solutions?

ASSESSMENT CHECKPOINT At this point in the lesson, children should be able to recognise that there are 24 hours in a day. They should be able to recognise and explain how they can find 24 hours from a given time by counting 24 hours from the start time. Children should recognise that while the finish time matches the start time, the day has changed.

ANSWERS

Question ❶: The postal worker will be back by 15 minutes past 9 on Wednesday morning.

Question ❷: No, the bandage should not be taken off yet. The bandage can be taken off at 25 minutes past 1 on Tuesday, which is 10 minutes later than the time shown.

Question ❸ a): There is not enough information to tell if the explorer finished the climb in time.

Question ❸ b): The children need to know what day it was and whether it was morning or night when the explorer started the climb. They also need to know what day it was and whether it was morning or night when she finished the climb.

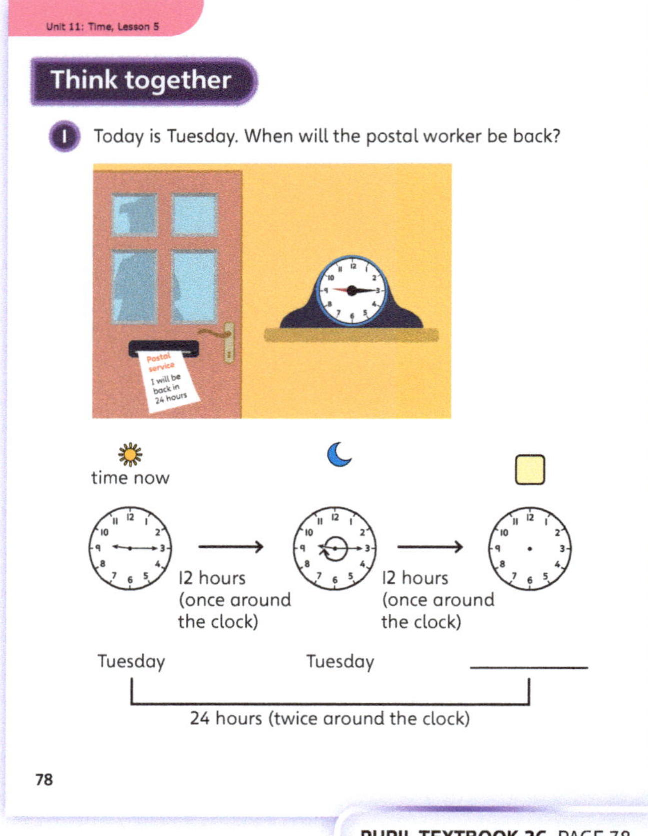

PUPIL TEXTBOOK 2C PAGE 78

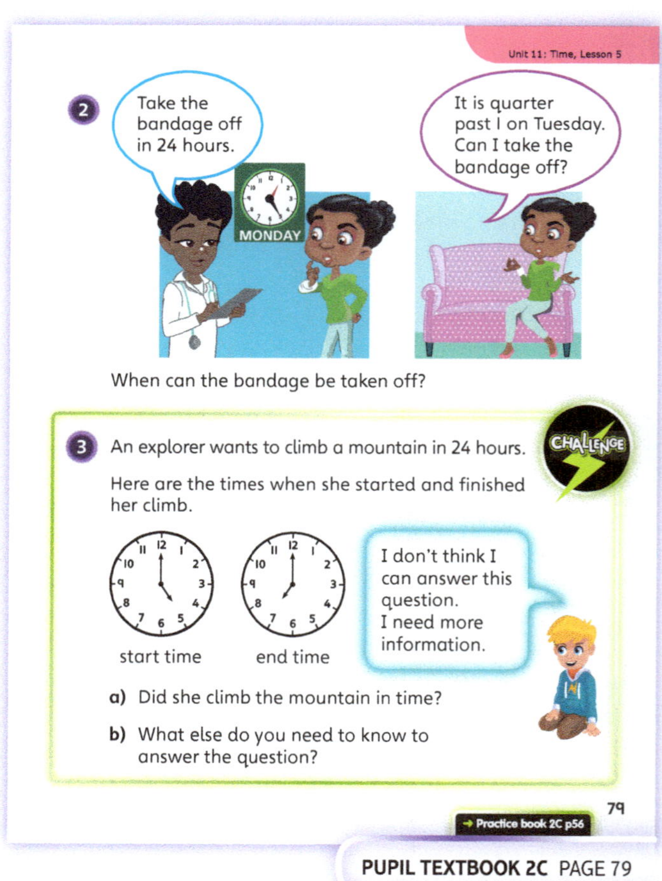

PUPIL TEXTBOOK 2C PAGE 79

Practice

WAYS OF WORKING Pair work

IN FOCUS Questions **1**, **2** and **3** give children pictorial representations of the time to scaffold their independent learning. Provide plastic clocks or laminated clock pictures to support children's work.

STRENGTHEN If children are struggling to remember to move to a new day when counting 24 hours, offer them a timeline with the hours and weekdays recorded on it. Ask children to find the day and time on the timeline and then count on another 24 hours.

DEEPEN When solving question **4**, deepen children's reasoning by asking them to find two ways to prove when Ella can or cannot eat a biscuit.

ASSESSMENT CHECKPOINT At this point in the lesson, children should be able to confidently recognise and explain that there are 24 hours in a day. Given any starting time, they should be able to find 24 hours in the future and explain how the time stays the same and it is the day that changes. Children should be able to use this understanding to solve simple problems confidently.

ANSWERS Answers for the **Practice** part of the lesson can be found in the *Power Maths* online subscription.

Reflect

WAYS OF WORKING Pair work

IN FOCUS Allow children to discuss, with a partner, what Maya should have said. Ask them what advice they would give Maya to correct her mistake. How would they prove to her that there are 24 hours in a day, not 12?

ASSESSMENT CHECKPOINT Look for children recognising how the misconception of 12 hours in a day could be made. Children should be able to explain that Maya should have travelled around the clock twice and give ways of explaining this, based on the models and representations used in the lesson.

ANSWERS Answers for the **Reflect** part of the lesson can be found in the *Power Maths* online subscription.

After the lesson ⏸

- Are children confident that there are 24 hours in a day?
- How did you challenge children's assumptions about the times they were not familiar with?

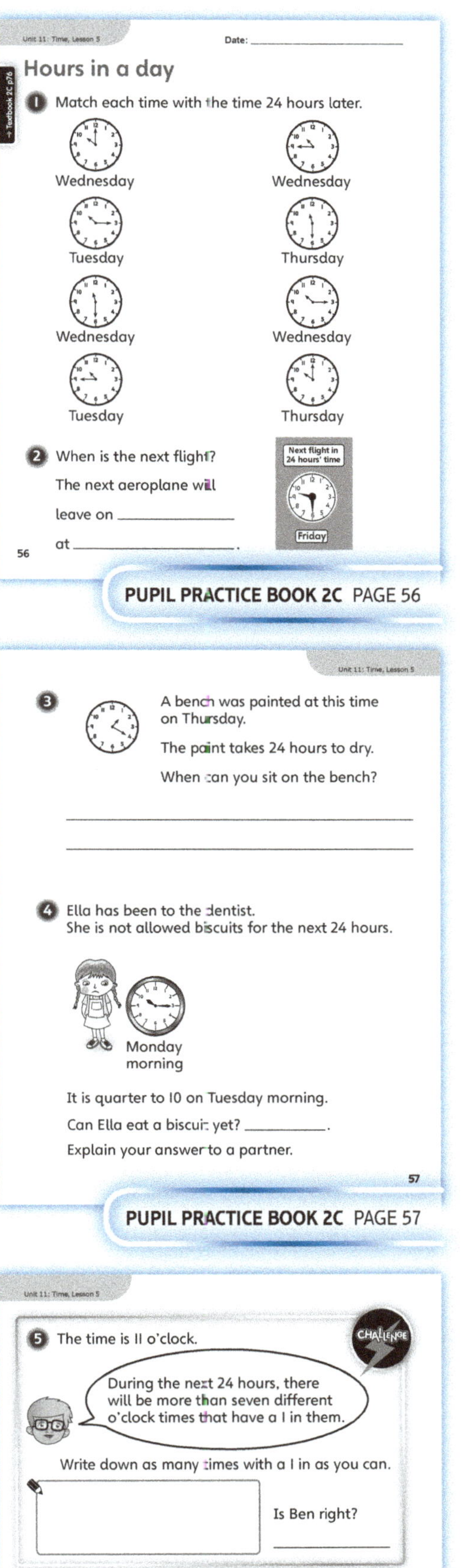

PUPIL PRACTICE BOOK 2C PAGE 56

PUPIL PRACTICE BOOK 2C PAGE 57

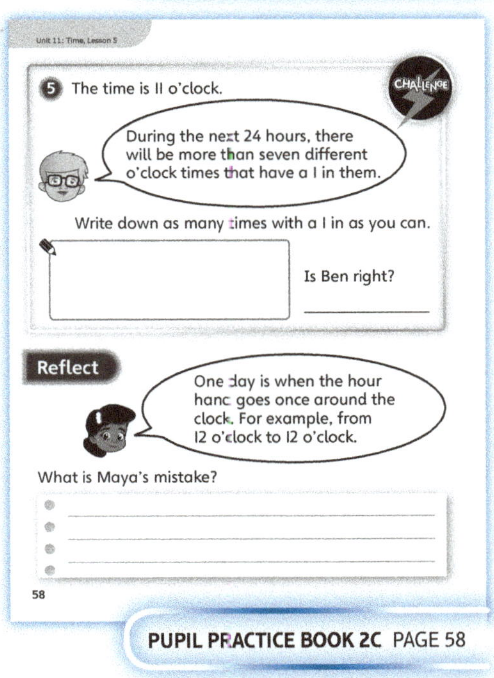

PUPIL PRACTICE BOOK 2C PAGE 58

End of unit check

Don't forget the unit assessment grid in your *Power Maths* online subscription.

 Group work adult led

- Question **1** assesses children's ability to identify a half-past time on an analogue clock.
- Question **2** assesses children's ability to identify a quarter-past time on an analogue clock.
- Question **3** assesses children's ability to recognise times to 5 minutes on an analogue clock and use the terminology of 'minutes past' and 'minutes to'.
- Question **4** assesses children's ability to recognise, measure, convert and compare durations of time.
- Question **5** assesses children's ability to count through 24 hours and recognise that the day has changed.

Think!

 Pair work

  This question assesses children's ability to read time on an analogue clock and in written form and to explain why the time in the picture matches the written time beneath. Ask: *What are the important words in the written times? Which hand shows the minutes? The hours? Can you match the hours in words to those in the picture?*

This question provides an opportunity to discuss the similarities and differences between times such as 'twenty minutes to 3' and 'forty minutes past 2'. Ask: *How are these times the same? How are they different?*

Encourage children, in pairs, to think through or discuss how they know that the times are right before writing their answer in **My journal**.

  Children will demonstrate mastery by counting in steps of five minutes and recognising at what times 'past' and 'to' should be used. They will be able to explain how the hour and minute hands show the written time.

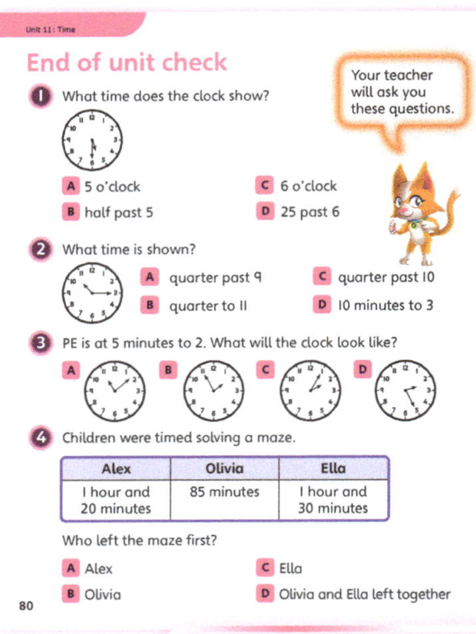

PUPIL TEXTBOOK 2C PAGE 80

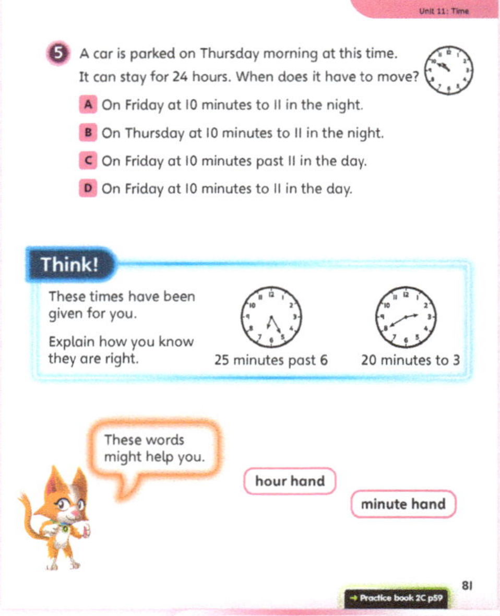

PUPIL TEXTBOOK 2C PAGE 81

Q	A	WRONG ANSWERS AND MISCONCEPTIONS	STRENGTHENING UNDERSTANDING
1	B	A or C suggests that children have mistaken a half-past time for o'clock. D may suggest that children have confused the hour and minute hand	Make sure any representations of analogue clocks are labelled clearly. Tell the story of your day using times. Children should show the times using analogue clocks. Depending on the fluency of the children, you can adapt this activity by: using only o'clock times; including 'half past' and 'quarter past/to'; giving start and end times and asking for the duration.
2	C	A suggests children have misread the hour hand. B suggests children do not understand the difference between quarter past and quarter to. D suggests children have confused the hour hand and the minute hand	
3	B	A or C suggests that children have misread the hands on the clock. D may suggest that children lack understanding of how the numbers on the clock face represent minutes.	
4	A	B suggests children have assumed that the time shown in minutes is less than that shown in hours. It could also indicate that they have converted the times inaccurately.	
5	D	A indicates children have only counted 12 hours. B suggests children have not moved through to the next day. C suggests children have misread the clock's hands.	

My journal

WAYS OF WORKING   Independent thinking

ANSWERS AND COMMENTARY

For each time, children may record answers such as those shown below.

I know the time is twenty-five minutes past 6 because:
- The hour has just passed 6.
- The minute hand is pointing at 5 which means 25 minutes.
- The minute hand is in the 'past' section of the clock.

I know the time is twenty minutes to 3 because:
- The hour hand is almost at 3.
- The minute hand is pointing to 8 which means 'twenty to' in minutes.
- The minute hand is in the 'to' section of the clock.

If children are struggling to give reasons, ask:
- *Where is the minute hand pointing? Can you count the minutes?*
- *What part of the clock is the minute hand in: 'to' or 'past'? How do you know?*
- *What do the hour hands tell you on each clock face?*

Power check

WAYS OF WORKING   Independent thinking

ASK

- *What did you know about reading times before you started this unit?*
- *What new things have you learnt?*
- *Do you think you could look at a clock at home and tell the time by yourself?*

Power puzzle

WAYS OF WORKING   Independent thinking

IN FOCUS   This puzzle will assess children's recognition of written time and the start and end times of 20-minute durations. Children should be able to recognise where 20 minutes have passed between their start time and end time, although it may help them to have a selection of the representations used in this unit at hand to support their thinking.

ANSWERS AND COMMENTARY   If children are unable to follow the route accurately, it may be beneficial to give them opportunities to practise:
- matching written times with pictorial and concrete representations of an analogue clock
- moving forwards and backwards through time, using the analogue clock
- recognising and using the key vocabulary appropriate to measuring time.

After the unit ⏸

- Measuring time is such a commonplace activity that it can be taken for granted. How will you encourage children to regularly measure time outside of this unit of work? Some ideas might include: using time within children's role play, giving children time-keeping responsibilities, asking children what the time is, and so on.
- Is your classroom a 'time-rich' environment? How many reminders or opportunities are there for children to engage with time around the school environment? Is there a clock in each room? How could the resources be improved to deepen children's understanding of and fluency with time?

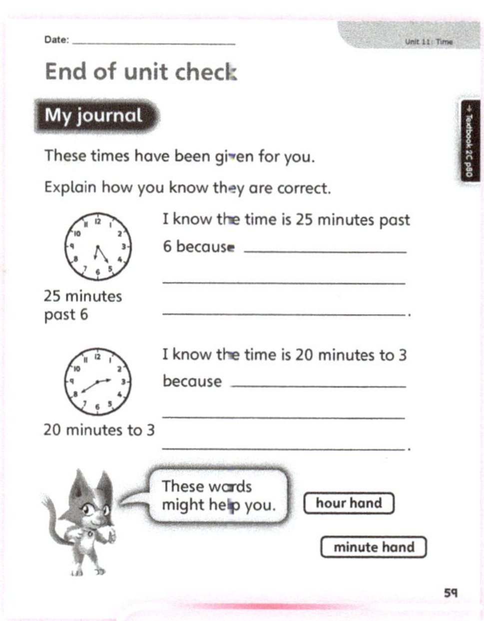

PUPIL PRACTICE BOOK 2C PAGE 59

START 4 o'clock	twenty-five past 4	ten past 5	twenty past 7
twenty past 4	twenty to 5	five past 5	twenty-five to 8
ten past 4	5 o'clock	twenty past 5	quarter to 7
half past 4	ten to 5	40 minutes past 5	half past 6
quarter to 5	quarter past 5	6 o'clock	20 minutes past 6
five to 5	twenty-five past 6	FINISH 7 o'clock	20 minutes to 7

PUPIL PRACTICE BOOK 2C PAGE 60

Strengthen and **Deepen** activities for this unit can be found in the *Power Maths* online subscription.

Unit 12
Problem solving and efficient methods

WHY THIS UNIT IS IMPORTANT

This unit brings together the key ideas of number that have been addressed in previous units and provides an opportunity for children to practise all four operations.

The use of the bar model is consistent throughout as this method allows children to clearly see which operation is needed to complete parts of a question at any given time.

A focus throughout the unit is to ensure that the methods that children use are as efficient as possible. Children must justify their choice of methods.

WHERE THIS UNIT FITS

→ Unit 11: Time
→ **Unit 12: Problem solving and efficient methods**
→ Unit 13: Position and direction

This unit mainly builds on work from Units 1, 2 and 3, focusing in particular on addition and subtraction, but also using the context of money (Unit 5) and touching on multiplication and division (Units 6 and 7) towards the end of the unit.

Before they start this unit, it is expected that children:
- know how to use the bar model to represent information given in a word problem
- understand how to distinguish between the four operations
- know key number facts to use within mental calculations.

ASSESSING MASTERY

Children who have mastered this unit will be able to fluently select the appropriate operation that is needed to complete a given step of a problem and will be able to use the bar model to represent these problems. They will also be able to select an efficient method to solve the calculation.

COMMON MISCONCEPTIONS	STRENGTHENING UNDERSTANDING	GOING DEEPER
Children may only focus on the numbers in a question and not consider the context of the problem. In doing so, they may select the wrong operation to complete the question.	Encourage children to use the bar model to represent a problem that is given in a context. Then, ask children to use their bar model to decide what operation they need to do.	Ask children to find all possible solutions that satisfy a problem and work in a way that ensures all solutions are found.
Children may use inefficient methods rather than carefully considering the numbers in the question.	Use resources to help children work in more efficient ways, such as base 10 equipment, to count in 10s rather than 1s.	Create maths stories for all four operations, ensuring they contain more than one step using different operations.

Unit I2: Problem solving and efficient methods

UNIT STARTER PAGES

Use these pages to introduce the problem-solving unit, ensuring that children understand what this is and what being efficient means.

STRUCTURES AND REPRESENTATIONS

Bar model: This model is crucial for this unit as it represents the questions and problems that children are presented with and enables them to identify what operation is required to solve different stages of the problem.

40	
I7	?

Part-whole model: Like bar models, children can use the part-whole model to model an addition or subtraction problem and identify what operation is needed.

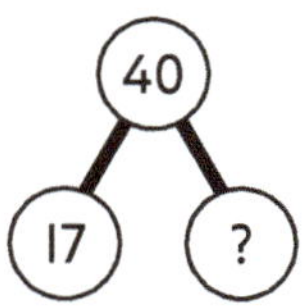

Number line: This model helps children work in an efficient way, such as counting in steps of 10 rather than 1.

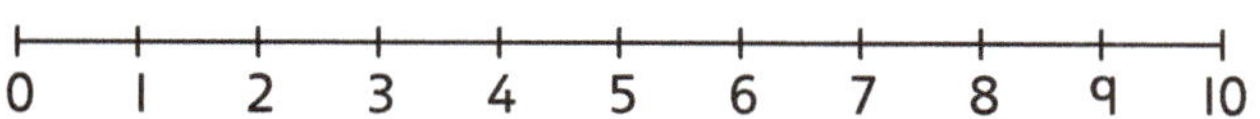

100 square: This model helps children see the links between numbers and again helps children work in a more efficient way.

I	2	3	4	5	6	7	8	9	10
II	12	13	14	15	16	17	18	19	20
21	22	23	24	25	26	27	28	29	30
31	32	33	34	35	36	37	38	39	40
41	42	43	44	45	46	47	48	49	50
51	52	53	54	55	56	57	58	59	60
61	62	63	64	65	66	67	68	69	70
71	72	73	74	75	76	77	78	79	80
81	82	83	84	85	86	87	88	89	90
91	92	93	94	95	96	97	98	99	100

KEY LANGUAGE

There is some key language that children will need to know as part of the learning in this unit:

➜ part, whole, part-whole model, partition

➜ add (+), addition, more than

➜ subtract (−), subtraction, difference, change, take away, less than

➜ divide (÷), division, share

➜ multiply (×), multiplication, lots of

➜ calculate mentally

➜ altogether, groups of, total, sum, total cost

➜ representation, bar model, efficient

➜ number fact, number line, 100 square

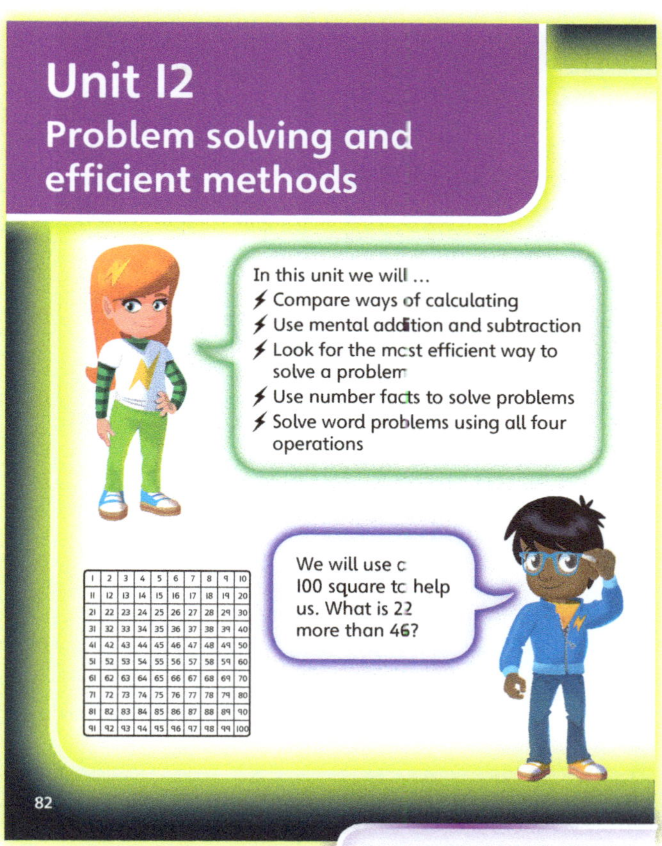

PUPIL TEXTBOOK 2C PAGE 82

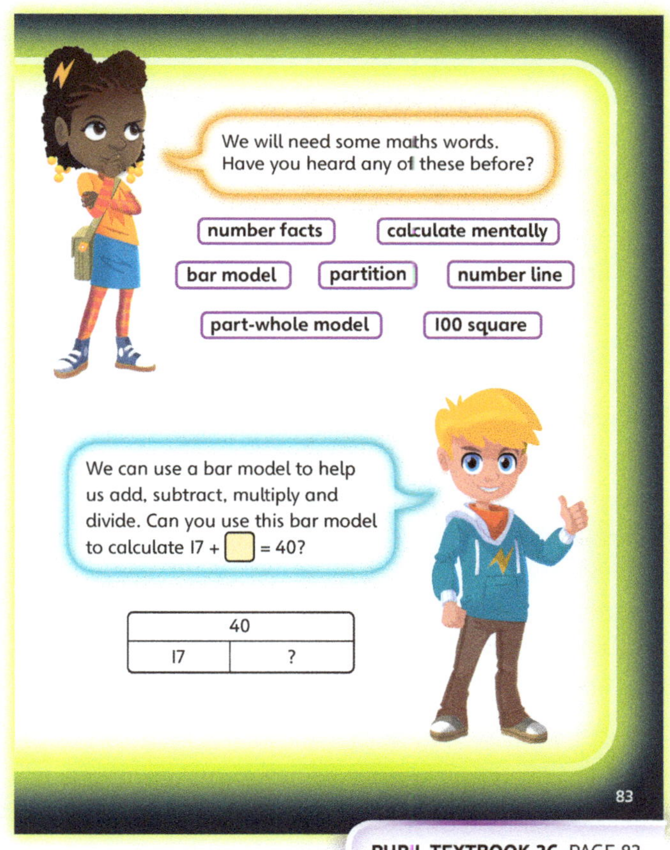

PUPIL TEXTBOOK 2C PAGE 83

My way, your way!

Learning focus

In this lesson, children will solve money problems using a variety of addition and subtraction strategies.

Before you teach 

- Are children confident interpreting word problems?
- How will you support children during the lesson?

NATIONAL CURRICULUM LINKS

Year 2 Number – number and place value

Use place value and number facts to solve problems.

Year 2 Number – addition and subtraction

Recognise and use the inverse relationship between addition and subtraction and use this to check calculations and solve missing number problems.

ASSESSING MASTERY

Children can identify different ways to solve the same problem and identify the efficiency of different methods, counting on or back during calculations, and switching between counting on in 10s and 1s as appropriate. Children can use the bar model to represent word problems and use this to identify the correct operations to solve each step.

COMMON MISCONCEPTIONS

Children may focus on the numbers within questions, ignoring the words, and therefore choose the incorrect operation or the wrong calculation to solve the problem. Ask:
- *Can the bar model be used to identify the operation to be used?*

STRENGTHENING UNDERSTANDING

If children are finding it difficult to interpret the calculation needed to solve each problem, provide a completed bar model to help with this process. Alternatively, provide opportunities for children to create their own bar models from word problems using strips of paper or coloured rods of different lengths.

GOING DEEPER

Challenge children to identify different methods to solve the same problem and identify strengths and weaknesses of each method. Can children identify the most efficient method and explain why? Children should record all steps of their working and use mathematical signs to increase the complexity of their answers.

KEY LANGUAGE

In lesson: bar model, strategy, method, operation, addition, +, subtraction, –, change

Other language to be used by the teacher: efficient

STRUCTURES AND REPRESENTATIONS

Bar model, number line

RESOURCES

Mandatory: coins, completed number line, blank number line

Optional: completed bar model, coloured rods

 In the eTextbook of this lesson, you will find interactive links to a selection of teaching tools.

Quick recap 

Ask children to count how many tables there are in the classroom, how many chairs, and how many children.

Discover

 Pair work

• Question **1** a): *Do you know the price to post the letter? Do you know the price to post the parcel?*
• Question **1** b): *What model could be used to find the total cost of posting the parcel and letter?*

 In question **1** a), children are required to interpret the picture shown and identify how to solve the problem. Children should observe the numbers 35 and 15 and the words 'more than' and realise that they must complete 35 + 15 to solve the problem.

In question **1** b), children are required to add the cost of posting the letter to the cost of posting the parcel. Children may recognise that they could also double the cost of posting the letter and then add the extra 15p. Discuss which method children prefer and which model will be the most helpful.

 Children can role-play the post office setting, identifying which coins are needed to pay for each item in the **Discover** scenario and adding these to find the total cost.

Question **1** a): The parcel costs 50p to post.

Question **1** b): The cost of posting the letter and the parcel is 85p.

PUPIL TEXTBOOK 2C PAGE 84

Share

 Whole class teacher led

• Question **1** a): *How can you calculate the total price to post the parcel?*
• Question **1** b): *Which part of the number line shows the cost of the parcel? Which part shows the cost of the letter?*
• Question **1** b): *What method did you choose? Why?*

 In this part of the lesson it is important for children to understand that there may be different ways to answer the same question. Children may use different mental strategies or representations to calculate the answer to the problem. Children should be able to count in 10s and 5s as an efficient way to solve the problem and should understand the benefits of working in this way, as opposed to counting in 1s.

 Encourage children to role-play being a shopkeeper and a customer to further strengthen their understanding of money questions. This should strengthen the real-life context that the problem presents and increase the likelihood of the correct cost being calculated.

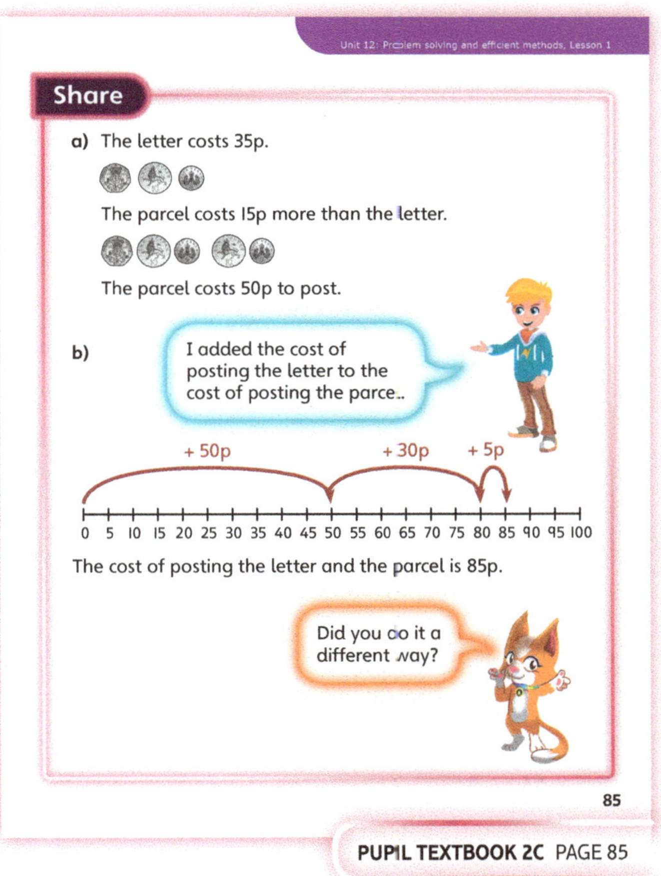

PUPIL TEXTBOOK 2C PAGE 85

Think together

WAYS OF WORKING Whole class teacher led (I do, We do, You do)

ASK

- *What is the unknown in each question?*
- *Which operation is needed at each stage of the problem?*

IN FOCUS Challenge children to consider whether the same method is appropriate for all questions. Each different method should be celebrated, as 'my way is not better than your way, it's just different'. Encourage children to record all steps of their working, rather than completing multiple steps mentally, which will increase the likelihood of calculation errors. Children may need support in understanding how this does not reduce the efficiency of calculation, but increases the reliability of their calculations.

STRENGTHEN Provide children with physical resources to create the bar models represented in each question. Manipulating the bars and creating the bar model will strengthen their understanding of which operation is required at different stages of the problem.

DEEPEN Challenge children to record the answers to questions ❶ and ❸ using the signs < and > to mathematically show if Sam and Seth have enough money in both situations.

ASSESSMENT CHECKPOINT Check to see that children can explain what the different parts of the bar model represent and how these help to determine the correct operation to solve the problem.

ANSWERS

Question ❶: Sam has enough. The cost is 90p.

Question ❷: Sam will get 30p change.

Question ❸: It is enough, 65p + 35p = £1 exactly.

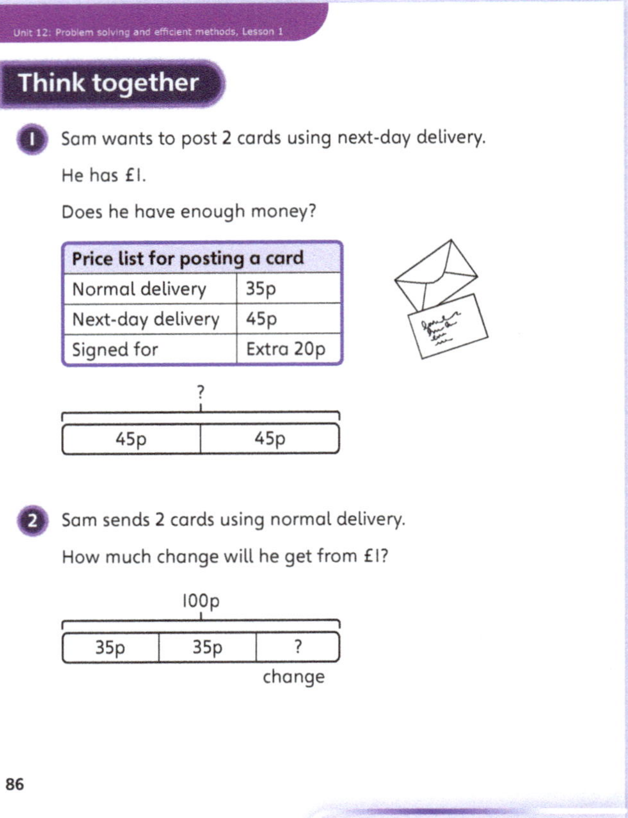

PUPIL TEXTBOOK 2C PAGE 86

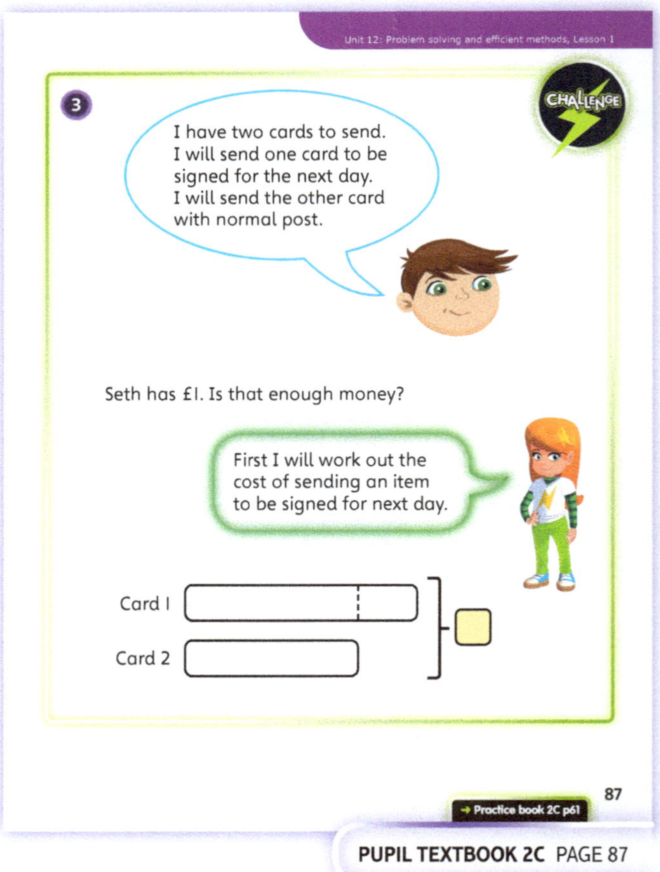

PUPIL TEXTBOOK 2C PAGE 87

Practice

WAYS OF WORKING Independent thinking

IN FOCUS Questions **1** to **4** are all one-step word problems. Only one operation is required to solve each one, but children are not told which operation they need to use. Encourage children to read the question carefully and think about what operation they need to do before they begin trying to solve the problem. Although a bar model has been provided in question **1** to help children understand the structure of the problem, allow them to use whatever calculation method they want to solve the problems. Encourage them to draw their own bar model for each one and to use their bar model to justify why they are using that operation.

STRENGTHEN Continue to encourage any children who find it difficult to interpret each problem to sketch bar models to represent each problem. These will increase the likelihood that children will identify the steps within the question and the appropriate operation required to solve the problem.

DEEPEN Question **5** is a two-step word problem. Firstly, emphasise the need to work in the same unit of currency, so change £1 to 100p. Then, discuss what steps children need to take, and what different methods are possible. Children might prefer to first add 31p and 37p together and secondly subtract the total from 100p. Alternatively, they might prefer to first subtract 31p from 100p and then subtract 37p from 69p. If children have a sound understanding of both methods, challenge them to create their own two-step word problems using the context of money.

ASSESSMENT CHECKPOINT Check what mental strategies children are using in calculations. Encourage children to not count on and back in 1s as this is an inefficient method and is more likely to lead to calculation errors.

ANSWERS Answers for the **Practice** part of the lesson can be found in the *Power Maths* online subscription.

Reflect

WAYS OF WORKING Independent thinking

IN FOCUS In this section of the lesson, children are presented with a money context using amounts in pounds. Some children may think that this increases the difficulty of the problem and this should be discussed as a class. Children should solve the problem independently and then discuss the different methods that have been used. It is important for children to understand that a variety of methods, such as a number line, a mental strategy or counting in 10s, may be used if they are efficient.

ASSESSMENT CHECKPOINT Check if children can verbalise the method they used to solve the problem. Further assessments can be made to see if children can identify the efficiency of their method compared to others' following class discussions.

ANSWERS Answers for the **Reflect** part of the lesson can be found in the *Power Maths* online subscription.

After the lesson ⏸

- Are children able to use more than one method to solve different styles of questions?
- Is there a strategy that the whole class finds more difficult to use that may require additional practice?

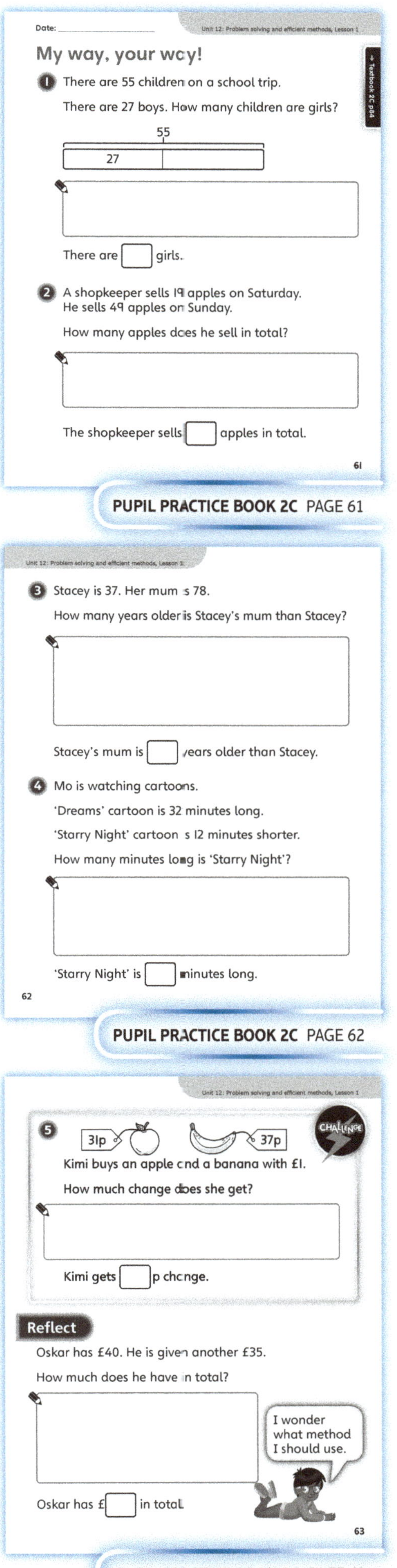

PUPIL PRACTICE BOOK 2C PAGE 61

PUPIL PRACTICE BOOK 2C PAGE 62

PUPIL PRACTICE BOOK 2C PAGE 63

Use number facts

Learning focus

In this lesson, children will make links between calculations to calculate unknown quantities, based on similarities and differences between the parts and the wholes.

Before you teach ▐▐

- How will you encourage children to use the most efficient calculation strategy?
- How will you ensure that children are exposed to the methods and approaches used by others?

NATIONAL CURRICULUM LINKS

Year 2 Number – number and place value

Use place value and number facts to solve problems.

ASSESSING MASTERY

Children can use known or given number facts as an efficient way to solve additional unknowns.

COMMON MISCONCEPTIONS

Children may see each calculation as a new question and therefore use an inefficient strategy to find the solution rather than build on what they already know. Similarly, children may only refer back to the initial fact that they have been given, rather than make links to their subsequent findings. Ask:
- *What is the same and what is different to what you already know?*

STRENGTHENING UNDERSTANDING

Using concrete manipulatives to model the different facts that are known and what needs to be calculated will increase the likelihood of children being able to identify what is the same and what can be used to solve the subsequent problem. Alternatively, allowing children to highlight or colour the different parts of the calculations that are the same will help them make stronger links.

GOING DEEPER

Children should be able to spot patterns within questions and, as a result, continue the pattern by creating subsequent calculations of their own, or create their own series of calculations that link in a similar way.

KEY LANGUAGE

In lesson: addition, part, whole, pattern

Other language to be used by the teacher: related, similar, different, link

STRUCTURES AND REPRESENTATIONS

Part-whole model

RESOURCES

Mandatory: base 10 equipment, blank part-whole models

Optional: coloured pencils

 In the eTextbook of this lesson, you will find interactive links to a selection of teaching tools.

Quick recap 

Ask children to complete this list of additions that total 11:

1 + ?   2 + ?   3 + ?   4 + ?   5 + ?

Talk about commutativity and then challenge children to use their first list to complete these addition facts:

6 + ?   7 + ?   8 + ?   9 + ?   10 + ?

Discover

 Pair work

- Questions ❶ a) and b): *How do the different questions link to each other?*
- Questions ❶ a) and b): *What is the same and what is different in each question?*

 In this part of the lesson, children can use a given fact, 5 + 8 = 13, to work out unknown quantities. Encourage children to use this fact in the subsequent calculations, rather than calculating each answer separately.

 Children can use base 10 equipment to model the calculations and explore the place value patterns and relationships.

Question ❶ a): 5 + 8 = 13
15 is 10 more than 5.
15 + 8 must be 10 more than 5 + 8.
15 + 8 = 23
Arun added the 1s first, then added the 10.

Question ❶ b): 25 + 8 = 33
35 + 8 = 43

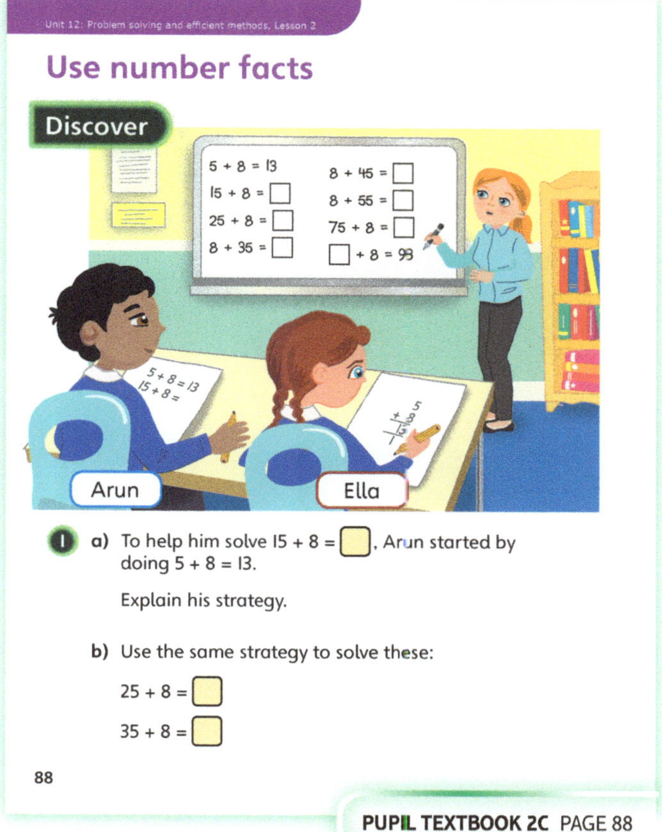

PUPIL TEXTBOOK 2C PAGE 88

Share

 Whole class teacher led

- Question ❶ b): *What is changing? What is staying the same?*
- Question ❶ b): *Where can you see 5 + 8 in each calculation?*

 In this part of the lesson, it is important for children to see the links between the different calculations presented. The use of base 10 equipment helps to show that only the number of 10s changes in each calculation, as opposed to the number of 1s. Therefore, the number of 1s in the whole will not change.

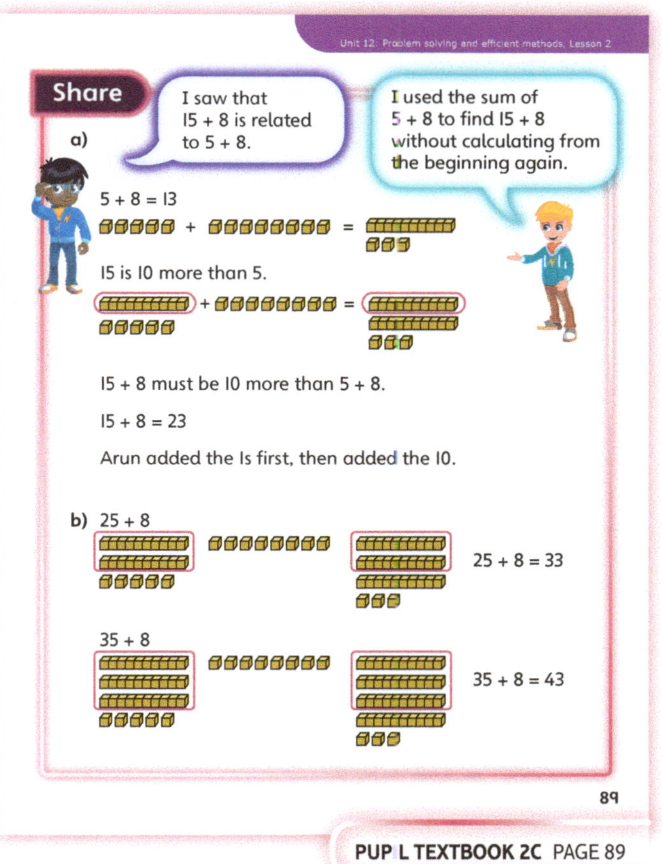

PUPIL TEXTBOOK 2C PAGE 89

Think together

WAYS OF WORKING Whole class teacher led (I do, We do, You do)

ASK

- Question **1**: *Look at the calculations 59 + 6, 79 + 6 and 29 + 6. What digit changes between the three number sentences? What digit stays the same?*
- Question **3**: *What changes between 28 + 36 and 76 + 28? What stays the same?*
- Question **3**: *Can you partition each 2-digit number into 10s and 1s to help you?*

IN FOCUS In this part of the lesson, children look at how they can use the answer to one calculation to work out another. They must look at what has stayed the same and what has changed in order to relate the calculations to each other. Questions **1** and **2** differ to previous questions as one of the parts, and therefore the whole, does not simply increase by 10 each time, but instead the parts may be 10, 20 or 30 more or less than those in the original calculation.

In question **3**, children will explore what happens when the 10s digit in one number increases by one 10, but the 10s digit in the other number decreases by one 10. Ask: *Can you predict the answer?* Do any children recognise that the answer will not change without needing to calculate?

STRENGTHEN Allowing children to continue to use base 10 equipment will help them see links between the different parts in the calculations. Alternatively, encourage children to highlight or colour the parts of the calculations that are the same to help them make stronger links.

DEEPEN Ask children to identify the pattern that links the different calculations within a question. Children can continue this pattern by making their own calculations, or create their own series of calculations that link together in a particular way.

ASSESSMENT CHECKPOINT Assess whether children can verbalise the connections between the calculations within a question. Are children using known facts to calculate new unknowns or are they treating each calculation separately?

If children are spotting, continuing and creating patterns, check to see if they are comfortable at reordering the parts of the addition calculations and if they are using more complex patterns, such as increasing both or alternate numbers by 10.

ANSWERS

Question **1** a): 79 is 20 more than 59.
Therefore 79 + 6 = 85.

Question **1** b): 29 is 30 less than 59.
Therefore 29 + 6 = 35.

Question **2**: 24 + 67 = 61 + 30 = 91

Question **3**: 28 + 36 = 64
76 + 28 = 64 + 40 = 104
38 + 26 = 64

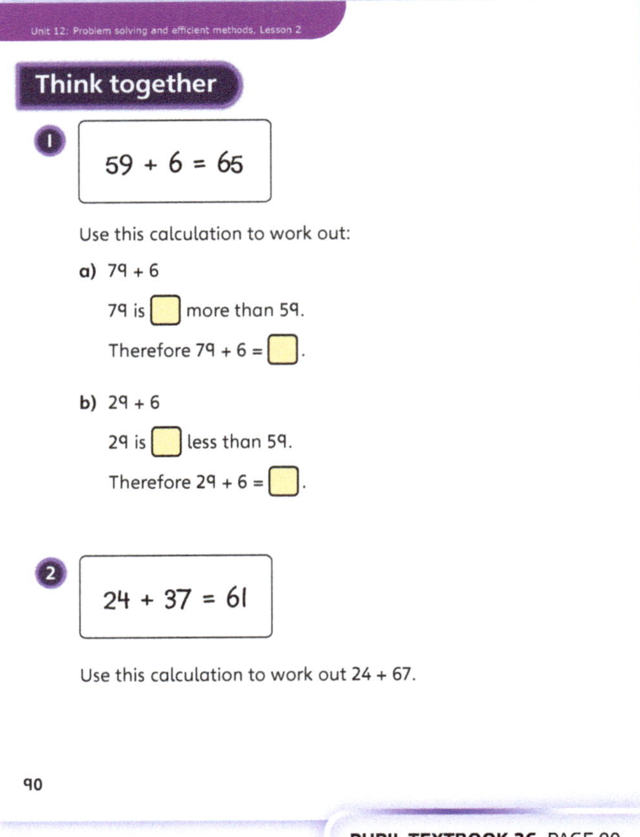

PUPIL TEXTBOOK 2C PAGE 90

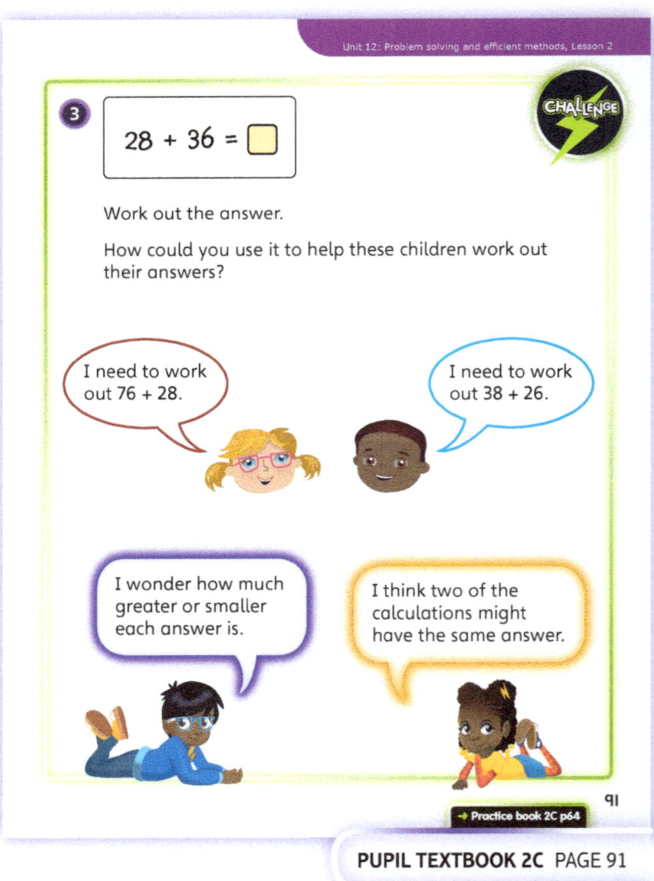

PUPIL TEXTBOOK 2C PAGE 91

Practice

WAYS OF WORKING Independent thinking

IN FOCUS In this part of the lesson, children are required to make links between calculations to find unknown quantities. In question ②, children's understanding of place value is tested. The focus should not be on finding the total of each calculation, but again on what is the same and different between the calculations. Question ⑤ shows children they can work out an answer without actually doing any calculation; they just need to be able to compare numbers. It is important that children can explain their reasoning verbally and communicate it on paper too.

STRENGTHEN Allow children to continue to work with concrete manipulatives to make the initial calculation. They must then adapt this calculation to reflect the new calculation and describe the changes they are making to one of the parts and how this will affect the whole.

DEEPEN Ask children to make a prediction of the number of different ways that ▢5 + ▢ = 65 can be completed and then work systematically to find all the possible solutions.

THINK DIFFERENTLY Question ④ is different as one of the parts increases or decreases by 1 for the first time as opposed to by a multiple of 10.

ASSESSMENT CHECKPOINT Check children's understanding by asking them to explain the links between the different calculations that they are presented with in each question. Children should be able to verbalise these similarities and differences and how they can be used to calculate subsequent unknowns.

ANSWERS Answers for the **Practice** part of the lesson can be found in the *Power Maths* online subscription.

Reflect

WAYS OF WORKING Independent thinking

IN FOCUS In this part of the lesson, for the first time children are presented with one of the parts and the whole for both calculations. To answer the written question, children do not need to solve each calculation and work out each missing number. They should instead be able to verbalise how knowing that the whole of the second calculation is 20 more than in the first calculation will affect the missing parts in both calculations.

ASSESSMENT CHECKPOINT Check to see which children solve each calculation to answer the question as opposed to using their understanding of the links between the numbers within the calculations.

ANSWERS Answers for the **Reflect** part of the lesson can be found in the *Power Maths* online subscription.

After the lesson 

- Did children use efficient methods to solve questions and make links between calculations?
- How will you encourage children to work efficiently?

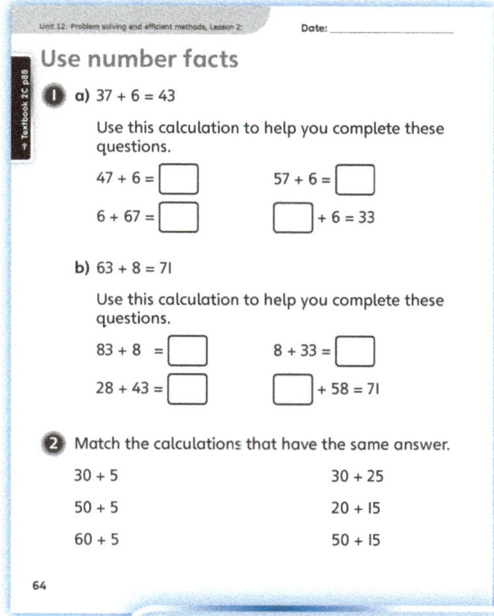

PUPIL PRACTICE BOOK 2C PAGE 64

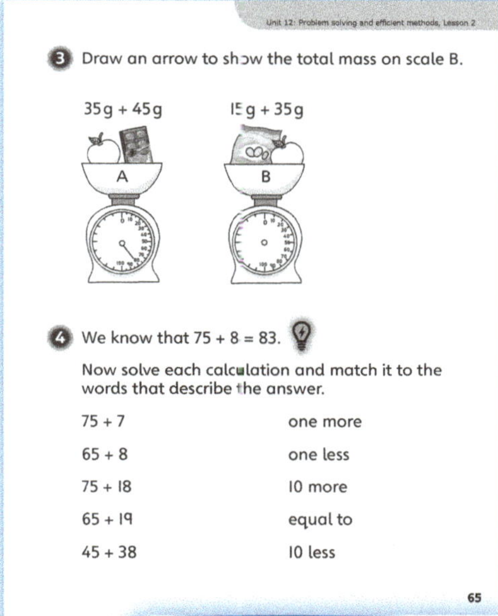

PUPIL PRACTICE BOOK 2C PAGE 65

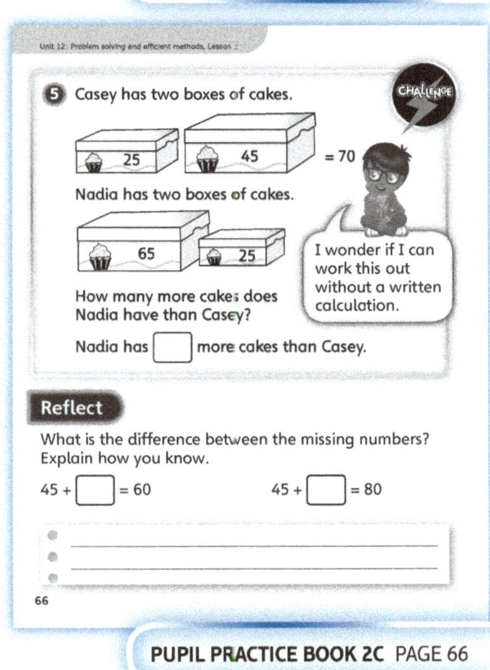

PUPIL PRACTICE BOOK 2C PAGE 66

129

Use a 100 square

Learning focus

In this lesson, children will become more familiar with the 100 square and use it to confidently count on and back in steps of 10 and 1 in addition and subtraction problems.

Before you teach

- Have children used the 100 square before?
- Do they find it easiest for calculating within 100?
- Can other resources supplement the 100 square?

NATIONAL CURRICULUM LINKS

Year 2 Number – number and place value

Use place value and number facts to solve problems.

Year 2 Number – addition and subtraction

Recognise and use the inverse relationship between addition and subtraction and use this to check calculations and solve missing number problems.

ASSESSING MASTERY

Children can add and subtract in steps of 10 and 1 and represent these visually on a 100 square. They can link moving on a 100 square to addition and subtraction and make steps within a calculation in different orders.

COMMON MISCONCEPTIONS

Children often do not know where to go at the end of a row. Sometimes they go back on the same row and get 45 + 8 = 43. They think of each row as a number line, not about the numbers themselves. Children may have seen a 100 square with 1 in the bottom left corner, and so move up a row to find 10 more. Ask:
- *What is 1 more than 50? Is 10 more than 45, 35?*

STRENGTHENING UNDERSTANDING

To increase familiarity with the 100 square, practise counting from 1 to 100 and 100 to 1 using the 100 square. To aid understanding of adding and subtracting 10 with the 100 square, base 10 equipment can also be used.

GOING DEEPER

Ask children to describe journeys from one number to another in different ways using the 100 square. For example, describe moving from 13 to 36: 13 + 10 = 23 (down one row), 23 + 10 = 33 (down one row), 33 + 3 = 36 (along 3 squares to the right), or 13 + 3 = 16 (3 squares to the right), 16 + 20 = 36 (down 2 rows).

KEY LANGUAGE

In lesson: square, row, column, more than, less than, across, down, right, left

Other language to be used by the teacher: support

STRUCTURES AND REPRESENTATIONS

100 square, number line

RESOURCES

Mandatory: laminated 100 square, 1–100 number line

Optional: counters, base 10 equipment, part-whole model

 In the eTextbook of this lesson, you will find interactive links to a selection of teaching tools.

Quick recap 

Display a large 100 square and cover some of the numbers. Challenge children to identify the hidden numbers.

Discover

WAYS OF WORKING Pair work

ASK

- Question **1** a) and b): *How many numbers are there in each row? How many rows are there?*
- Question **1** a) and b): *What is 10 more than 45?*
- Question **1** a) and b): *What is 20 more than 45?*
- Question **1** a) and b): *How can the 100 square be used to show this?*

IN FOCUS In this part of the lesson, children refamiliarise themselves with the 100 square. Remind children that there are 10 numbers on each row and that there are 10 rows. Recap how to use the 100 square when children reach the end of a row. Question **1** a) helps with this process. Remind children that when they count on from 50, they then need to move to 51 on the next row down.

PRACTICAL TIPS Children can move a counter on their own 100 square, saying each number aloud as they count on and back.

ANSWERS

Question **1** a): Count on 8 from 45.
$$45 + 8 = 53$$

Question **1** b): Count on in 10s from 45. Then count on in 1s from 75.
$$45 + 32 = 77$$

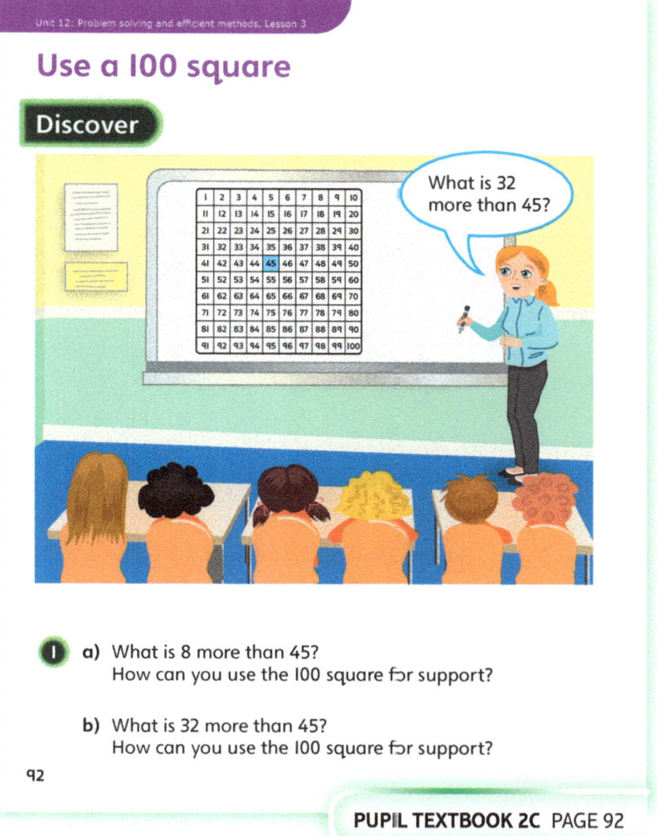

PUPIL TEXTBOOK 2C PAGE 92

Share

WAYS OF WORKING Whole class teacher led

ASK

- Question **1** a): *Is there more than one way to calculate 45 + 8?*
- Questions **1** a) and b): *In which direction do you move when counting backwards?*
- Questions **1** a) and b): *How does moving along the rows and columns link to addition and subtraction?*

IN FOCUS In this part of the lesson, children are exposed to different ways that the 100 square could be used to calculate the same question. For example, in question **1** a), you could count on 8 jumps of 1, or you could add 10 by moving down one row and then count back 2.

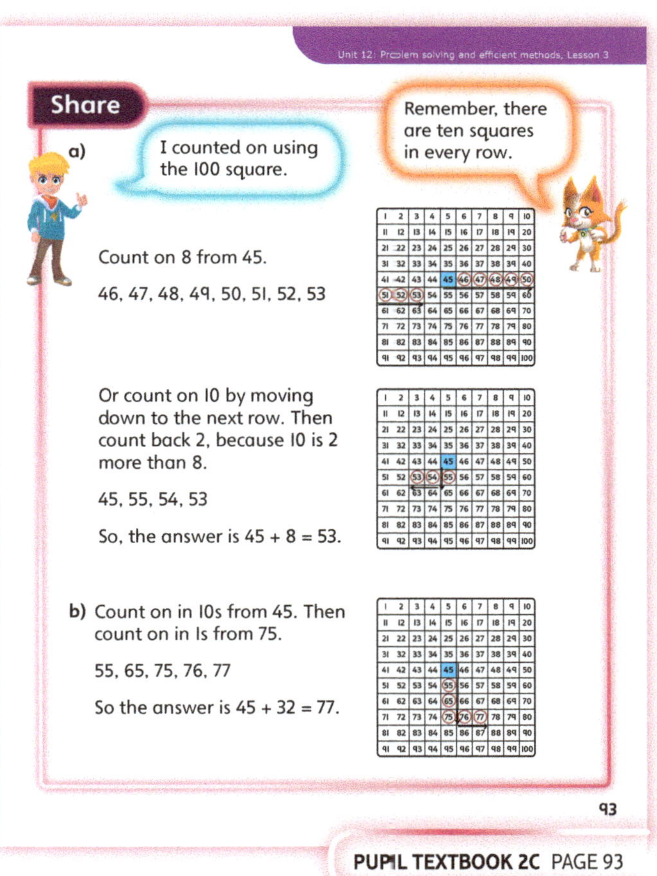

PUPIL TEXTBOOK 2C PAGE 93

Think together

WAYS OF WORKING Whole class teacher led (I do, We do, You do)

ASK

- Question **1**: *Which direction do you move on the 100 square if you add 10? Which direction do you move if you add 1?*
- Question **2**: *Which direction do you move on the 100 square if you subtract 10? Which direction do you move if you subtract 1?*
- Questions **1** to **3**: *Does the order that you add or subtract in matter? Why is having ten numbers in a row useful? How does the 100 square help with adding and subtracting in 10s and 1s?*

IN FOCUS This part of the lesson provides an opportunity for children to count on and back, completing addition and subtraction problems using the 100 square. It also requires the number line to be used to strengthen links between how these jumps are shown in different ways when using different resources.

In question **3**, children could record the jumps on the number line in different ways. To follow Millie's working, children could jump back in two steps of 10 or one jump of 20 then one jump of 6 or six jumps of 1. Highlighting the different ways that children record this on their number line will highlight the different mental calculation strategies they are using.

STRENGTHEN Provide children with part-whole models to help them record the partitioned number in 10s and 1s. This could be partitioned further to show how many jumps of 10 are required.

Children can record the steps that they take on a laminated 100 square, once they have established the number of jumps of 10 and 1 they need to take. This allows children to check that the number of jumps they have made matches the question.

DEEPEN How many different journeys can you take with the same starting and finishing number? How are these different journeys shown differently on a number line?

ASSESSMENT CHECKPOINT Check that children can link the jumps they are making on the 100 square to the question they are solving. See which children are using inefficient methods such as only counting in 1s, rather than counting in 10s and 1s.

ANSWERS

Question **1**: 88

Question **2**: 31

Question **3**:

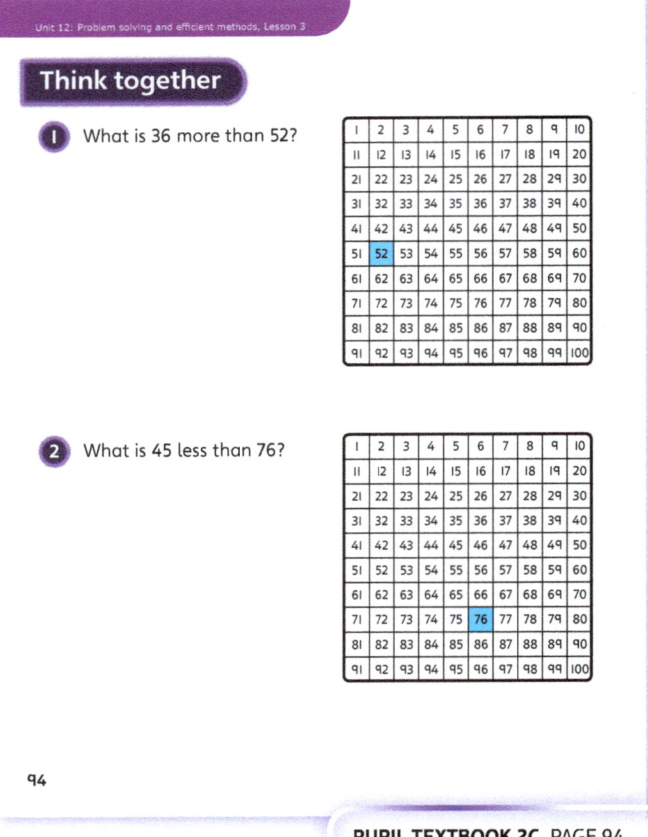

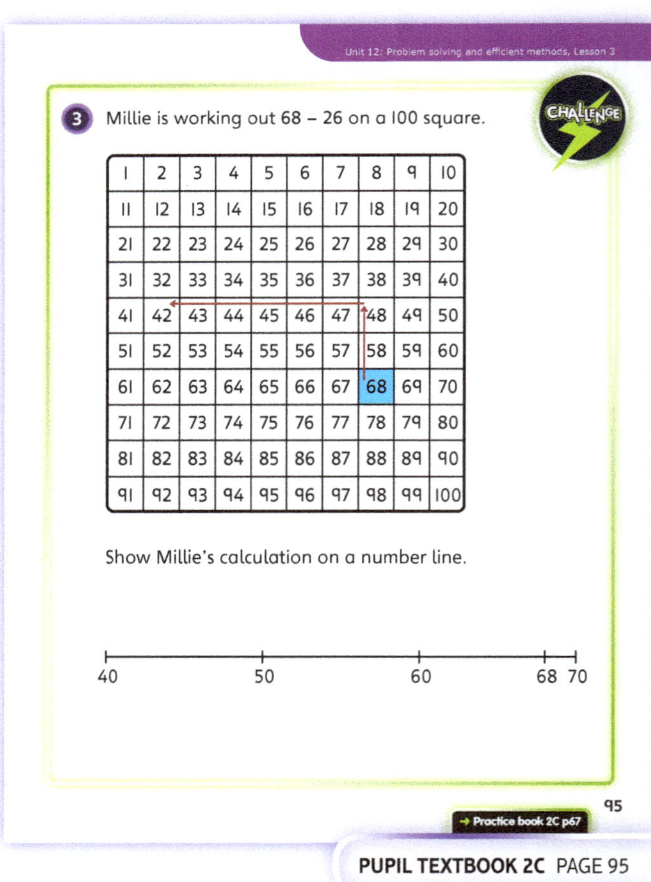

Practice

WAYS OF WORKING Independent thinking

IN FOCUS These questions give children the opportunity to work independently using the 100 square and a number line to support and visually show addition and subtraction calculations. Throughout the questions, children are required to move forwards and backwards in 10s and 1s and relate these jumps to addition and subtraction. Children may work in different ways and these differences should be highlighted and celebrated to expose children to different methods that can be used to answer the same questions.

STRENGTHEN Allowing children to work with base 10 equipment alongside the 100 square will help them understand how to move around and link movements to addition, subtraction, counting on and counting back in jumps of 10.

DEEPEN Ask children to represent 42 – 24 in as many different ways as they can on a number line and on the 100 square, ensuring that the representations on each resource match each other for each possible way.

THINK DIFFERENTLY Question ④ requires children to count on or back to solve an addition and a subtraction and to represent these calculations on a 100 square and a number line.

ASSESSMENT CHECKPOINT Check to see that children can explain how the jumps that they have recorded on the 100 square and the number line link to each other and to addition and subtraction. Children should be expected to work in jumps of 10s and 1s, rather than only in 1s.

ANSWERS Answers for the **Practice** part of the lesson can be found in the *Power Maths* online subscription.

Reflect

WAYS OF WORKING Independent thinking

IN FOCUS This **Reflect** improves children's understanding of commutativity. Encourage children to explain how the jumps they have shown on the 100 square show what has been added and why both methods result in the same total.

ASSESSMENT CHECKPOINT Check to see if children can explain what is the same and what is different about the two different calculations that they have shown on the same 100 square. Check if children can use the correct mathematical vocabulary during these explanations. Reflect also on whether it is easier or more efficient to start with the bigger number.

ANSWERS Answers for the **Reflect** part of the lesson can be found in the *Power Maths* online subscription.

After the lesson ⏸

- Are children comfortable counting in 10s and 1s rather than only in 1s?
- Did children who began by counting in 1s recognise how this was an inefficient method of calculation?

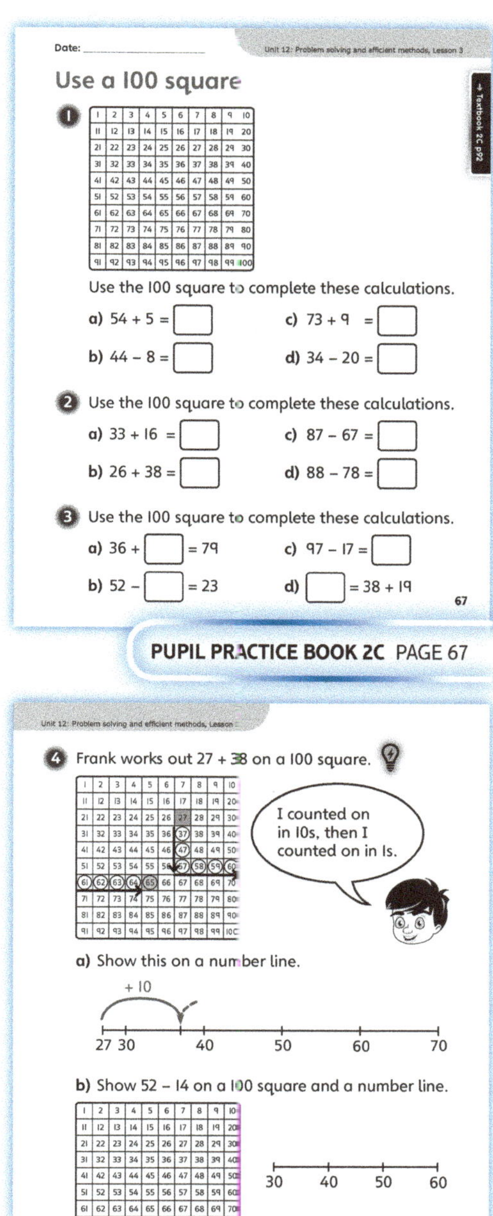

PUPIL PRACTICE BOOK 2C PAGE 67

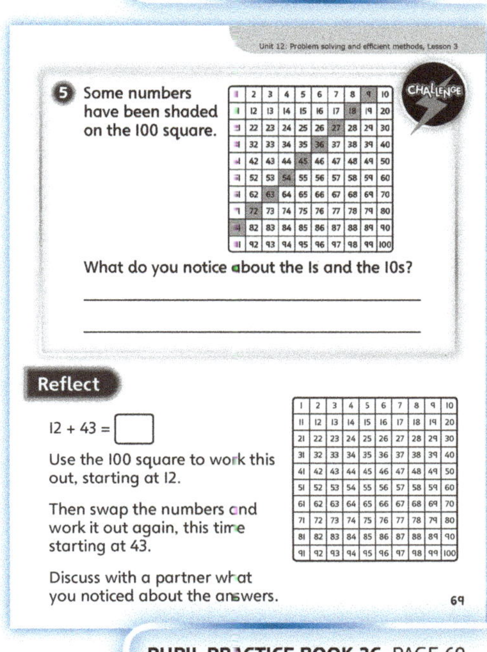

PUPIL PRACTICE BOOK 2C PAGE 68

PUPIL PRACTICE BOOK 2C PAGE 69

Getting started

Learning focus

In this lesson, children will find multiple answers to the same questions and increase their confidence in choosing a starting point and using trial and improvement to work from it.

Before you teach 

- Are children confident in problem solving?
- Can children find more than one answer?

NATIONAL CURRICULUM LINKS

Year 2 Number – number and place value

Use place value and number facts to solve problems.

Year 2 Number – addition and subtraction

Solve problems with addition and subtraction, using concrete objects and pictorial representations, including those involving numbers, quantities and measures.

ASSESSING MASTERY

Children can find multiple answers to the same problem and are confident at choosing a starting point and using trial and improvement to work from it.

COMMON MISCONCEPTIONS

In the questions provided, it is likely that children will consider all of the different ways to make the target number and forget that there is a limitation in place where each number can only be used once. Ask:

- *Is each solution possible using the number cards?*

Children may also repeat solutions or work in a way where they miss possible solutions if they are trying to find all the ways to satisfy the problem. Ask:

- *Have you worked in a systematic way to find all the solutions?*

STRENGTHENING UNDERSTANDING

Some children may find the addition and subtraction in the problems a barrier to accessing the problems. Provide children with numbers bonds within 20 to increase the ease of access to the problem.

GOING DEEPER

Encourage children to find multiple solutions to the same problem and develop the way they work to find all possible solutions. Challenge children to find all the possible ways to solve each question. Encourage children to work in a systematic way to prove they have found all possible solutions.

KEY LANGUAGE

In lesson: partition, trial and improvement, solutions, possible, impossible, sum, total, digit

Other language to be used by the teacher: strategy

STRUCTURES AND REPRESENTATIONS

Part-whole model

RESOURCES

Mandatory: number cards 1–9, number sentence scaffolds

Optional: cubes, number bonds within 20, starting point prompts

 In the eTextbook of this lesson, you will find interactive links to a selection of teaching tools.

Quick recap

Practise listing addition facts together as a class. Which facts do children know best? For example, doubles or near doubles.

Discover

ASK

- Question **1** a): *What does sum mean?*
- Question **1** a): *How many ways can you make 14 using two cards? How can you list your answers to make sure you have them all?*
- Question **1** b): *How many different ways can you make 14 using three cards if one of them has to be the same as one of Filip's numbers? How can you list your answers to make sure you have them all?*

IN FOCUS For both questions **1** a) and b), ensure children understand that there is only one of each digit card and as a result the same number cannot be used more than once. Encourage children to work. systematically – perhaps using a table – to make sure they have found all possible answers.

PRACTICAL TIPS Children can use cubes with part-whole models to represent and partition Filip and Kat's numbers.

ANSWERS

Question **1** a): Filip could have 9 and 5 or 8 and 6.

Question **1** b): **Share** shows just two possibilities:
 If Filip has 9 and 5, Kat could have:
 1, 4, 9;   2, 3, 9;   5, 2, 7;   5, 1, 8;   5, 3, 6.
 If Filip has 8 and 6, Kat could have:
 1, 5, 8;   2, 4, 8;   1, 7, 6;   3, 5, 6.

Share

ASK

- Question **1** a): *If you choose 3 as the first card, what do you then need to think about?*
- Question **1** a): *Could any number greater than 4 be one of Filip's cards?*

IN FOCUS This section of the lesson emphasises the trial and improvement thinking process. In question **1** a), ensure children understand that if 2 did not work, 1 would not work either. Similarly, both 3 and 4 would need to be paired with a number larger than 9. Starting with 9 and working down, you can have 9 and 5, or 8 and 6.

A systematic process is very important for question **1** b). Children should consider Filip's two possible pairings in turn, and try to partition each number into two other numbers. A table similar to this one might help.

If Filip has 9 and 5		If Filip has 8 and 6	
Partitioning 9 gives:	1+8+5 2+7+5 3+6+5	Partitioning 8 gives:	1+7+6 3+5+6
Partitioning 5 gives:	9+1+4 9+2+3	Partitioning 6 gives:	8+1+5 8+2+4

One further combination, 3 + 7 + 4, is not valid because it doesn't include a digit that Filip could have. Encourage children to think about which cards they have already used and which cards they have not.

PUPIL TEXTBOOK 2C PAGE 96

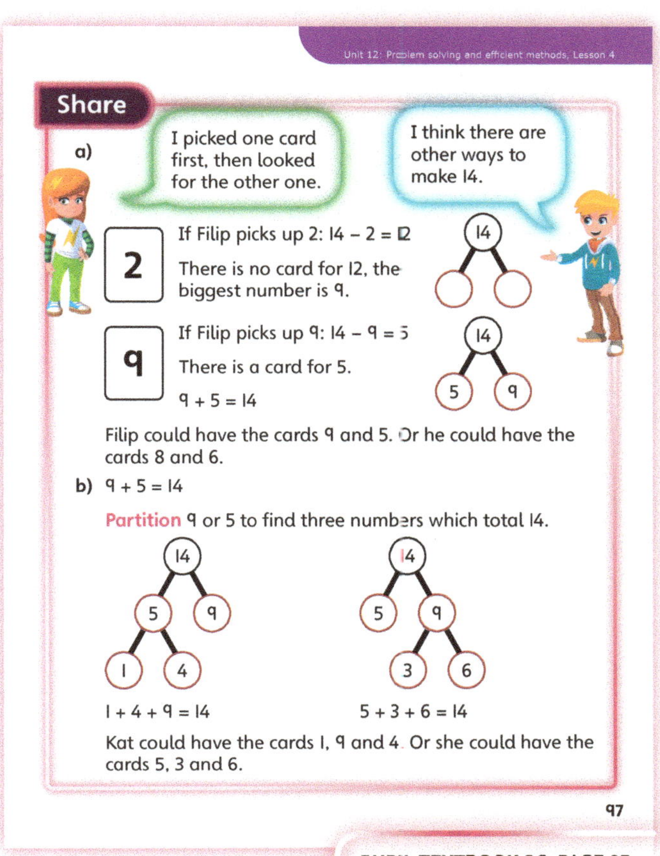

PUPIL TEXTBOOK 2C PAGE 97

Think together

 Whole class teacher led (I do, We do, You do)

- Question **1**: *Why has a table been used to record the different possible ways to make 14?*
- Question **1**: *Why have 13 and 12 been crossed out? Will you need to cross out any other numbers?*

 This part of the lesson helps children understand that, instead of working randomly and choosing any starting card, they can work in a more systematic way, starting with the largest or smallest card and working through all possibilities. This is demonstrated in the table in question **1**.

 Provide children with a set of number cards so they can play the game or explore for themselves. This will help children to see that they cannot use $7 + 7 = 14$ because they only have one of each card.

 Challenge children to find all of the possible ways to make 14 using two and three cards. This will provide an opportunity for children to realise that trial and improvement is a good strategy to get you started, but you need to move to working systematically to find all the possible solutions.

 Assess whether children are working systematically or randomly when choosing a starting point.

Question **1**: 5 and 9, 6 and 8

Question **2**: Answers can vary. For example:
    1, 4, 9
    2, 3, 9
    5, 1, 8
    5, 2, 7
    5, 3, 6

Question **3**: If Filip has 5 and 9, Kat could have:
    8, 2 and 4; or
    7, 6 and 1; or
    7, 4 and 3.
    If Filip has 6 and 8, Kat could have:
    9, 1 and 4; or
    9, 2 and 3; or
    7, 4 and 3; or
    7, 5 and 2.

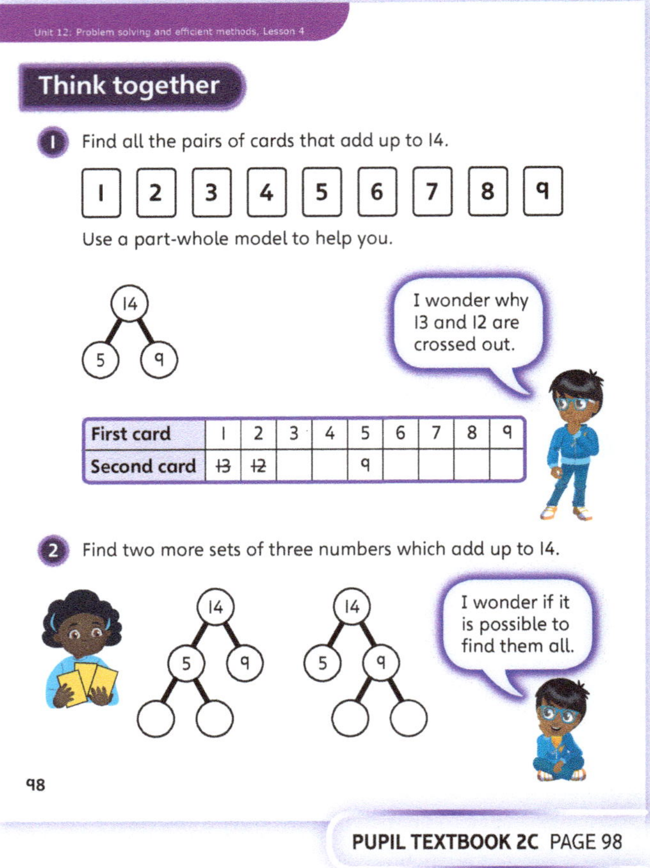

First card / Second card table:

First card	1	2	3	4	5	6	7	8	9
Second card	13	12			9				

PUPIL TEXTBOOK 2C PAGE 98

PUPIL TEXTBOOK 2C PAGE 99

Practice

WAYS OF WORKING Independent thinking

IN FOCUS In this part of the lesson, children apply what they have learnt about working systematically to questions with new totals. Question ③ requires children to think slightly differently as their place value understanding is tested.

STRENGTHEN Children who find these questions or the concept of only using a digit once difficult should continue to be provided with number cards to 'act out' the problem. If children find choosing a number to start from difficult, provide them with a starting point.

DEEPEN Children who can complete the questions should be encouraged to find all the possible answers systematically. Questions ② and ⑤ can be completed in many different ways. Question ② is similar to the example from earlier in the lesson and the same restriction could be applied where each digit can only be used once. Question ⑤ also builds on work from the previous lesson.

THINK DIFFERENTLY Question ③ requires children to select pairs of 1-digit numbers and put them together to make 2-digit numbers that meet given criteria. Instruct children that they are putting the numbers together, *not* adding them (for example, 9 and 6 can be put together to make 96 or 69).

ASSESSMENT CHECKPOINT For all questions, children should be able to explain how they have chosen to start the question, what they have learnt from this starting point and how they can justify that their answer meets the criteria in the question. This will allow their understanding of addition, commutativity, place value and problem solving to be assessed.

ANSWERS Answers for the **Practice** part of the lesson can be found in the *Power Maths* online subscription.

Reflect

WAYS OF WORKING Pair work

IN FOCUS This question again requires children to choose their own starting point. From this point they can use trial and improvement to try and complete the problem. Encourage children to find as many different ways to complete the problem as possible.

ASSESSMENT CHECKPOINT Children should be able to explain where they chose to start, why they chose this point and what they learnt from this choice.

ANSWERS Answers for the **Reflect** part of the lesson can be found in the *Power Maths* online subscription.

After the lesson ⏸

- Were children comfortable with choosing their own starting point and making mistakes?
- What was a successful way to encourage children to take risks during the lesson?

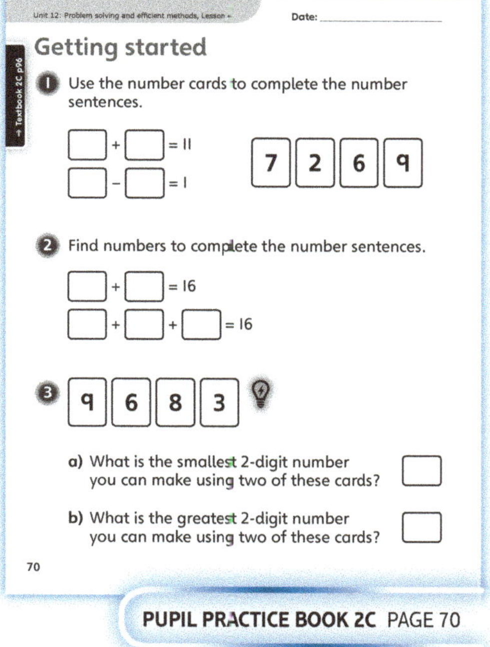

PUPIL PRACTICE BOOK 2C PAGE 70

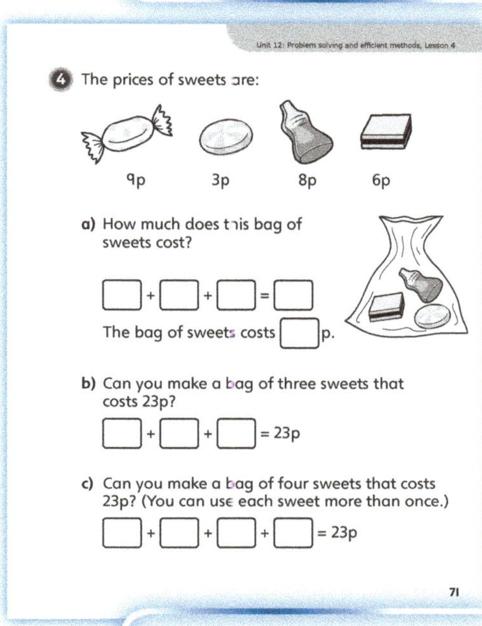

PUPIL PRACTICE BOOK 2C PAGE 71

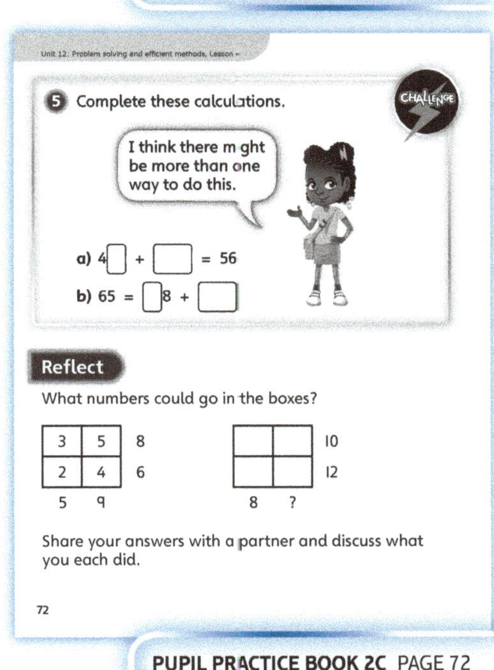

PUPIL PRACTICE BOOK 2C PAGE 72

Missing numbers

Learning focus

In this lesson, children will identify what they know from a question and use it to work out unknowns, rearranging number sentences as appropriate.

Before you teach 

- Will children find interpreting questions difficult?
- How will you support children during the lesson?

NATIONAL CURRICULUM LINKS

Year 2 Number – addition and subtraction

Recognise and use the inverse relationship between addition and subtraction and use this to check calculations and solve missing number problems.

ASSESSING MASTERY

Children can identify what is known and not known from a given problem and use their understanding of the inverse to rearrange a calculation to confirm what they need to calculate and which operation will be needed to achieve this. Children can use the bar model to visually represent this process.

COMMON MISCONCEPTIONS

Children are likely to see the numbers, but may ignore the context and so use the wrong operation. Ask:
- *Does using a bar model help work out which operation you need to find the answer?*

Children who use inefficient methods to calculate, such as counting in 1s, are likely to make errors. Ask:
- *What resources would help you to use a more efficient calculation method?*

STRENGTHENING UNDERSTANDING

Allowing children to work with resources to support mental calculations will increase efficiency and reduce errors from inefficient methods. Encourage children who find interpreting the question difficult to use the bar model to represent it. If they find this difficult, provide resources to manipulate the bars, such as coloured rods. If this is still too difficult, provide completed bar models to help children identify the necessary operation.

GOING DEEPER

Challenge children to explore how a whole that has been split into three parts can be represented in different ways, completing the fact family for these calculations.

KEY LANGUAGE

In lesson: missing number, unknown, altogether

Other language to be used by the teacher: represent, donate

STRUCTURES AND REPRESENTATIONS

Part-whole model, bar model, number line

RESOURCES

Mandatory: base 10 equipment

Optional: coloured rods or strips of paper to create bar models, completed bar models, number line, blank part-whole models, bead strings, place value counters, money

 In the eTextbook of this lesson, you will find interactive links to a selection of teaching tools.

Quick recap 

Ask children to identify the parts and the whole in the number sentence 5 + 7 = 12 and to show it on a part-whole model. Repeat with the number sentence 20 − 7 = 13.

Discover

 Pair work

ASK

- Questions **1** a) and b): *How could you write these as missing number sentences?*
- Questions **1** a) and b): *Would using a bar model help with this process?*

IN FOCUS In this part of the lesson, children need to interpret what the question is asking them to calculate. Encourage them to write this as a missing number sentence which they can manipulate to show what they are trying to find out. For example, in question **1** a), £55 = £20 + ☐ could be rearranged as £55 − £20 = ☐. Support children with this by encouraging their use of the bar model to represent the amounts Marta has.

PRACTICAL TIPS Children can use base 10 equipment to represent and partition the numbers in the scenario.

ANSWERS

Question **1** a): Marta had £35.

Question **1** b): Marta needs to save another £45.

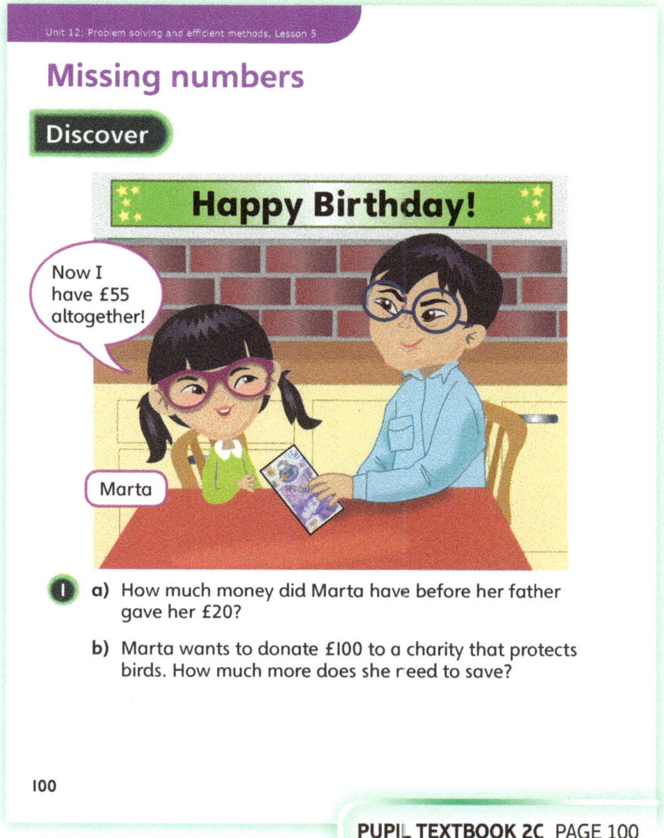

PUPIL TEXTBOOK 2C PAGE 100

Share

 Whole class teacher led

ASK

- Questions **1** a) and b): *How are the pictures similar? How are the pictures different?*
- Questions **1** a) and b): *How do the number sentences relate to the models?*

IN FOCUS In this part of the lesson, children are exposed to different representations that show the problem and help them to calculate the answer in an efficient way. The use of base 10 equipment, for example, is an effective way to move children on to counting more efficiently than in 1s.

STRENGTHEN Some children may find other resources more useful to interpret the problem. Providing children with a choice of resources to use may increase the effectiveness of the resource. Other resources that might be useful include bead strings, place value counters and real money.

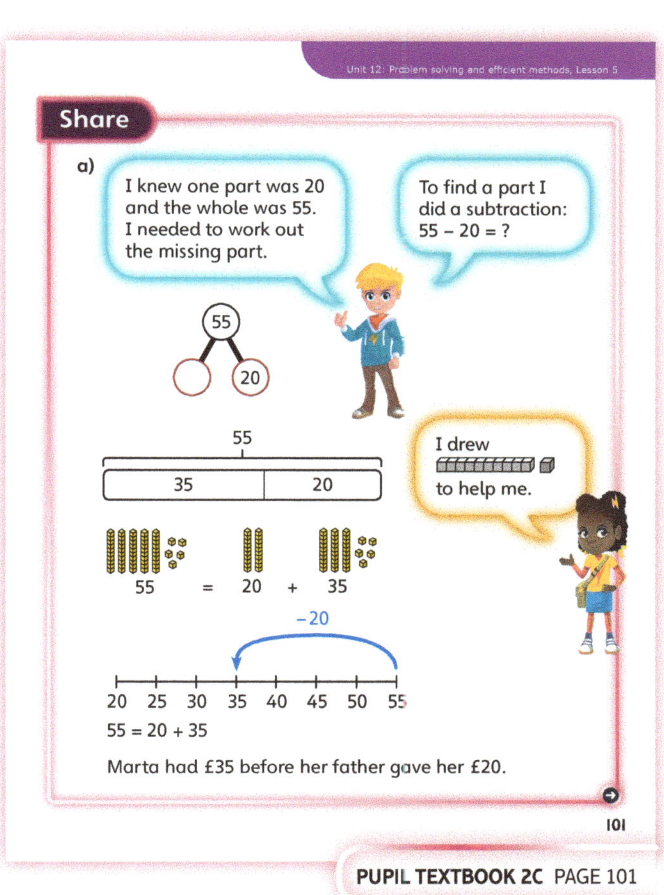

PUPIL TEXTBOOK 2C PAGE 101

Think together

WAYS OF WORKING Whole class teacher led (I do, We do, You do)

ASK

- Question ❶ and ❷: *Can the bar model help to represent the problem?*
- Question ❸: *What method do you find the easiest to calculate each answer?*

IN FOCUS This section of the lesson allows children to use different methods to complete the same question. Some children may find some methods difficult to use and as a result may use methods of their own.

In question ❸, it may not be obvious that the value represented by the star (34) is the same for all calculations.

STRENGTHEN If children are unsure of what is known and unknown and, as a result, of what operation is needed to complete the question, allow them to use the bar model for support. The use of coloured rods or strips of paper will help with this process as children are able to move the bars around to understand the problem.

DEEPEN Children can continue the sequence of calculations found in question ❸ by either adding more calculations before or after the first and final calculations. Alternatively, they could create their own sequence of calculations that follows a similar pattern.

ASSESSMENT CHECKPOINT Use questions ❶ and ❷ to check whether children can identify what is known and what is unknown. Can they then use what is known to calculate the unknown? Check to see if children can explain the different parts of the bar model and how this helps to identify the operation needed to complete the problem.

ANSWERS

Question ❶: 22 more cups are needed.

Question ❷: Joe needs 30p more.

Question ❸: 22 + 34 = 56
23 + 34 = 57
24 + 34 = 58
25 + 34 = 59
26 + 34 = 60

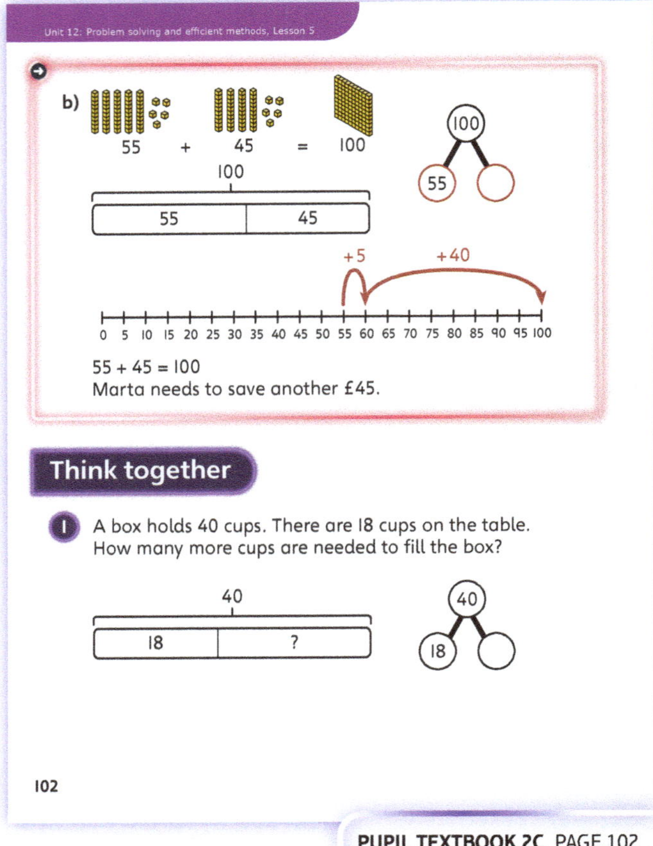

PUPIL TEXTBOOK 2C PAGE 102

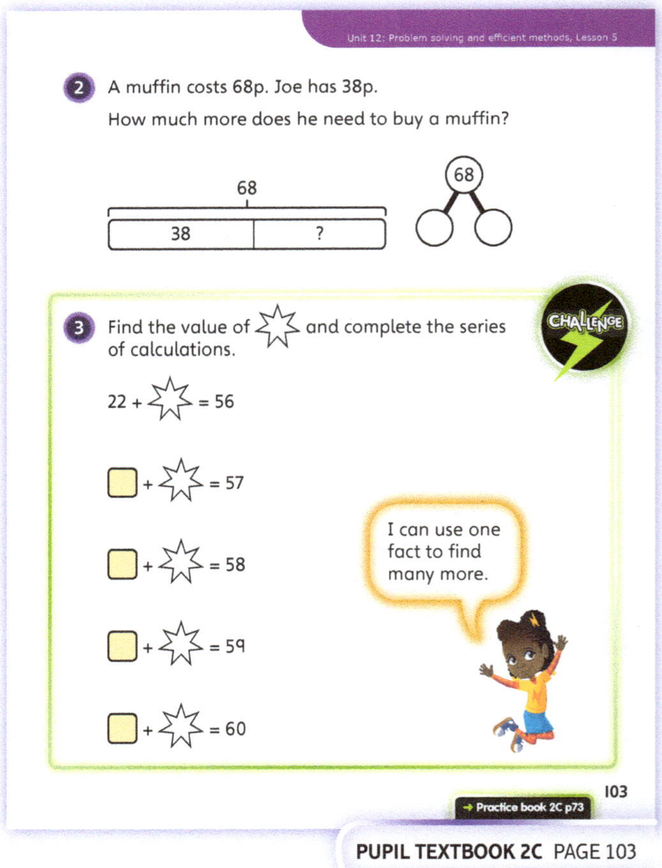

PUPIL TEXTBOOK 2C PAGE 103

Practice

WAYS OF WORKING Independent thinking

IN FOCUS These questions test children's understanding of the part-whole model, and how the parts and whole relate to each other and can be used to calculate unknowns. This is scaffolded in question **1** as there are no unknowns and children can use number bonds to 20 to help recognise if a calculation would not make sense. Children's answers to question **1** and how the parts and the whole relate can be used to support other questions. Question **4** is open-ended and children can use their strategy from the lesson to work out different ways that it can be completed.

STRENGTHEN Children who find the part-whole model too abstract should continue to create bar models. Alternatively, provide a completed bar model for each question, although physically making the model will allow children to manipulate and move the parts to model the calculations they are recording.

DEEPEN Provide children with a part-whole model where the whole has been split into three parts. Challenge children to record all possible addition and subtraction calculations for this fact family.

ASSESSMENT CHECKPOINT Children should be able to verbalise what the parts and the whole are in each question and how knowledge of this helped them to solve each question.

ANSWERS Answers for the **Practice** part of the lesson can be found in the *Power Maths* online subscription.

Reflect

WAYS OF WORKING Independent thinking

IN FOCUS In this part of the lesson, all support has been removed and children must apply what they have learnt throughout the lesson to the abstract calculations containing a missing number. At this stage, children may have different methods to solve the same problem and these should be shared with the class.

ASSESSMENT CHECKPOINT Check children's explanations of how they solved each question to assess their understanding of the relationship between the parts and wholes in additions and subtractions.

ANSWERS Answers for the **Reflect** part of the lesson can be found in the *Power Maths* online subscription.

After the lesson ⏸

- Have children mastered how the parts and the whole relate to each other in additions and subtractions?
- Did children learn different ways to solve the same questions from class discussions?

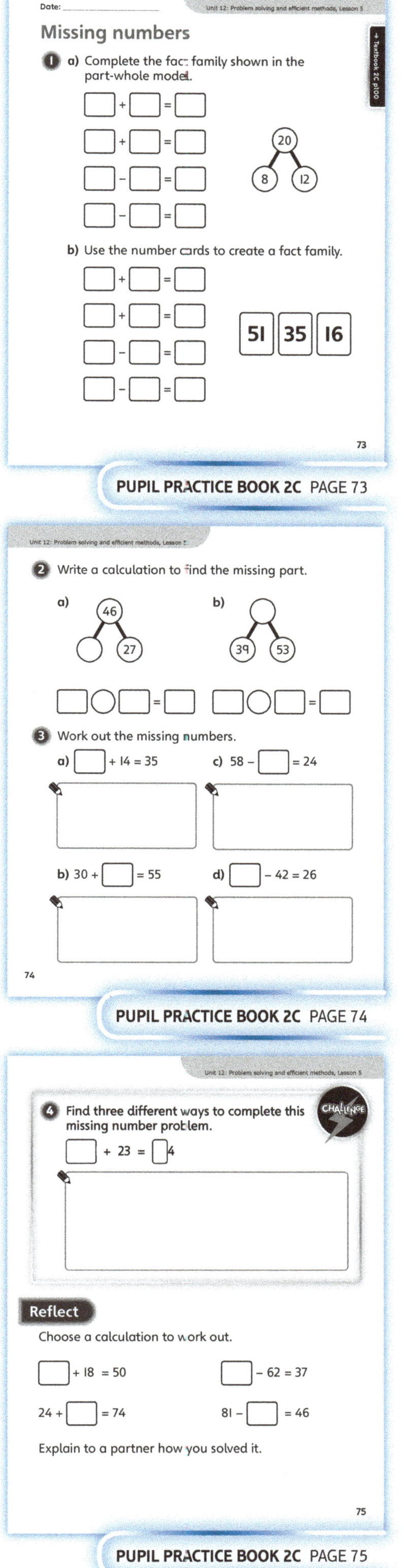

PUPIL PRACTICE BOOK 2C PAGE 73

PUPIL PRACTICE BOOK 2C PAGE 74

PUPIL PRACTICE BOOK 2C PAGE 75

Mental addition and subtraction ❶

Learning focus

In this lesson, children will use what they know to calculate unknown quantities. Children will apply methods and strategies that they have learnt in previous lessons.

Before you teach 

- How will you support children who found it difficult to identify the unknown values in the previous lesson?
- How can we refine the methods used by children to be more efficient?

NATIONAL CURRICULUM LINKS

Year 2 Number – number and place value

Use place value and number facts to solve problems.

Year 2 Number – addition and subtraction

Solve problems with addition and subtraction, applying their increasing knowledge of mental and written methods.

ASSESSING MASTERY

Children can make links between questions and use these links in order to work in a more efficient manner. Children can apply calculation strategies learnt from previous lessons and are flexible with the method that they use, based on the question they are presented with.

COMMON MISCONCEPTIONS

Children may not spot relationships between facts so they calculate from the beginning each time. Ask:
- *Can you use any known facts to help with the calculation?*

STRENGTHENING UNDERSTANDING

The use of manipulatives such as base 10 equipment will help children be more efficient in their mental calculations.

GOING DEEPER

Encourage children to visualise the steps that they are completing during mental calculations. Being able to visualise numbers in this way and describe what they can see pictorially or in writing will help them work more effectively in this way.

KEY LANGUAGE

In lesson: calculate mentally, efficient, incorrect, addition, +, subtraction, –

Other language to be used by the teacher: method

STRUCTURES AND REPRESENTATIONS

Part-whole model, number line

RESOURCES

Mandatory: base 10 equipment

Optional: bead string, number line

 In the eTextbook of this lesson, you will find interactive links to a selection of teaching tools.

Quick recap 

Challenge children to complete these additions. What do they notice?

12 + 5   12 + 7   12 + 2   12 + 4   6 + 12

Discover

 Pair work

- Questions **1** a) and b): *What do you know?*
- Questions **1** a) and b): *What do you need to find out?*
- Questions **1** a) and b): *Can you represent the information as a bar model?*

 Children must look at the information given about Tim and his family and work out what they know and what they don't know. A bar model could help to determine this and how the known and unknown quantities relate to each other.

 Children can use base 10 equipment to represent, partition, manipulate and calculate with the numbers in the scenario.

Question **1** a): Tim's brother is 17 years old.

Question **1** b): Tim's mum is 37 years old.

PUPIL TEXTBOOK 2C PAGE 104

Share

 Whole class teacher led

- Question **1** a): *What is the most efficient method that you can use?*
- Question **1** b): *Can you use what you calculated in part a) to help you with this calculation?*

 In this part of the lesson, it is important to discuss the different methods that children use to calculate the answers to the same question. Draw children's attention to Flo's statement in question **1** a) and discuss an efficient mental method that could be used.

Question **1** b) could be answered by returning to Tim's age and calculating 31 years more than this, or by considering how many more years older Tim's mum is than Tim's brother, and working from the previous answer. It is likely that this will need to be modelled for children to understand.

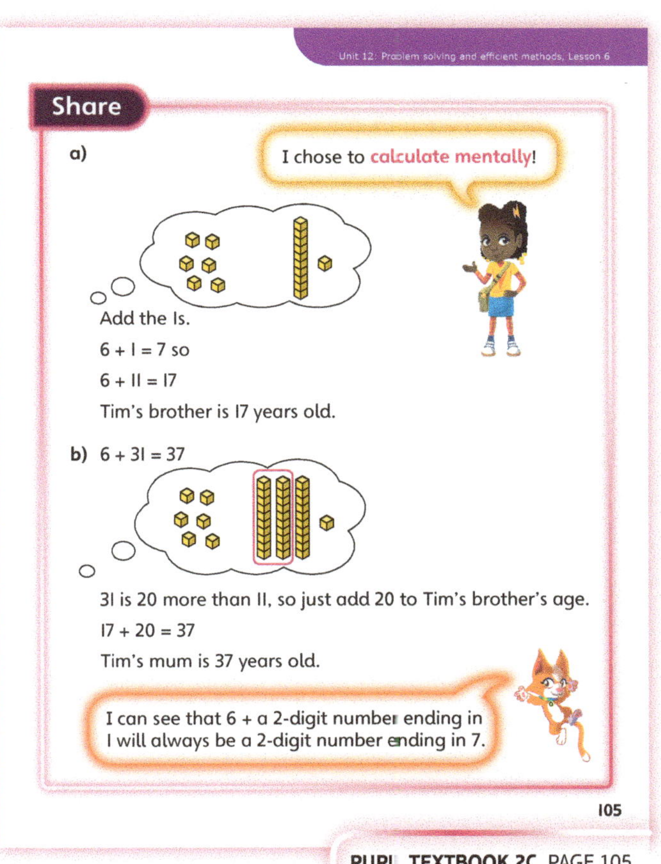

PUPIL TEXTBOOK 2C PAGE 105

Think together

WAYS OF WORKING Whole class teacher led (I do, We do, You do)

ASK

- Question **1**: *How can base 10 equipment be used to help you work efficiently?*
- Question **2**: *Do you have to return to Tim's age to solve this question? Why not?*

IN FOCUS This section of the lesson again allows children to discuss the different methods that can be used to solve the same problem. Question **2** highlights that it is not always the most efficient method to return to the original information. If children were to do this, they would then have to recalculate what they have already worked out, rather than simply working from a previous answer.

STRENGTHEN Using base 10 equipment will help children discover different mental methods that they can use to calculate the age of the different family members. It will also make it easier for children to count on in 10s and 1s, rather than just 1s.

DEEPEN Encourage children to visualise the calculations that they complete mentally. Ask them to draw pictorially what they can see, or write down what they can see at different stages of their mental calculation.

ASSESSMENT CHECKPOINT Check to see if children are working in an efficient way, or if they are still counting on in 1s. Also, check to see that children are building on what they have already calculated, rather than returning to the original information each time. Question **3** provides an opportunity to assess the different mental calculation strategies that children are comfortable using.

ANSWERS

Question **1**: Tim's grandma is 67 years old.

Question **2**: 37 + 40 = 77
Tim's grandpa is 77 years old.

Question **3**:
25 + 4 = 29	25 − 4 = 21
42 + 30 = 72	42 − 30 = 12
36 + 7 = 43	36 − 7 = 29
28 + 12 = 40	38 − 18 = 20

Children may explain how to use place value.

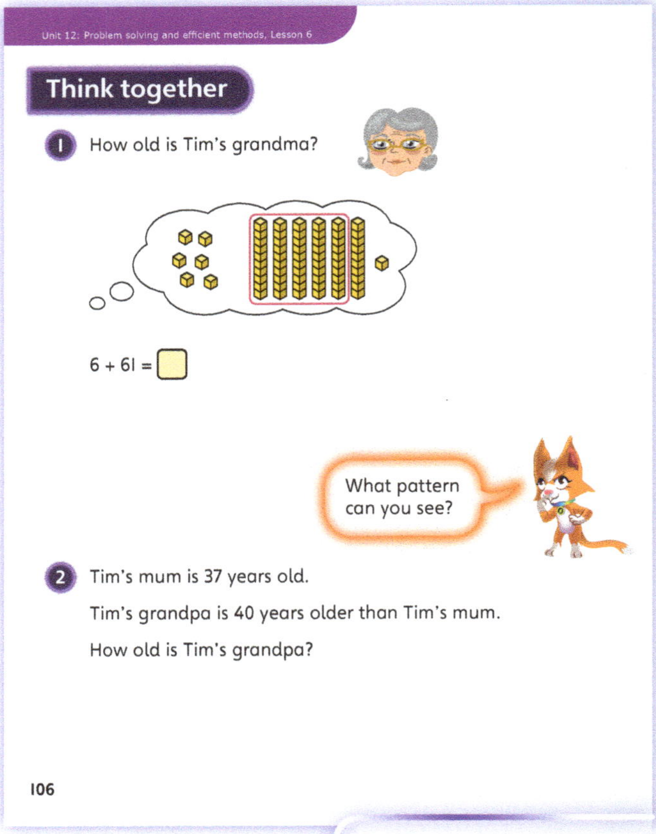

PUPIL TEXTBOOK 2C PAGE 106

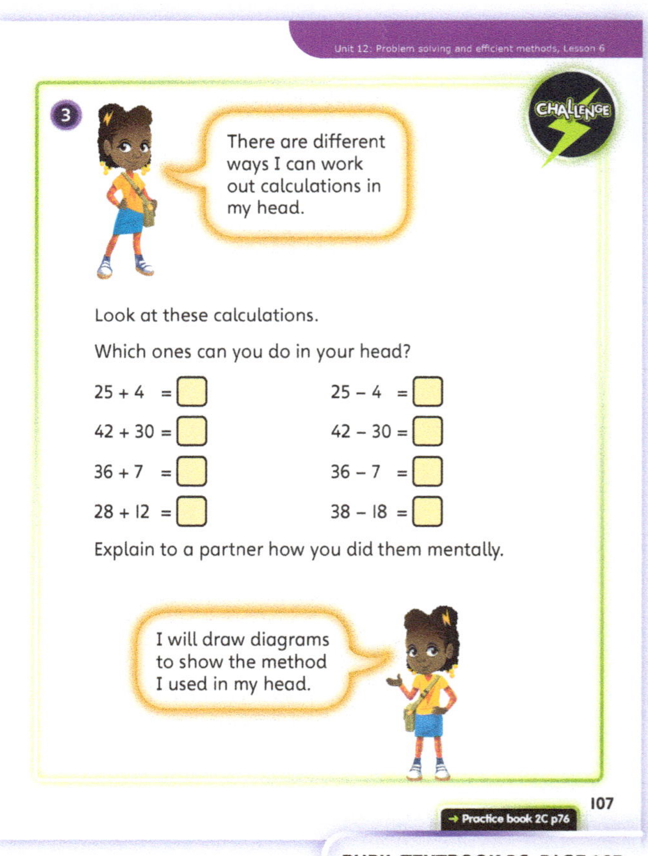

PUPIL TEXTBOOK 2C PAGE 107

Practice

WAYS OF WORKING Independent thinking

IN FOCUS The **Practice** questions allow children to apply what they have learnt in this and previous lessons. Children should be making links between numbers in different calculations and using known facts to efficiently calculate unknowns. They are also required to identify errors within calculations and should explain how they know calculations are incorrect. Encourage their use of mental methods as they work through the problems.

STRENGTHEN Using resources, for example, a bead string for question ❶, will help children make links between questions. If two different bead strings are used, children will be more likely to be able to comment on what is the same and what is different in the different calculations.

DEEPEN Encourage children to continue to visualise the mental calculations they complete. Explaining these to each other or drawing a pictorial representation will again strengthen this process and expose children to different mental strategies.

ASSESSMENT CHECKPOINT Children should be able to explain how they know that calculations are incorrect or how they have used information effectively in order to be efficient. Check to see which children are treating each question as a new calculation and as a result are not being efficient.

ANSWERS Answers for the **Practice** part of the lesson can be found in the *Power Maths* online subscription.

Reflect

WAYS OF WORKING Pair work

IN FOCUS In this part of the lesson, the different methods that children have used throughout the lesson should be discussed. These can then be applied to the different questions presented. Children may have different methods to complete the same question and these again should be celebrated.

ASSESSMENT CHECKPOINT Check to see if children can apply an appropriate mental calculation for each calculation given. The method that they use should be an efficient way to solve the calculation, rather than used blindly.

ANSWERS Answers for the **Reflect** part of the lesson can be found in the *Power Maths* online subscription.

After the lesson ⏸

- Do children have a range of strategies that they can use appropriately according to the question type?
- Are children working in an efficient way for the majority of calculations?
- Can children visualise the maths that they are completing and explain what they see to their peers?

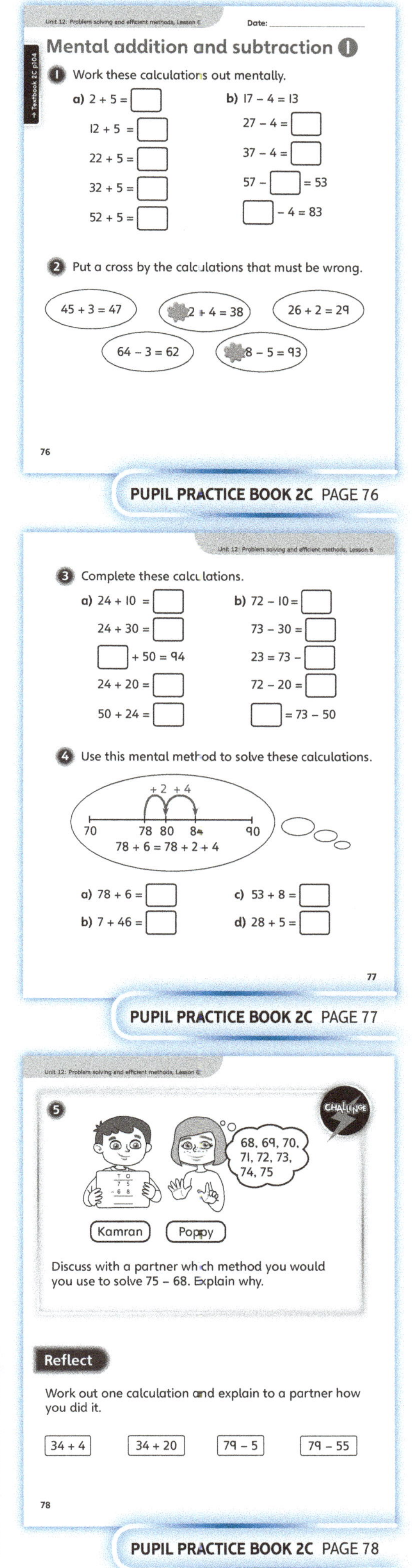

PUPIL PRACTICE BOOK 2C PAGE 76

PUPIL PRACTICE BOOK 2C PAGE 77

PUPIL PRACTICE BOOK 2C PAGE 78

Mental addition and subtraction ②

Learning focus

In this lesson, children will learn how to add or subtract a multiple of 10 to or from a number and then adjust to reflect the amount that should have been added or subtracted.

Before you teach ⏸

- Are children confident using a number line to show addition and subtraction?
- What are the barriers that may prevent children accessing the strategy taught in this lesson?

NATIONAL CURRICULUM LINKS

Year 2 Number – number and place value

Use place value and number facts to solve problems.

Year 2 Number – addition and subtraction

Solve problems with addition and subtraction, applying their increasing knowledge of mental and written methods.

ASSESSING MASTERY

Children can add or subtract a multiple of 10 to or from a number to represent a near multiple of 10 and then adjust the total appropriately to reflect the actual calculation.

COMMON MISCONCEPTIONS

Children are likely to confuse whether they should add or subtract 1 or 2 in the final stage of a mental calculation. For example, they may add 20 rather than 19 but then add another 1 to the total, rather than subtracting. Ask:
- *Can you show your calculation on a number line to show what needs to be added or subtracted?*

STRENGTHENING UNDERSTANDING

Using a number line will help children see the total amount that has been added or subtracted to or from a number and whether the final stage of the mental calculation requires addition or subtraction.

GOING DEEPER

Children should have the opportunity to work forwards and backwards through problems to gain a deeper understanding of the method. Working in pairs and explaining stages that they complete and then asking their partner to tell them the starting number, final number or whole calculation will also deepen their understanding.

KEY LANGUAGE

In lesson: adjust, addition, +, subtraction, –, difference, change

Other language to be used by the teacher: extra, count on

STRUCTURES AND REPRESENTATIONS

Number line, 100 square

RESOURCES

Mandatory: 1–100 number line, laminated 100 square

Optional: counters, base 10 equipment

 In the eTextbook of this lesson, you will find interactive links to a selection of teaching tools.

Quick recap 

Ask children to solve these additions using a 100 square:

23 + 10   46 + 20   38 + 30   79 + 20

Discover

 Pair work

- Questions **1** a) and b): *Is it easier to add (or subtract) multiples of 10, or to use numbers like 19, 28 or 37?*
- Questions **1** a) and b): *How could you use multiples of 10 to help with these calculations?*

 Use this part of the lesson to introduce children to the strategy of adding or subtracting a multiple of 10 to or from a number. Explain that this is an easier mental step and they can then adjust the answer by an appropriate amount to reflect the calculation that they are completing.

 Children can move a counter on their own 100 square, saying each number aloud as they count on and back.

Question **1** a): A skateboard and a pair of knee-pads cost £45.

Question **1** b): Ben's skateboard costs £37.

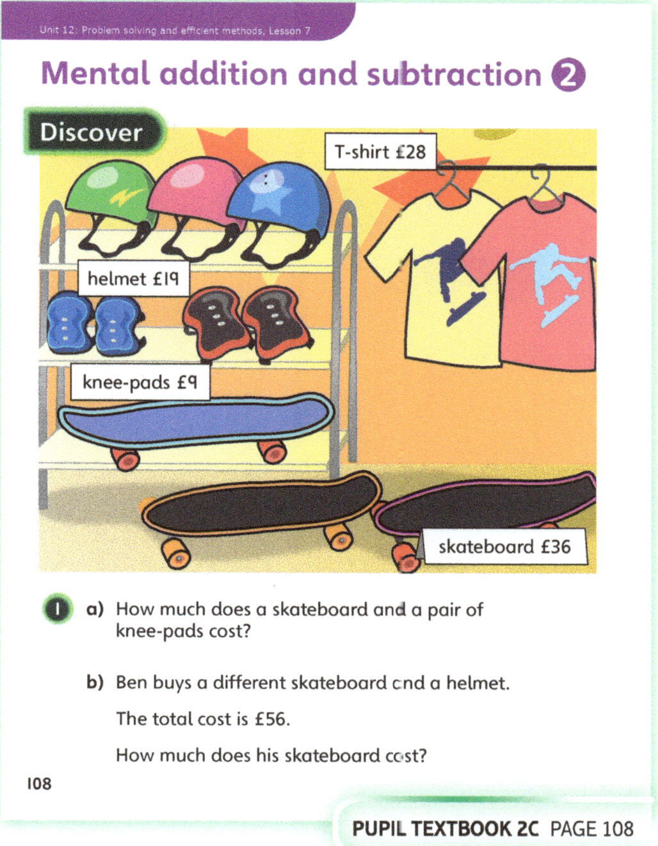

PUPIL TEXTBOOK 2C PAGE 108

Share

 Whole class teacher led

- Questions **1** a) and b): *What is the same and what is different about the methods used?*
- Questions **1** a) and b): *Why don't you subtract as the final stage for both questions?*

 This section of the lesson visually demonstrates to children the process of making the next 10 and adjusting, using a 100 square and a number line. It will be useful for children to model the stages of the calculations on these representations in order to understand the process of adjusting and the differences between addition and subtraction.

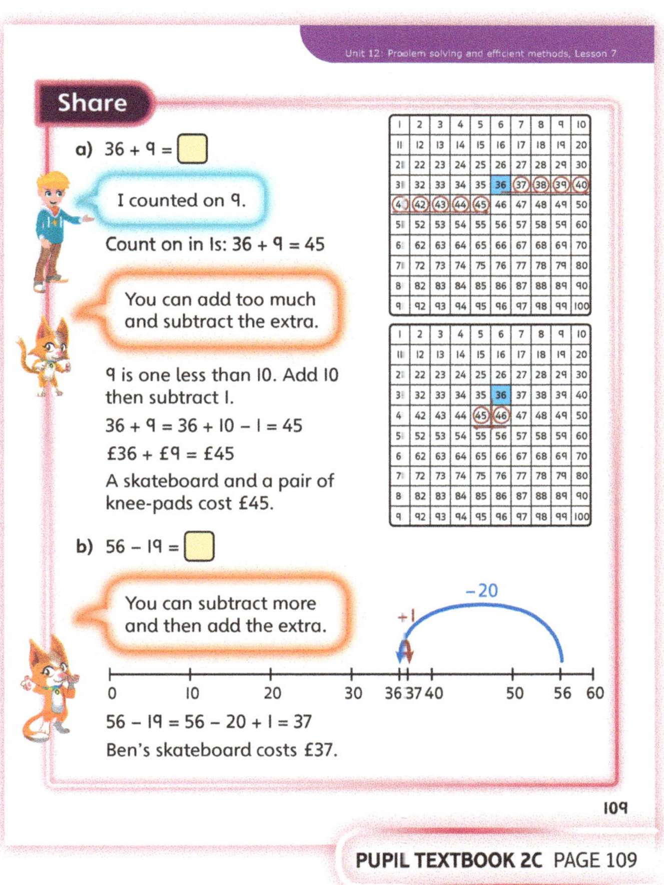

PUPIL TEXTBOOK 2C PAGE 109

Think together

 Whole class teacher led (I do, We do, You do)

- *Why do you sometimes add and sometimes take away during the final step in the calculation?*
- *Can this method only be used for numbers that are 1 away from 10?*

 These questions model adding and subtracting near multiples of 10 again using the 100 square and a number line. Questions ❶ and ❷ do not show the final adjusting stage and children must work out whether it is necessary to adjust by adding or subtracting 1 at the final stage.

 Drawing the stages of the calculation on a number line will help to show what has been added or subtracted at different stages in the process. Using actual money alongside this will also help children to understand how much has been added or subtracted at different points.

 Deepen understanding by challenging children to answer questions such as: *I start from a number, I add 20 to this number and then add another 1. The number that I finish on is 37. What calculation did I do?*

 Check that children can verbalise why they either add or subtract within the final stage of the calculation process. If they can explain which operation is necessary then they understand the method.

Question ❶: $54 + 19 = 54 + 20 - 1 = £73$

Question ❷: $50 - 29 = 50 - 30 + 1 = £21$

Question ❸: Kara: $56 - 29 = 27$
David: $37 + 18 = 55$

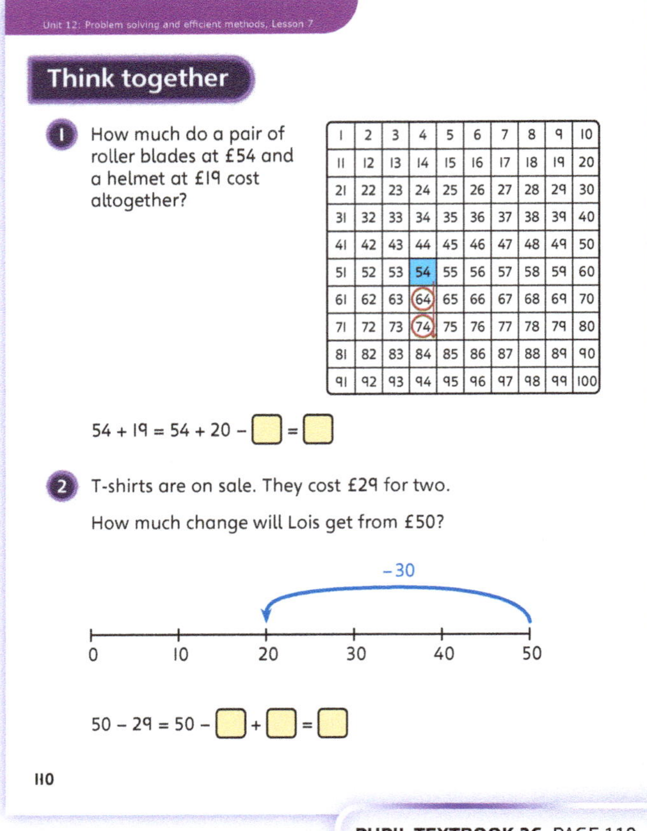

PUPIL TEXTBOOK 2C PAGE 110

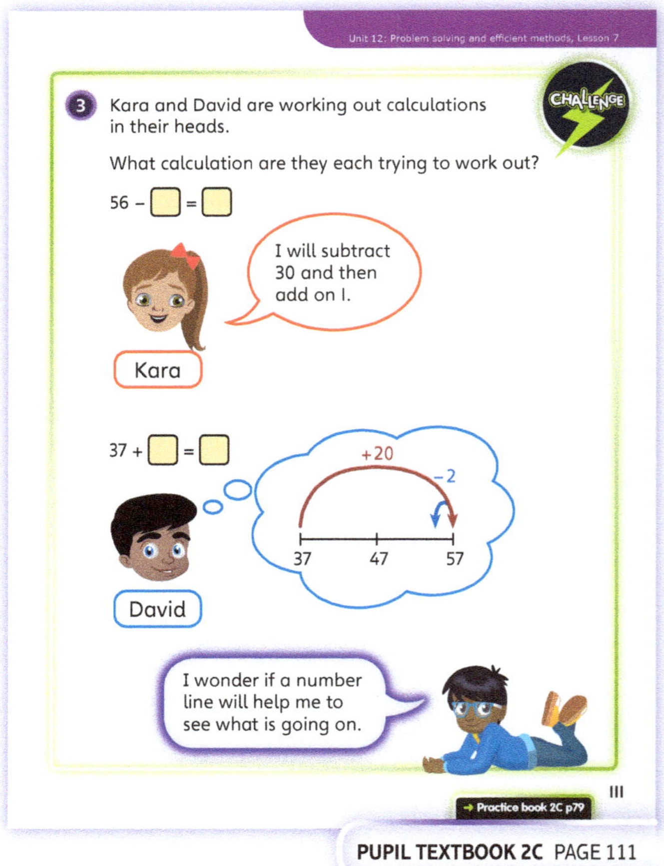

PUPIL TEXTBOOK 2C PAGE 111

Practice

WAYS OF WORKING Independent thinking

IN FOCUS Questions ❶ and ❷ practise the method that has been introduced in the earlier parts of the lesson. Questions ❸ and ❹ practise a slight variation of the method as both numbers have been changed by the same amount to make the mental calculation easier. Children may need additional guidance for this method as they will need to be clear on what makes a calculation easier to complete mentally (i.e. not having to make an exchange, but simply subtracting tens and then subtracting ones).

STRENGTHEN Allow children to continue to work with a 100 square or number line to reinforce the use of the different methods: counting on, subtracting too much and adding the extra, or adjusting both numbers so that no exchange is needed.

DEEPEN Children can work in pairs to create questions for a partner to solve that are similar to question ❺ using either of the methods practised within this lesson.

ASSESSMENT CHECKPOINT Check to see if children can explain the mental calculations that they are completing and if they are using a particular method, such as partitioning or counting on or back.

ANSWERS Answers for the **Practice** part of the lesson can be found in the *Power Maths* online subscription.

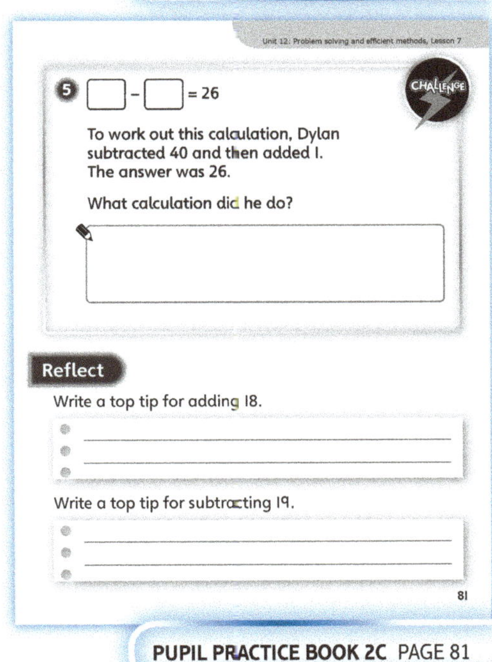

PUPIL PRACTICE BOOK 2C PAGE 79

PUPIL PRACTICE BOOK 2C PAGE 80

Reflect

WAYS OF WORKING Independent thinking

IN FOCUS This section of the lesson allows children to demonstrate their understanding of the strategy that they have learnt during the lesson. Allow children to complete this independently and then share and critique the accuracy of each other's explanations.

ASSESSMENT CHECKPOINT Check children's explanation to assess their understanding of the method learnt in this lesson.

ANSWERS Answers for the **Reflect** part of the lesson can be found in the *Power Maths* online subscription.

After the lesson ⏸

- Are children able to effectively use the strategy of adding or subtracting near multiples of 10 to and from numbers?
- Would children benefit from additional practice to strengthen their understanding?

PUPIL PRACTICE BOOK 2C PAGE 81

Efficient subtraction

Learning focus

In this lesson, children will look at efficient methods for subtracting. They will be challenged to choose an appropriate method based on the question they are presented with.

Before you teach 

- Do children always rely on the same method to calculate subtraction problems?
- How will you offer extra support to make them confident in an additional method?

NATIONAL CURRICULUM LINKS

Year 2 Number – addition and subtraction

Solve problems with addition and subtraction using concrete objects and pictorial representations, including those involving numbers, quantities and measures.

ASSESSING MASTERY

Children can identify an appropriate and efficient subtraction method based on the question that they are presented with. Children start to spot patterns to help them with subtraction.

COMMON MISCONCEPTIONS

Children may feel that they always have to use the same method for subtraction. Ask:
- *Is there a more efficient method you can use?*

STRENGTHENING UNDERSTANDING

Using a number line will help children to see alternative methods that could be used for some questions, such as 97 – 89. If children identify where 89 and 97 are on the number line, they can see that counting on is an efficient strategy to choose.

GOING DEEPER

Ask children to create a list of their own calculations that best suit taking away, finding the difference, counting on and making 10. Children should be able to justify why each calculation best suits each method.

KEY LANGUAGE

In lesson: efficient, near, count on, take away, subtract, difference

Other language to be used by the teacher: break apart

STRUCTURES AND REPRESENTATIONS

Number line

RESOURCES

Mandatory: blank number line

Optional: counters, 100 square

 In the eTextbook of this lesson, you will find interactive links to a selection of teaching tools.

Quick recap 

Ask children to complete these subtractions:

23 – 10   46 – 20   38 – 30   79 – 20

Discover

WAYS OF WORKING Pair work

ASK

- Questions ❶ a) and b): *What methods can we use for subtraction?*
- Questions ❶ a) and b): *Are you more likely to make mistakes with some methods than others?*

IN FOCUS In this part of the lesson, it is important to discuss the different methods that children in the class feel comfortable using for subtraction. It is also important to establish that at different times, different methods are more efficient.

PRACTICAL TIPS Children can move a counter on a 100 square, saying each number aloud as they count on and back.

ANSWERS

Question ❶ a): 61 − 18 = 43

Question ❶ b): 61 − 56 = 5

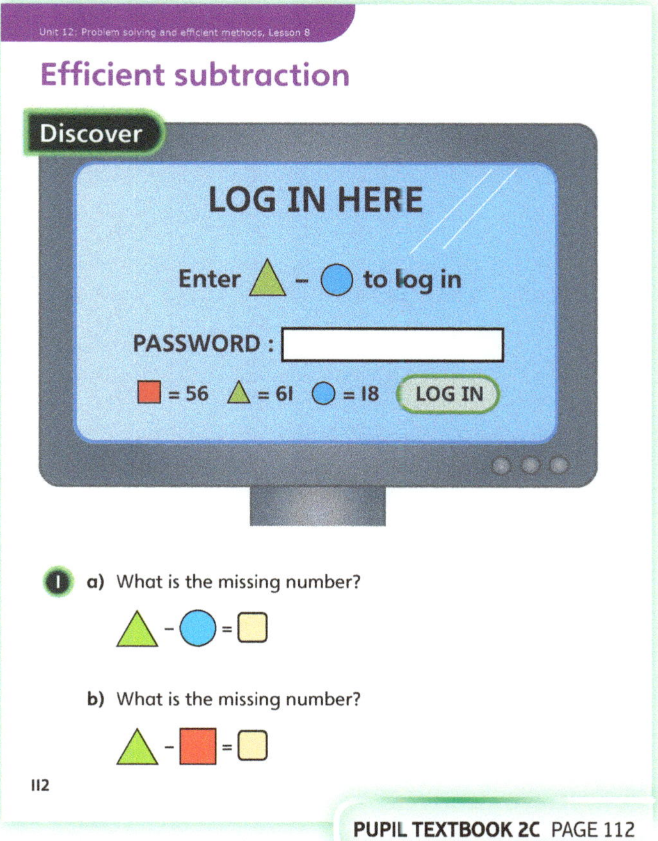

PUPIL TEXTBOOK 2C PAGE 112

Share

WAYS OF WORKING Whole class teacher led

ASK

- Question ❶ a): *Why is Astrid's method not efficient? What method would be more efficient to use?*

IN FOCUS Question ❶ a) models using a number line to first subtract the correct multiple of 10, and then the correct number of 1s. Discuss why this is an efficient method to calculate 61 − 18.

Question ❶ b) can be approached in two ways. Children could first subtract the correct multiple of 10 (50), and then the correct number of 1s (6). Alternatively, they could find the difference by counting on from 56 to 61.

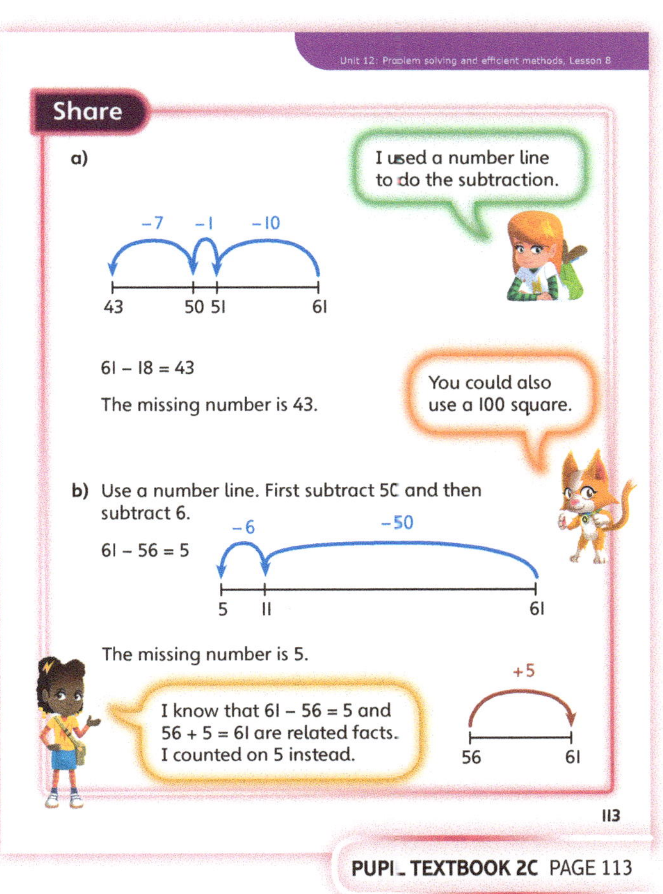

PUPIL TEXTBOOK 2C PAGE 113

Think together

WAYS OF WORKING Whole class teacher led (I do, We do, You do)

ASK

- *Is taking away the most efficient method to solve these problems?*
- *What resource could you use to show a more efficient strategy?*

IN FOCUS This part of the lesson requires children to consider what the most efficient method for subtraction is. For some calculations, counting on might be more efficient than taking away.

STRENGTHEN Provide children with a number line that they can identify both numbers on to clearly show if taking away is the most efficient way to solve the problem.

DEEPEN Challenge children to create their own questions that best suit taking away, counting on and rounding and adjusting (from the previous lesson).

ASSESSMENT CHECKPOINT Check to see which method children are choosing to use and if they can justify why this is the best method to use for each question.

ANSWERS

Question ❶: 56 − 18 = 38

Question ❷: 56 − 49 = 7   56 − 47 = 9
56 − 48 = 8   56 − 46 = 10
The answers are 1 more each time.

Question ❸: 81 − 72 = 9
48 − 8 = 40
81 − 8 = 73
72 − 24 = 48

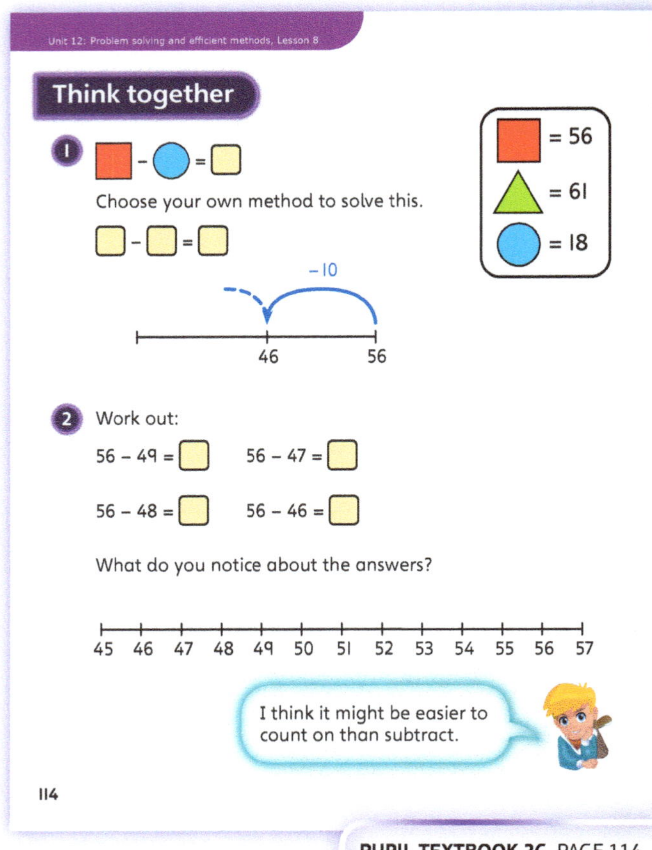

PUPIL TEXTBOOK 2C PAGE 114

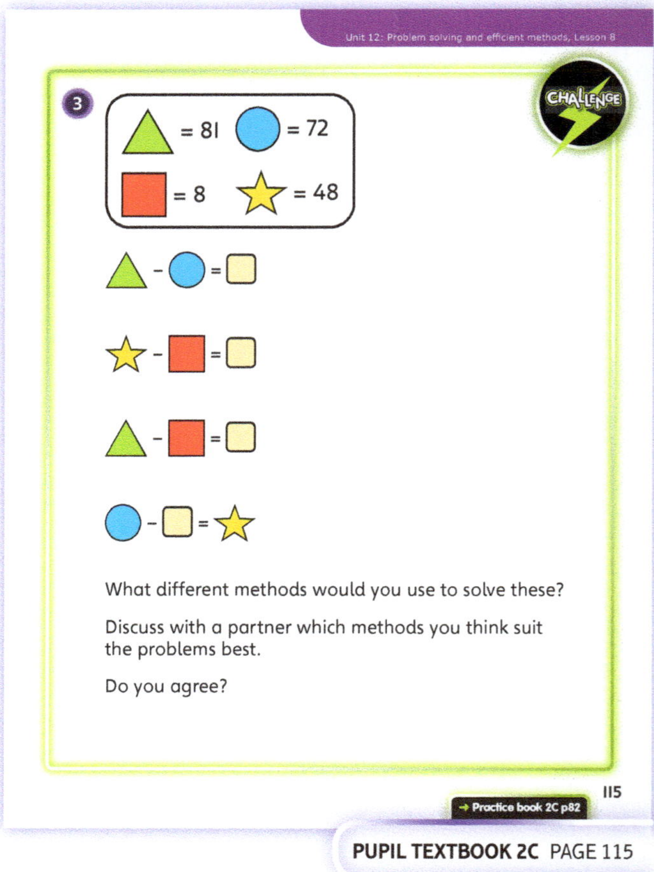

PUPIL TEXTBOOK 2C PAGE 115

Practice

WAYS OF WORKING Independent thinking

IN FOCUS This part of the lesson provides an opportunity for children to work independently to choose the most efficient calculation method for each question.

STRENGTHEN Recording the information that they are given as a bar model may help children identify the most efficient method to use if they can see the relative size of the bars and relate this to the most appropriate strategy.

DEEPEN Ask children to draw a pictorial representation to demonstrate how the method they have used to solve each question is the most efficient.

ASSESSMENT CHECKPOINT Check to see if children can justify the method that they have chosen for each question and why they think this is the most efficient choice.

ANSWERS Answers for the **Practice** part of the lesson can be found in the *Power Maths* online subscription.

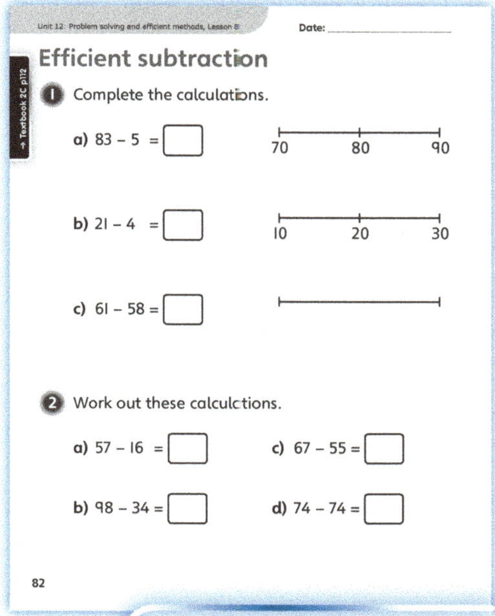

PUPIL PRACTICE BOOK 2C PAGE 82

PUPIL PRACTICE BOOK 2C PAGE 83

Reflect

WAYS OF WORKING Pair work

IN FOCUS Allow children to work with a partner to choose the method that they believe is the most efficient to solve each calculation. Once the class have done this, allow them to share, critique and build on each other's opinions.

ASSESSMENT CHECKPOINT The quality of children's verbal or written explanations will allow you to make assessments of the clarity of their understanding of the different methods.

ANSWERS Answers for the **Reflect** part of the lesson can be found in the *Power Maths* online subscription.

After the lesson ⏸

- How did children respond to the methods presented in the lesson?
- Which method were children least comfortable using?
- How could children become more confident at using this method?

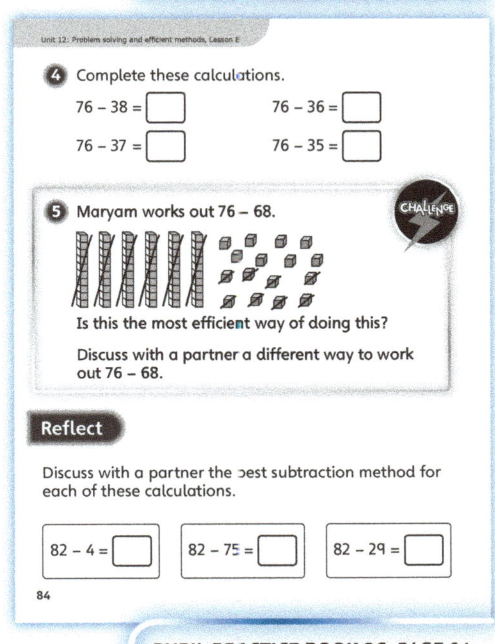

PUPIL PRACTICE BOOK 2C PAGE 84

Solve problems – addition and subtraction

Learning focus

In this lesson, children will solve a variety of different 1- and 2-step problems that will require appropriate calculation strategies.

Before you teach 

- Will children find it difficult to interpret the problems?
- How will you provide additional support?

NATIONAL CURRICULUM LINKS

Year 2 Number – number and place value

Use place value and number facts to solve problems.

Year 2 Number – addition and subtraction

Solve problems with addition and subtraction, applying their increasing knowledge of mental and written methods.

ASSESSING MASTERY

Children can decide fluently whether a problem involves addition, subtraction, or both operations. As a result of this understanding, they can choose appropriate strategies to solve the problem in an efficient manner.

COMMON MISCONCEPTIONS

Children may not use an appropriate method to solve the problem. Ask:
- *What method is the most efficient to solve the problem? What other methods have you considered?*

STRENGTHENING UNDERSTANDING

Support children to build bar models and manipulate the parts and the whole to increase their understanding of the problem and the likelihood that they choose the correct operation to solve the problem.

GOING DEEPER

Children should find the most efficient method for each problem. Ask children to explain why they think each method is the most efficient. Alternatively, challenge children to draw what they are visualising with their mental calculations.

KEY LANGUAGE

In lesson: bar model, total cost, change, left over, addition, +, subtraction, –, whole, part

Other language to be used by the teacher: represent, method

STRUCTURES AND REPRESENTATIONS

Bar model, part-whole model

RESOURCES

Mandatory: blank bar models

Optional: coloured rods or strips of paper to create the bar model, laminated part-whole diagrams

 In the eTextbook of this lesson, you will find interactive links to a selection of teaching tools.

Quick recap 

Ask children to solve these additions:

23 + 11   46 + 22   38 + 31   79 + 19

Discover

WAYS OF WORKING Pair work

ASK

- Question 1 a): *What does altogether mean?*
- Questions 1 a) and b): *Do you need to add or subtract for each question?*
- Questions 1 a) and b): *How could the bar model be used to represent each question?*

IN FOCUS This part of the lesson begins with a simple addition problem. Children should be able to complete this question as a result of their place value knowledge, or by counting on from 58 in four jumps of 10.

The second part of the question is more complex and it is likely that children will need to draw the bar model in order to understand how to solve it. To solve the question, children will also have to remember that £1 = 100p.

PRACTICAL TIPS Children can make their own bar models based on the scenario using strips of paper or coloured rods of different lengths.

ANSWERS

Question 1 a): A cup of tea and a teacake cost 98p altogether.

Question 1 b): One egg costs 26p.
One piece of toast costs 48p.

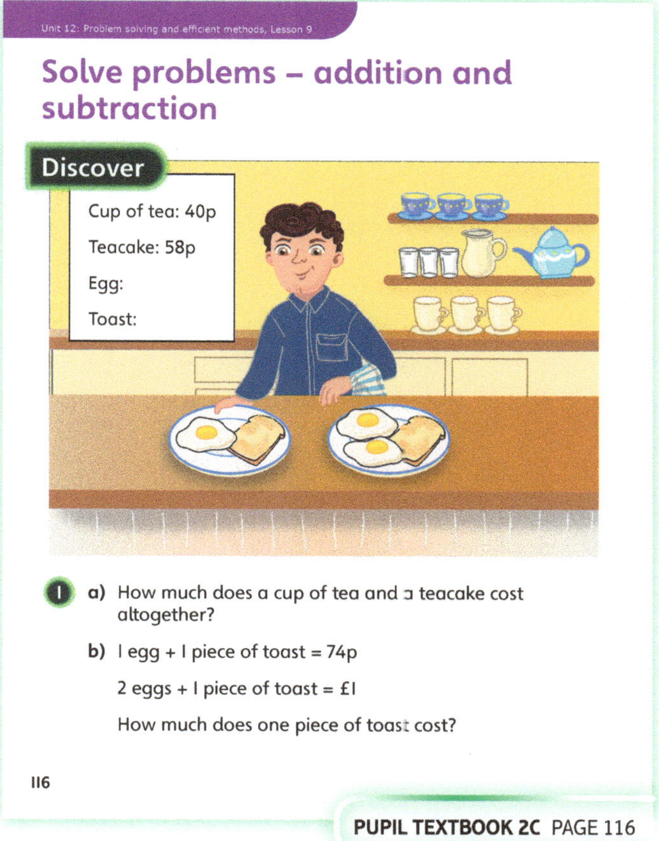

PUPIL TEXTBOOK 2C PAGE 116

Share

WAYS OF WORKING Whole class teacher led

ASK

- Question 1 a): *Why has addition been used for this question?*
- Question 1 b): *How can you calculate the price of one egg?*
- Question 1 b): *Can you represent all the information that you know as a bar model?*

IN FOCUS In question 1 b), children learn how the bar model can be used to represent all of the information that is known and what is required to be calculated. The part-whole model can then additionally be used to highlight that subtraction should be used to calculate the information that the question requires you to find: the price of the toast. After drawing the two bar models, encourage children to see that they cannot immediately find the price of a piece of toast. Ask: *Is it possible to find the price of the toast straight away? Which item can you find the price of first? Can you explain how?*

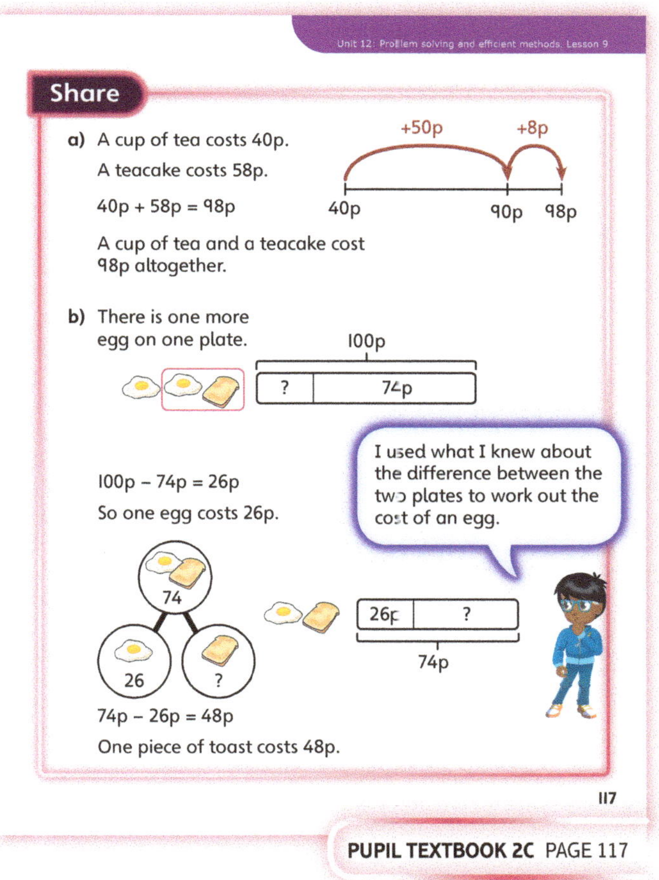

PUPIL TEXTBOOK 2C PAGE 117

Think together

WAYS OF WORKING Whole class teacher led (I do, We do, You do)

ASK

- *What part of each bar model shows the number that you need to find?*
- *What is the most efficient calculation strategy for each question?*

IN FOCUS In these questions, the bar model is provided to support children with the decision to use addition or subtraction. Children should be able to link the numbers shown on the bars to numbers found in the question and use this representation to establish whether addition or subtraction is needed to find the solution.

STRENGTHEN If children find skip counting or remembering division and multiplication facts difficult, provide times-table facts for them to refer to. This will help them with their calculations.

DEEPEN Challenge children to create their own story that involves a combination of addition and subtraction. To deepen understanding further, ask them to draw a bar model including a '?' to show what must be calculated.

ASSESSMENT CHECKPOINT Check to see if children can recognise whether a problem requires addition or subtraction. Assess whether children can explain how they know, with the use of a bar model to help, as opposed to guessing.

ANSWERS

Question **1**: A teacake costs 32p more than an egg.

Question **2**: Filip will get 4p change.

Question **3**: The red block is 30 cm.
The yellow block is 25 cm.

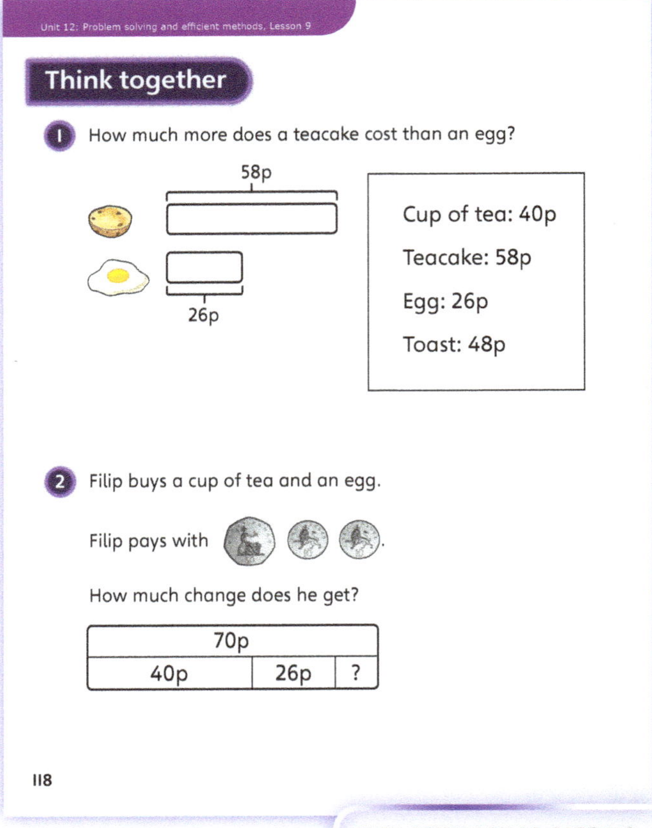

PUPIL TEXTBOOK 2C PAGE 118

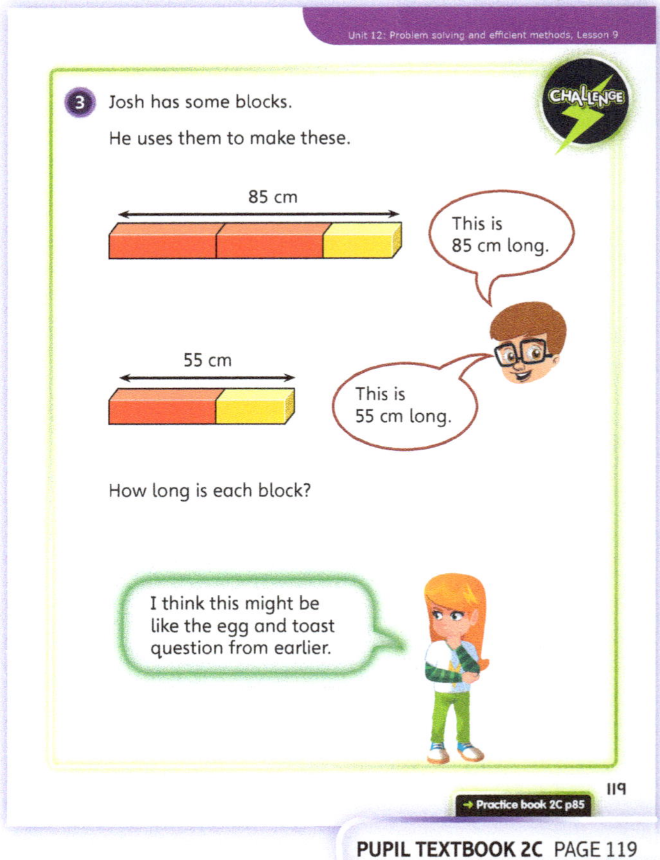

PUPIL TEXTBOOK 2C PAGE 119

Practice

WAYS OF WORKING Independent thinking

IN FOCUS In this part of the lesson, children are no longer provided with the bar model to assist them in making the decision about the appropriate operation to use. Encourage children to draw their own model to help them with this decision. Some questions are 2-step problems. Encourage children to keep a record of the information that they calculate as they go, rather than completing too many steps mentally, which may lead to errors.

STRENGTHEN Provide children with blank bar models with parts of appropriate sizes to match the questions. Children can then position values given to them in the question in appropriate parts to help them identify the operation they must use to solve the problem.

DEEPEN Challenge children to rank the difficulty of each question and justify why they think questions are easier or harder than each other. This will make children carefully consider the steps that they took to complete the question and the efficiency of these steps.

THINK DIFFERENTLY In question **4**, children work through a word problem. This could be solved by addition but children should identify that it is more efficient to do a subtraction.

ASSESSMENT CHECKPOINT Check to see that children have chosen the correct operation for each question, that their bar models match the information provided in the question, and that the bars are in the appropriate orientation to each other.

ANSWERS Answers for the **Practice** part of the lesson can be found in the *Power Maths* online subscription.

Reflect

WAYS OF WORKING Independent thinking, pair work

IN FOCUS In this part of the lesson, children are required to work backwards from the abstract calculation and create a maths story to match. Allow children to do this independently and then share their stories with a partner.

ASSESSMENT CHECKPOINT Check to see whether children use appropriate mathematical language within their maths story. Children can be assessed to see if they can identify which calculation matches the maths story of others in their class.

ANSWERS Answers for the **Reflect** part of the lesson can be found in the *Power Maths* online subscription.

After the lesson ⏸

- Did children use the bar model effectively to model the problems that they were presented with?
- What was the most common misconception that the use of the bar model elicited?

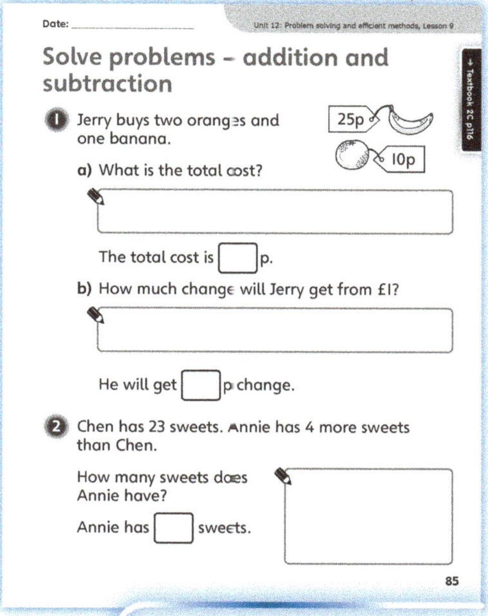

PUPIL PRACTICE BOOK 2C PAGE 85

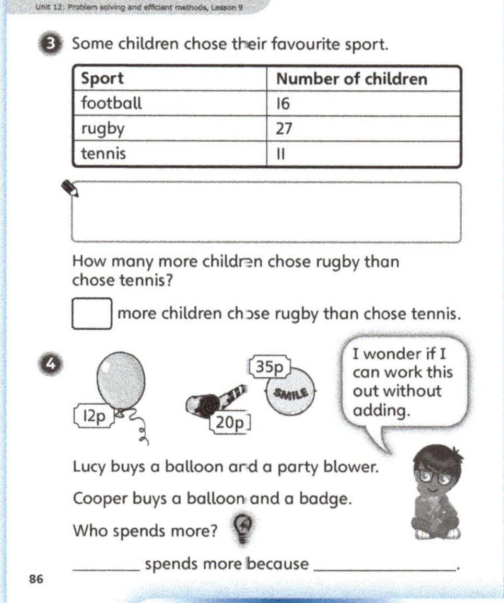

PUPIL PRACTICE BOOK 2C PAGE 86

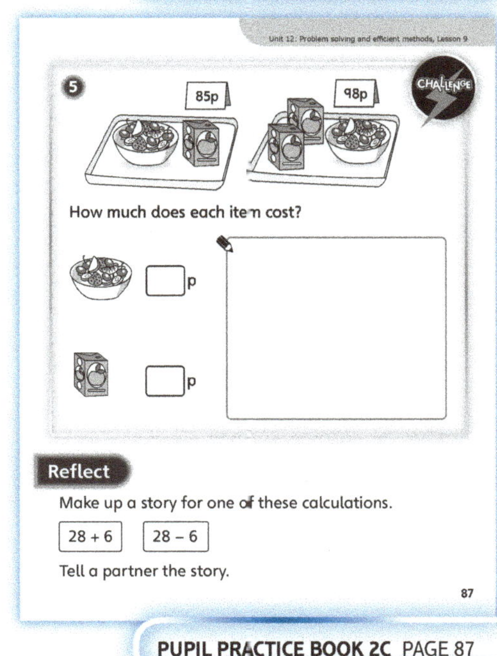

PUPIL PRACTICE BOOK 2C PAGE 87

Solve problems – multiplication and division

Learning focus

In this lesson, children will solve multiplication and division problems. They will decide if a problem requires multiplication or division to solve it, using the bar model to help make their choice.

Before you teach 

- Do children need to recap key multiplication and division learning points before this lesson?

NATIONAL CURRICULUM LINKS

Year 2 Number – multiplication and division

Solve problems involving multiplication and division, using materials, arrays, repeated addition, mental methods, and multiplication and division facts, including problems in contexts.

ASSESSING MASTERY

Children can decide fluently whether a problem requires multiplication or division. Children can choose appropriate strategies to solve the problem in an efficient manner.

COMMON MISCONCEPTIONS

Children may struggle to know which operation they need. They may simply look at the numbers and guess the operation, ignoring the context. Ask:
- *Can you use a bar model to show the operation you used to find the unknown?*

STRENGTHENING UNDERSTANDING

Give bar models to children who find it difficult to represent the problems, so they have a better chance of choosing the correct operation. Provide multiplication and division facts for children to select the appropriate fact from.

GOING DEEPER

Ask children to create their own maths story for multiplication or division, or both, in a multi-step problem, and draw their own bar model to represent the story. Children can then solve each other's problems in pairs.

KEY LANGUAGE

In lesson: multiply, ×, divide, ÷, bar model, total cost, share, equally

Other language to be used by the teacher: operation, equal parts

STRUCTURES AND REPRESENTATIONS

Bar model

RESOURCES

Mandatory: coloured rods to make the bar model (or pre-cut strips of card)

Optional: times-table muliplication facts

 In the eTextbook of this lesson, you will find interactive links to a selection of teaching tools.

Quick recap 

Practise counting on and back together in 5s as a class.

Discover

WAYS OF WORKING Pair work

ASK

- Questions **1** a) and b): *Can you draw a bar model to represent these problems?*
- Questions **1** a) and b): *What needs to be the same for all of the bars?*

IN FOCUS In this part of the lesson, children are expected to apply what they have been practising during recent lessons, using the bar model for addition, subtraction, multiplication and division word problems. It is important for children to remember that for these questions, it is vital that the bars are equal in length to accurately represent the question. Using repeated addition may help some children effectively make this jump.

PRACTICAL TIPS Children can use coloured rods or strips of paper to create a bar model for each calculation. Remind them that in multiplication and division each bar or group must be equal.

ANSWERS

Question **1** a): The total cost is £20.

Question **1** b): Three balls cost £6.

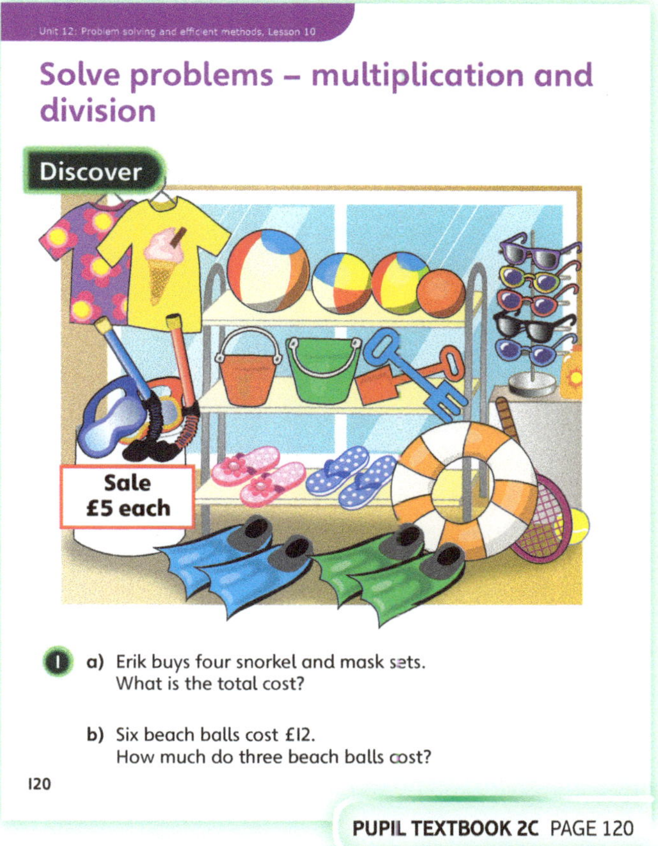

PUPIL TEXTBOOK 2C PAGE 120

Share

WAYS OF WORKING Whole class teacher led

ASK

- Question **1** b): *Do you always have to find the price of one item?*
- Question **1** b): *Is there a more efficient way to calculate the answer?*
- Question **1** b): *Who calculated the answer in the same way as Flo?*

IN FOCUS In this part of the lesson, children are shown each question using the bar model. This should help children to see how the bar model can be used for multiplication and division questions. Again, it is important to highlight to children that the bars in each question are equal in length.

Question **1** b) also discusses halving as an alternative way to calculate the answer. This is more efficient than calculating the price of one ball and then multiplying by 3.

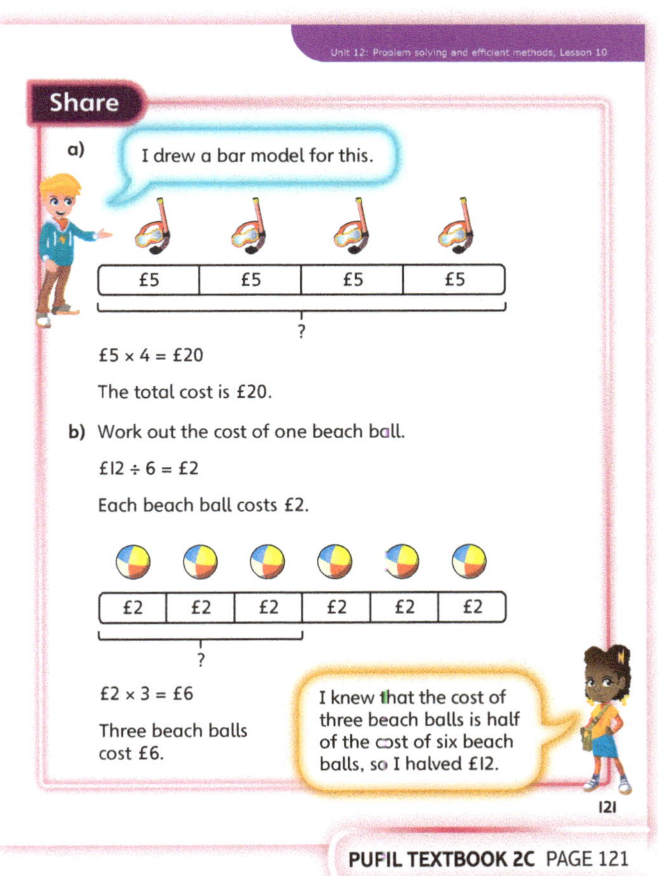

PUPIL TEXTBOOK 2C PAGE 121

Think together

WAYS OF WORKING Whole class teacher led (I do, We do, You do)

ASK

- Questions **1** and **2**: *How does the bar model help you to work out if the question needs multiplication or division?*
- Question **3**: *What is the most efficient way to calculate the answer?*

IN FOCUS In this section of the lesson, children are provided with bar models to help with their decision of whether each question requires multiplication or division. Question **3** shows a number of the same items grouped in different ways, for example, one box of 20 is the same as two boxes of 10. Some children may simply want to work out how many lollies there are altogether and then halve this number. Alternatively, some may want to halve each group, which they will find difficult for the boxes of 5 lollies, rather than creating groups of the same number of objects.

STRENGTHEN Provide children who find skip counting or remembering division and multiplication facts difficult times-table facts that they can refer to, to help with calculations.

DEEPEN Children can create their own story that involves either multiplication or division. To deepen understanding further, they can draw a bar model including a '?' to show what must be calculated.

ASSESSMENT CHECKPOINT Check to see which children are able to correctly identify which questions require multiplication or division. They should be able to justify how they know, rather than just guessing.

ANSWERS

Question **1**: 20

Question **2**: £6

Question **3**: They will each get 30 ice lollies.

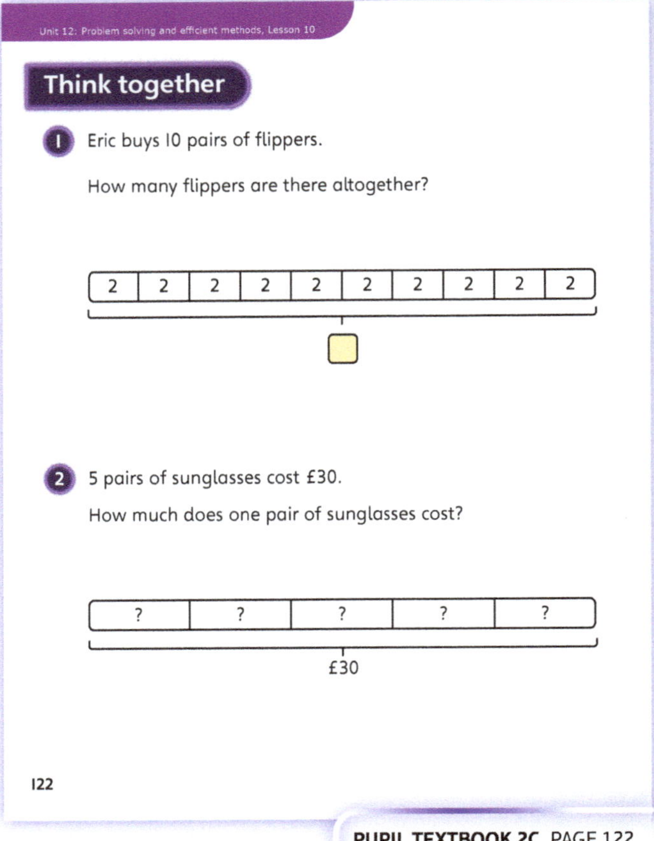

PUPIL TEXTBOOK 2C PAGE 122

PUPIL TEXTBOOK 2C PAGE 123

Practice

WAYS OF WORKING Independent thinking

IN FOCUS In the questions in this part of the lesson, children are no longer given bar models to help them. Encourage children to draw bar models in the space provided and record the calculation that they are completing.

STRENGTHEN Children who find drawing the bar model difficult should have the opportunity to make it with resources first, in order for mistakes to be made in a way that can easily be changed.

DEEPEN In question **4**, challenge children to find how many different ways the packs of balloons can be shared between the two children. Although the end result will be the same, it will tackle the misconception that items must be shared in 1s.

THINK DIFFERENTLY In question **4**, children need to choose the best way to share different amounts of balloons between two children. Some children may be tempted to share the balloons one by one but they should soon realise that this is not an efficient method when working with bigger numbers.

ASSESSMENT CHECKPOINT Check to see that children have identified the operation correctly and that the bar model reflects the information given in the question accurately.

ANSWERS Answers for the **Practice** part of the lesson can be found in the *Power Maths* online subscription.

Reflect

WAYS OF WORKING Independent thinking

IN FOCUS In this section of the lesson, children are required to work backwards from the abstract calculation and create a word problem to match. Allow children to do this independently and then share the different stories with the class.

ASSESSMENT CHECKPOINT Check to see whether children use appropriate mathematical language within their maths story. Assess whether children can identify which calculation matches the maths story of other children in the class.

ANSWERS Answers for the **Reflect** part of the lesson can be found in the *Power Maths* online subscription.

After the lesson ⏸

- Which operation did children find most difficult?
- Would children benefit from additional practice of this operation before moving on?

PUPIL PRACTICE BOOK 2C PAGE 88

PUPIL PRACTICE BOOK 2C PAGE 89

PUPIL PRACTICE BOOK 2C PAGE 90

Solve problems – using the four operations

Learning focus

In this lesson, children will practise the four operations. They will solve problems with multiple steps and use the bar model to represent these steps.

Before you teach 

- What misconceptions have previously been seen that are likely to occur again?
- How will these misconceptions be addressed?

NATIONAL CURRICULUM LINKS

Year 2 Number – number and place value

Use place value and number facts to solve problems.

ASSESSING MASTERY

Children can represent problems using the bar model and select the appropriate operation to solve the problem. Children carefully choose the most efficient method to solve the problem.

COMMON MISCONCEPTIONS

Children may mix up the operations needed to solve different problems. This is especially likely as this is the first time all four operations have been used in the same lesson. Ask:

- *Can you represent the problem using the bar model? How does this show which operation to use?*

STRENGTHENING UNDERSTANDING

Show examples of the four operations displayed in different ways using bar models. Children can make their bar model look the same by knowing the operation, or make a bar model based on information in the question and work out the operation from the examples. Making the bar model with rods before drawing it is easier.

GOING DEEPER

Ask children to identify where questions with more than one operation can be completed in different ways. Children should record the different ways of completing them and explain why others cannot be completed in this way.

KEY LANGUAGE

In lesson: bar model, representation, addition, +, subtraction, –, multiplication, ×, division, ÷, altogether, left over, difference, change

Other language to be used by the teacher: operation

STRUCTURES AND REPRESENTATIONS

Bar model, part-whole model, number line

RESOURCES

Mandatory: coloured rods

Optional: cubes or counters, bar models showing the four operations, base 10 equipment

 In the eTextbook of this lesson, you will find interactive links to a selection of teaching tools.

Quick recap 

Practise counting on and back together in 2s, then 5s and then 10s as a class.

Discover

 Pair work

- Question ❶ a): *How many apples are in the bags? How many apples are in the boxes?*
- Question ❶ b): *Can you draw a bar model to represent this information?*

 In this part of the lesson, children are required, for the first time, to choose one of the four operations to solve the problem. Encourage children to work efficiently, and not to use repeated addition to calculate the answer.

 Children can use cubes or counters to represent the apples. They can then make bar models with strips of paper or coloured rods of different lengths to show what calculation is needed.

Question ❶ a): There are 22 apples altogether.

Question ❶ b): There are 10 apples left.

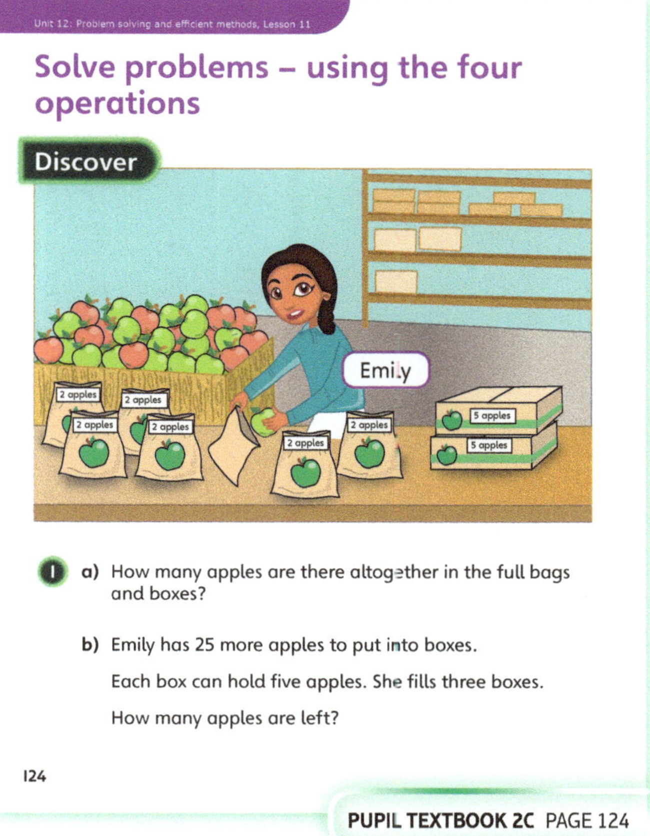

PUPIL TEXTBOOK 2C PAGE 124

Share

 Whole class teacher led

- Question ❶ a): *Why have the bars been arranged like this?*
- Question ❶ b): *Why are some of the bars the same length?*

 In this part of the lesson, children are presented with the information as a bar model. Discussions about these representations will help children to correctly make similar representations for later questions.

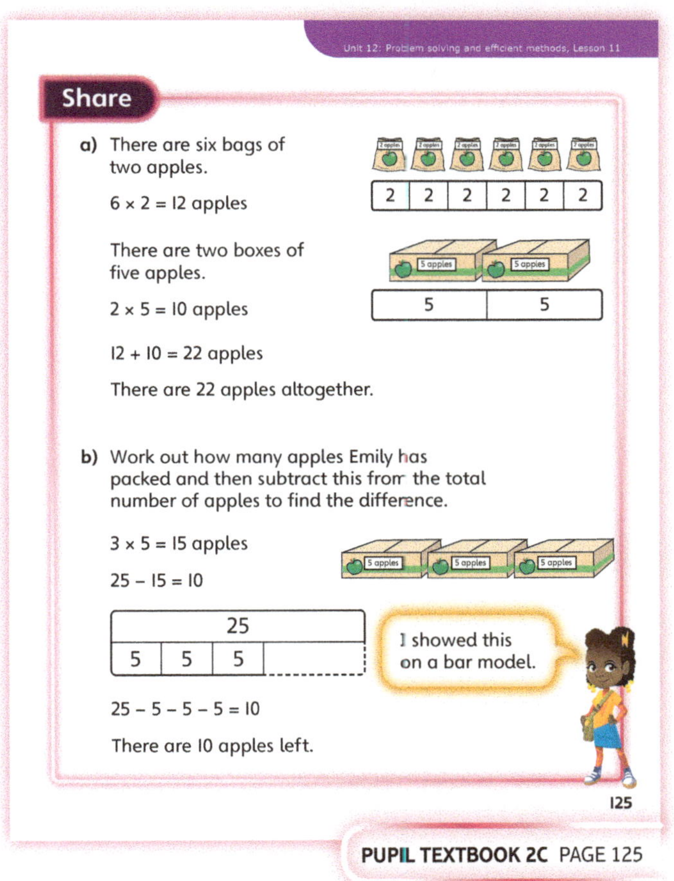

PUPIL TEXTBOOK 2C PAGE 125

Think together

WAYS OF WORKING Whole class teacher led (I do, We do, You do)

ASK

- Question **3**: *Why do some questions need more than one operation?*
- Question **3**: *Should all steps of the problem be done at the same time? Why not?*

IN FOCUS In this section of the lesson, children are initially provided with bar models to help them interpret the question. These are then removed for question **3** where children are required to apply what they have learnt from previous questions.

In question **3**, children are required to recognise three different stages within one calculation. It may support children to think: 'First I will …', 'Then I will …', and 'Finally I will …'.

STRENGTHEN Provide resources, such as base 10 equipment, to help children with their calculations. These are more favourable than leaving children to count in 1s to find the answer. Alternatively, provide key number facts that children could choose from to solve the questions.

DEEPEN Challenge children to make their own maths stories that use three operations at a time. Working in this open way will challenge children to find a context where this is possible and use numbers that will work for the three operations.

ASSESSMENT CHECKPOINT Check to see whether children can explain why, at different stages of each question, different operations are needed. Children should relate the answers that they give to both the information in the question and the bar model. Children should also be using known facts to calculate the answers rather than counting on or back in 1s. Check to see how children are calculating each answer.

ANSWERS

Question **1**: £30 + £8 = £38

Question **2**: Mantas will have £16 left.

Question **3**: Liam could buy 5 £2 bags with the £10 change.

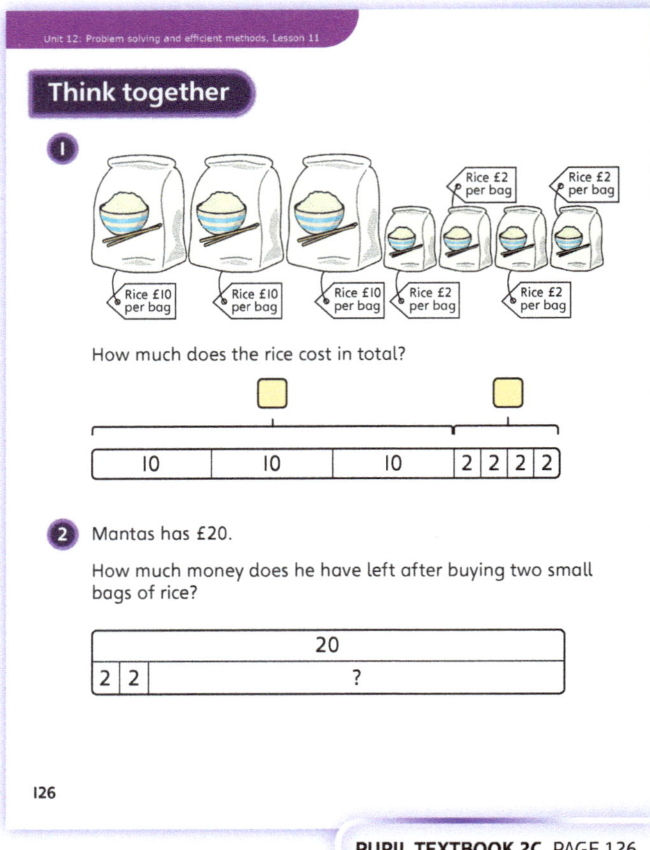

PUPIL TEXTBOOK 2C PAGE 126

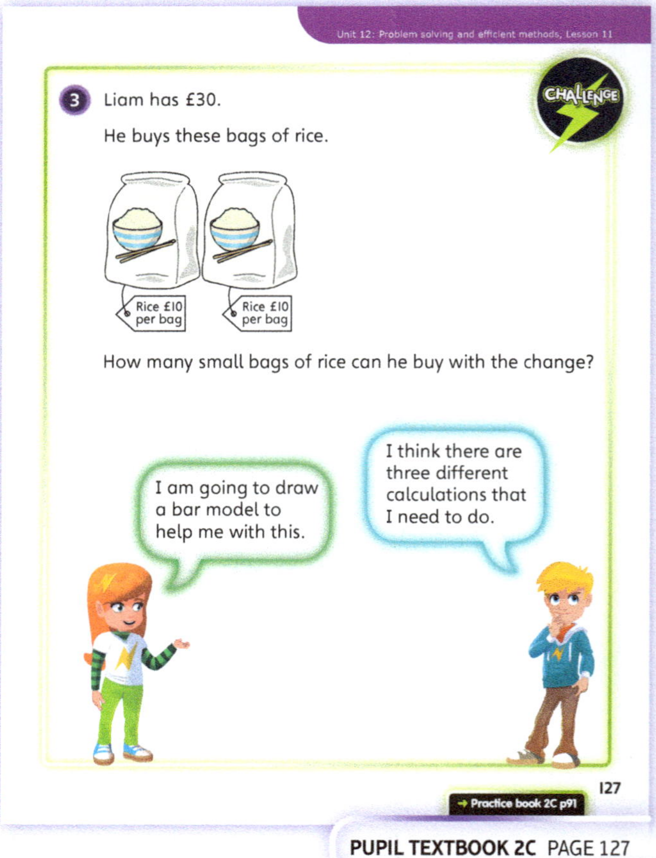

PUPIL TEXTBOOK 2C PAGE 127

Practice

WAYS OF WORKING Independent thinking

IN FOCUS Questions in this section of the lesson allow children to practise using the four operations in similar situations and will challenge children's number sense. The questions begin with the operations being treated separately and then combine the operations within different steps of the problem in an increasingly difficult way.

Children have not been provided with a bar model for questions in this part of the lesson. Encourage children to draw a bar model to help them understand the necessary operation at different stages of the calculation.

STRENGTHEN Children may need support to correctly draw or make bar models for the more complex, multi-step questions. If this is a step too far, provide bar models, highlighting the different steps of the question.

DEEPEN It is possible to answer some questions in different ways or orders. Challenge children to explore the possibilities, using different strategies, and record these as written calculations.

ASSESSMENT CHECKPOINT Ask children to explain why they have chosen to complete the different steps of a question in a particular order. These explanations will allow you to assess their understanding.

ANSWERS Answers for the **Practice** part of the lesson can be found in the *Power Maths* online subscription.

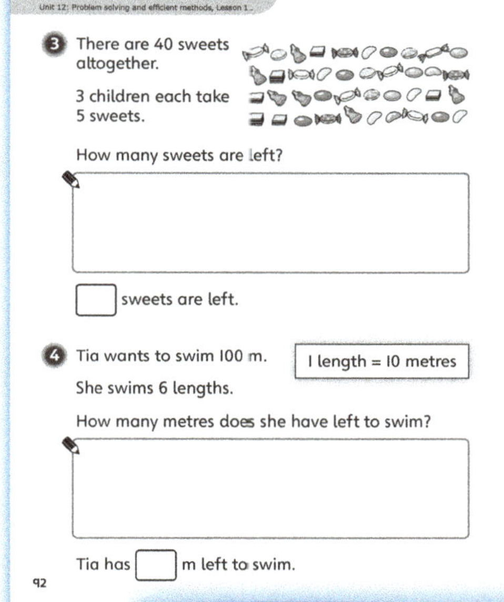

PUPIL PRACTICE BOOK 2C PAGE 91

PUPIL PRACTICE BOOK 2C PAGE 92

Reflect

WAYS OF WORKING Independent thinking

IN FOCUS Give children the opportunity to work independently to write their own maths story problems based on the calculations given. Children should then have the opportunity to share their stories with others in the class.

ASSESSMENT CHECKPOINT Check to see whether the language that children use for the different stages of the story matches the operation that they are required to use.

ANSWERS Answers for the **Reflect** part of the lesson can be found in the *Power Maths* online subscription.

After the lesson ⏸

- Do children recognise the effectiveness of the bar model?
- Are children using the bar model of their own accord?

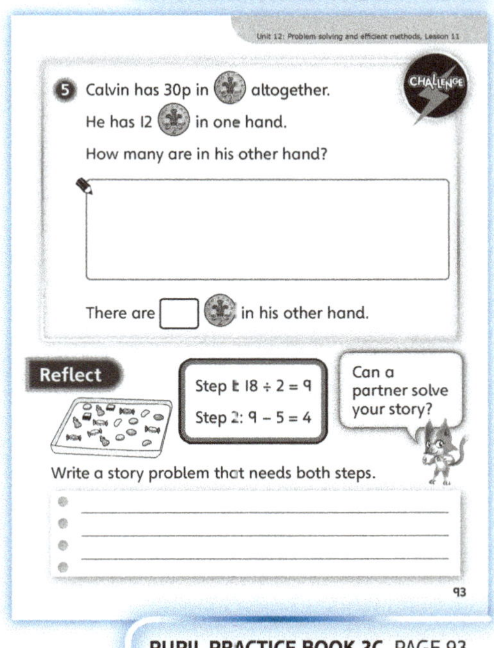

PUPIL PRACTICE BOOK 2C PAGE 93

End of unit check

> **Don't forget the unit assessment grid in your _Power Maths_ online subscription.**

WAYS OF WORKING Group work adult led

IN FOCUS The focus of these questions is whether children can select the appropriate operation as a result of the problems they are presented with, and solve these using an efficient strategy.

Think!

WAYS OF WORKING Pair work or small groups

IN FOCUS This question requires children to select an appropriate operation as a result of the information that they are provided with.

Children must work with multiplication and division at different stages of the question and self-identify where these stages occur.

Children should aim to work in an efficient way as a result of their mental calculations.

ANSWERS AND COMMENTARY Children who have mastered this unit will be able to fluently select the appropriate operation that is needed to complete a given step of a problem, and will be able to use the bar model to represent these problems. They will also be able to select an efficient method to solve the calculation.

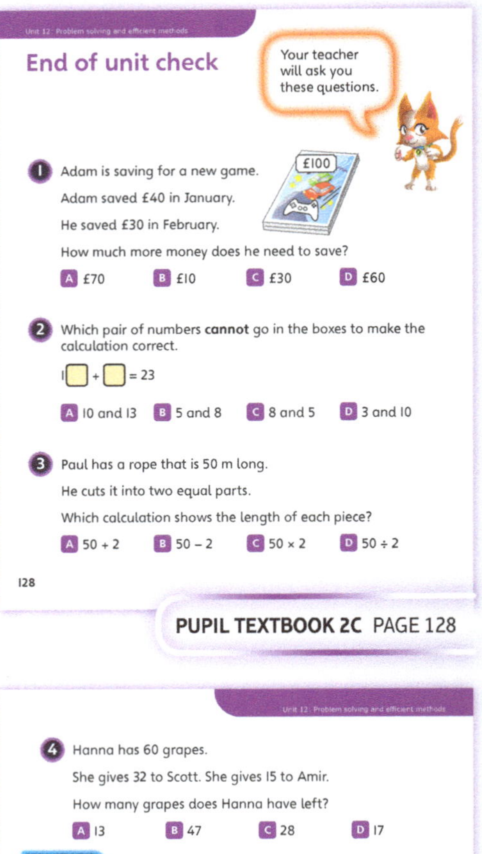

PUPIL TEXTBOOK 2C PAGE 128

PUPIL TEXTBOOK 2C PAGE 129

Q	A	WRONG ANSWERS AND MISCONCEPTIONS	STRENGTHENING UNDERSTANDING
1	C	A suggests children have simply added the two numbers. B suggests children have simply subtracted one number from the other. D suggests children have calculated how much more money Adam needs just based on what he has saved in January.	Children who are still finding it difficult to interpret the word problems that they are given, and as a result are choosing the incorrect operation or guessing an operation based on the numbers that they see, should have additional exposure to the bar model.
2	A	If children have answered B, C or D they are likely to have ignored the 1 before the first blank box indicating one 10.	
3	D	A, B or C suggest that children do not fully understand 'equal parts' or do not fully understand the context of the problem.	Children should have the opportunity to build bar models as a result of reading or hearing word problems. They should also have the opportunity to create word problems of their own from completed bar models. Alternatively, they could match bar models to word problems and/or calculations.
4	A	B suggests children have simply added the number of grapes that Hanna has given away. C suggests children have only taken away the number of grapes that Hanna gave to Scott. D suggests they have calculated the difference between the number of grapes that Scott and Amir have been given.	

My journal

WAYS OF WORKING Independent thinking

ANSWERS AND COMMENTARY

First I multiplied 4 by 10.
$4 \times 10 = 40$. There are 40 oranges altogether.

Then I divided 40 by 5.
$40 \div 5 = 8$.

I got the answer 8. There are 8 bags of 5 oranges.

If children are finding it hard to identify the correct operation to complete the different stages of the task, ask: *What does each number represent? Could you use resources to represent the oranges, the boxes and the bags? Could you draw a bar model to represent the information that you are given?*

Power check

WAYS OF WORKING Independent thinking

ASK

- *What mental calculation strategies did you use for addition problems at the beginning of this unit?*
- *What mental calculation strategies do you now have for addition problems?*

Power play

WAYS OF WORKING Pair work or small groups

IN FOCUS Use this **Power play** to assess whether children are confident at identifying when to use addition, subtraction and division. Children should be able to use mental strategies to calculate all the answers. There may be different methods used for the second part as some children may relate what each character drops to the original amount they had, rather than considering the amount as a whole. Discussions about efficiency of their methods could be had at this point.

ANSWERS AND COMMENTARY $26 + 24 = 50$ (6 ones and 4 ones = 10 ones = 1 ten, 2 tens + 2 tens + 1 ten = 5 tens). There are 50 pieces of bread in total.

There are 30 pieces of bread left. Children should be able to recognise that this can be calculated by combining the 13 pieces Hansel dropped and the 7 pieces Gretel dropped. They should use the known number fact $13 + 7 = 20$ and mentally subtract this from 50, either by counting back in jumps of 10 or using the known number fact of 5 tens – 2 tens is 3 tens.

Each bird gets 6 pieces of bread. Three of the birds get 18 pieces altogether. Children should recognise that the problem requires division from the language 'share equally'. The most efficient method is to use the known fact $5 \times 6 = 30$ to calculate $30 \div 5 = 6$. Then use the number fact $3 \times 6 = 18$ to calculate 18 pieces altogether for 3 birds to share.

After the unit ❙❙

- Are children now confident at selecting the correct operation needed to solve a problem?
- Do children understand how to be efficient in their calculations, and do they have a variety of different strategies to use?

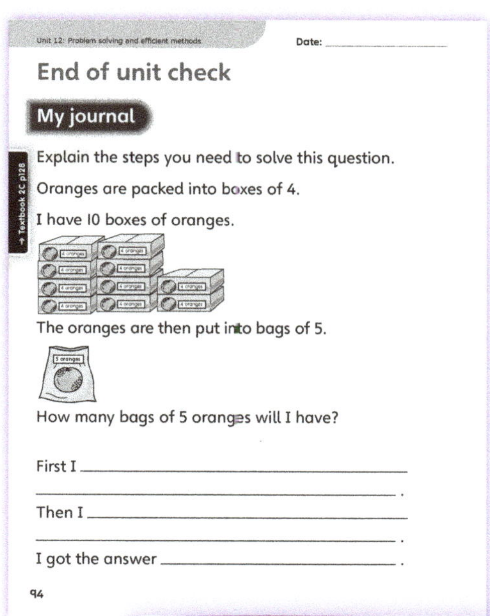

PUPIL PRACTICE BOOK 2C PAGE 94

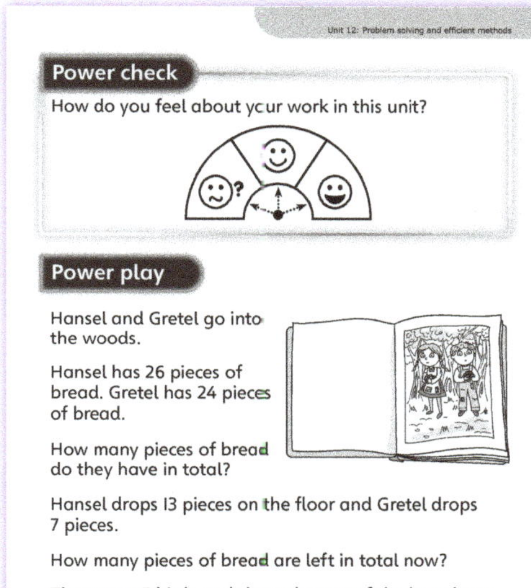

PUPIL PRACTICE BOOK 2C PAGE 95

Strengthen and **Deepen** activities for this unit can be found in the *Power Maths* online subscription.

Unit 13
Position and direction

WHY THIS UNIT IS IMPORTANT

This unit focuses on describing position in relation to other objects, describing lateral and rotational movement and describing and completing repeating patterns. Children will apply their previous learning about fractions and their knowledge of 2D shapes to describe degrees of turn.

This unit also helps children to develop their logical and computational thinking in order to follow and describe sequences relating to movement.

WHERE THIS UNIT FITS

→ Unit 12: Problem solving and efficient methods

→ **Unit 13: Position and direction**

→ Unit 14: Statistics

Before they start this unit, it is expected that children:
- know how to describe the position of an object in relation to one or more other objects
- understand halves and quarters and the relationship between them
- know positional and directional language such as forwards, backwards, left, right, between, above and below.

ASSESSING MASTERY

Children who have mastered this unit will be able to use correct mathematical language to describe position and lateral and rotational movement. Children will be able to follow a series of instructions related to movement and they will be able to plan a series of instructions to arrive at a desired goal.

COMMON MISCONCEPTIONS	STRENGTHENING UNDERSTANDING	GOING DEEPER
Children may confuse left and right, particularly when the orientation of the object is different to their own orientation (for example, when two children face each other, their left hands are not opposite each other).	Provide practical opportunities for children to either turn themselves or objects left and right, labelling the left and right sides if necessary. Ask children to raise their left or right hand and then point to the left or right hand of a partner who is facing them. What do they notice? Can they explain this?	Encourage children to create their own problems using various sizes of grids, such as on paper or outside in the playground. Children could design a treasure hunt with obstacles to avoid and ask a partner to work out a possible route. They could then guide a partner along this route using only verbal instructions, including information about the angle and direction of turn.
Children may struggle to identify quarter, half and three-quarter turns around a point.	Provide children with a paper circle and demonstrate how to fold it in half and then in half again to divide it into quarters. Children can label one quarter, one half and three quarters, and refer to this paper circle during the lessons in this unit.	Children could also use programmable toys to devise and solve their own problems.

Unit 13: Position and direction

UNIT STARTER PAGES

Use Astrid's comment to introduce the unit and ask children what they already know about these things. Discuss with children what Ash might mean and ask whether they can demonstrate a half turn. Question children to find out whether they know any of the key language that Flo introduces. Use Dexter's comment to discuss what clockwise and anticlockwise might mean. Does the picture of the clock help children to explain these terms?

STRUCTURES AND REPRESENTATIONS

Using a circle divided into quarters can help the children to visualise quarter, half and three-quarter turns around a point. Using the image of a clock face can help reinforce the direction of clockwise and anticlockwise turns. Curved arrows can also be used to illustrate the direction and fraction of a turn.

KEY LANGUAGE

There is some key language that children will need to know as part of the learning in this unit:

➔ quarter turn, half turn, three-quarter turn, whole turn
➔ clockwise, anticlockwise
➔ forwards, backwards
➔ left, right
➔ up, down
➔ turn
➔ middle
➔ position
➔ pattern
➔ above, below
➔ top, bottom
➔ between
➔ cube, cylinder
➔ circle, semicircle
➔ triangle, rectangle, square

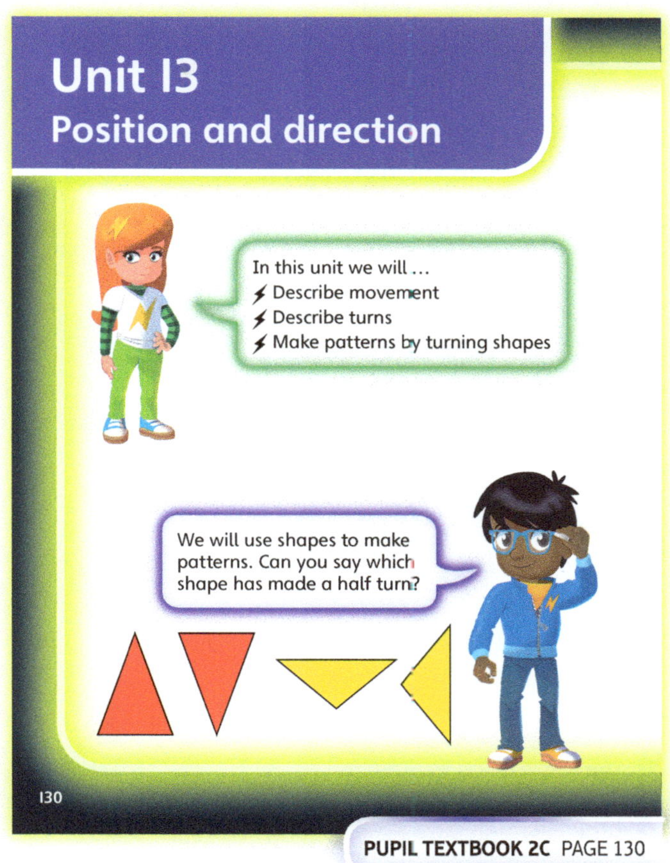

PUPIL TEXTBOOK 2C PAGE 130

PUPIL TEXTBOOK 2C PAGE 131

Language of position

Learning focus

In this lesson, children will use practical positional language to describe scenes or arrangements of items.

Before you teach 

- Can children show you their left and right hands?
- Can children make a quick sketch of an arrangement of items that they are shown, staying true to the relative position between objects?

NATIONAL CURRICULUM LINKS

Year 2 Geometry – position and direction

Use mathematical vocabulary to describe position, direction and movement, including movement in a straight line and distinguishing between rotation as a turn and in terms of right angles for quarter, half and three-quarter turns (clockwise and anticlockwise).

ASSESSING MASTERY

Children can use key vocabulary to accurately describe the position of items or shapes that are shown in a variety of different arrangements.

COMMON MISCONCEPTIONS

Children may confuse left and right. Ask:
- *Which hand do you use to hold a pencil when you write? What is the other hand called?*

STRENGTHENING UNDERSTANDING

Provide children with practical situations in which they can develop the key language of position.

GOING DEEPER

Connect this learning with other topics such as shape and space by making arrangements of 2D or 3D shapes for children to describe using positional language.

KEY LANGUAGE

In lesson: next to, over, under, left, right, above, below, beneath, behind, middle, top, bottom, between, beside

STRUCTURES AND REPRESENTATIONS

Arrangements, sometimes in a grid formation

RESOURCES

Optional: 2D or 3D shapes, small items or outdoor apparatus to practically model positioning

 In the eTextbook of this lesson, you will find interactive links to a selection of teaching tools.

Quick recap 

Introduce a class discussion about position by asking: *Who is sitting next to you? Behind you? In front of you? On your left? On your right?*

Discover

 Pair work

ASK

- Question **1** a): *What do you notice in this picture?*
- Question **1** a): *What can you tell a partner about the picture?*
- Question **1** b): *Can you describe where the boat/the van/ the sun is in the picture without pointing to it?*
- Question **1** b): *Where would you like to stand in this scene?*

IN FOCUS Children are introduced to the use of positional language as a meaningful way to describe a realistic scene showing an arrangement of several different items. Start by asking children to describe what items they can see in the scene, before entering into a discussion about the relative positions.

PRACTICAL TIPS Ask children to work with a partner. One child looks at the picture while the other looks away. The first child describes different aspects of the scene and their partner tries to imagine or sketch it.

ANSWERS

Question **1** a): Children's answers will vary, but they should use correct positional language, for example:
The lorry is above the boat.
The duck is next to a sheep.
The birds are over the tree.
The boat is under the bridge.

Question **1** b): Children's answers will vary, but they should use correct positional language, for example:
The sun is behind a cloud.
The duck is between the two sheep.
The tree is below the birds.
The boat is beneath the bridge.

Share

 Whole class teacher led

ASK

- Question **1** a): *Can you see any other things that are above the bridge?*
- Question **1** a): *Can you see anything else that is next to the duck?*
- Question **1** b): *Can you see anything else that is behind the van?*
- Question **1** b): *Can you show me what these words mean using small toys or objects?*

IN FOCUS Questions **1** a) and b) develop children's understanding of positional language as being relational between two items. For example, you can use the position of the boat in order to describe the position of the lorry. Look together at Ash's comment and discuss whether children agree that some of the position words have a similar meaning. Try using these words interchangeably to see if the sentences still make sense. Ask: *Are there any more words that mean the same as each other?*

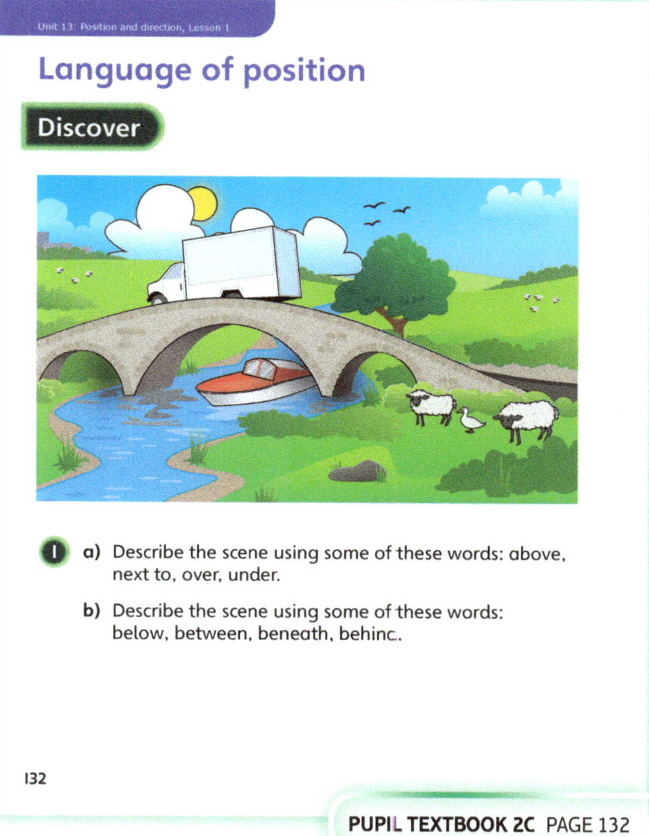

PUPIL TEXTBOOK 2C PAGE 132

PUPIL TEXTBOOK 2C PAGE 133

Think together

 Whole class teacher led (I do, We do, You do)

- Question **1**: *Do you know a way to remember left and right?*
- Question **2**: *Could you describe the position of an item (for example, the socks) in a different way?*
- Question **3**: *What different positions can you arrange the objects in? How is the best way to describe their positions to a partner?*

 Question **1** requires children to consider and describe left and right from the viewer's point of view. Children can also use left and right to describe the position of items in question **2**. However, due to the grid layout they will also need to use positional words to explain which row each item is on and where each sits in relation to other items. In question **3**, children use a range of positional language to describe their own arrangement of 3D shapes to a partner who then tries to copy it. Encourage children to try to place the objects in a way that they know how to describe using the words they have learnt in this lesson. They should avoid, for example, placing an object in an obscure diagonal position relative to the other objects, as this will be too difficult for them

 Provide small world items or outdoor apparatus for children to explore the concepts and vocabulary of position.

 Use question **1** to discuss why left and right can cause confusion. Ask: *What side is the walker standing for the driver of the yellow car? What side is the tree on for that driver? Is that the same for you? Why not?*

Challenge children to think of as many different ways they can to describe the position of one particular object. For example, if they say 'the bike is below the football', you could respond with 'Well done. Where is the bike in relation to the banana?'

 Question **2** assesses whether children can use a wide range of vocabulary to describe the position of an item, including the fact that it can be relational between two items.

Question **1**: The man walking the dog is on the left. The yellow car is in the middle of the road. The tree is on the right.

Question **2**: Children's answers will vary, but they should be able to correctly describe the position of an item, for example: The zebra is on the bottom row on the right of the banana.

Question **3**: Children's answers will vary, for example: My cube is to the left. The cone is balanced on top of the cube. The pyramid is to the right of the cube. The sphere is behind the pyramid.

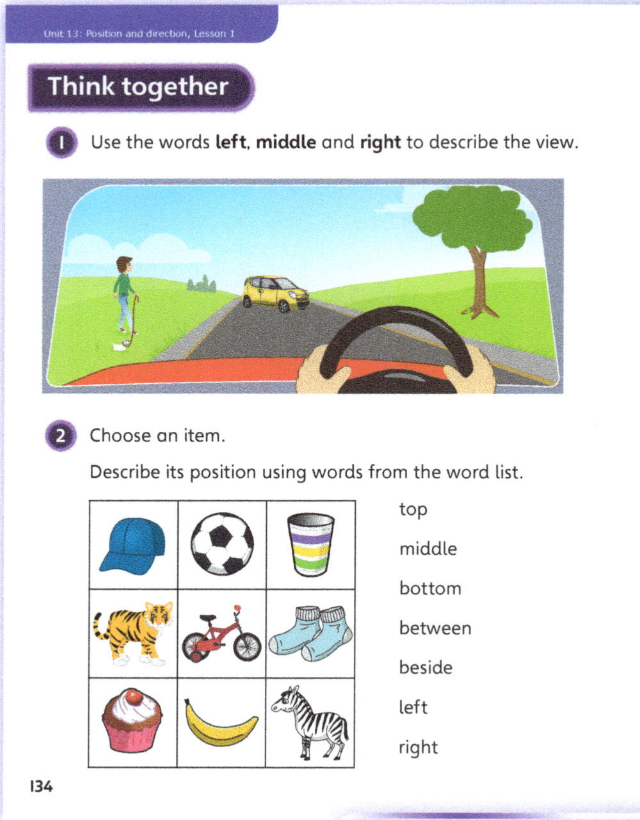

PUPIL TEXTBOOK 2C PAGE 134

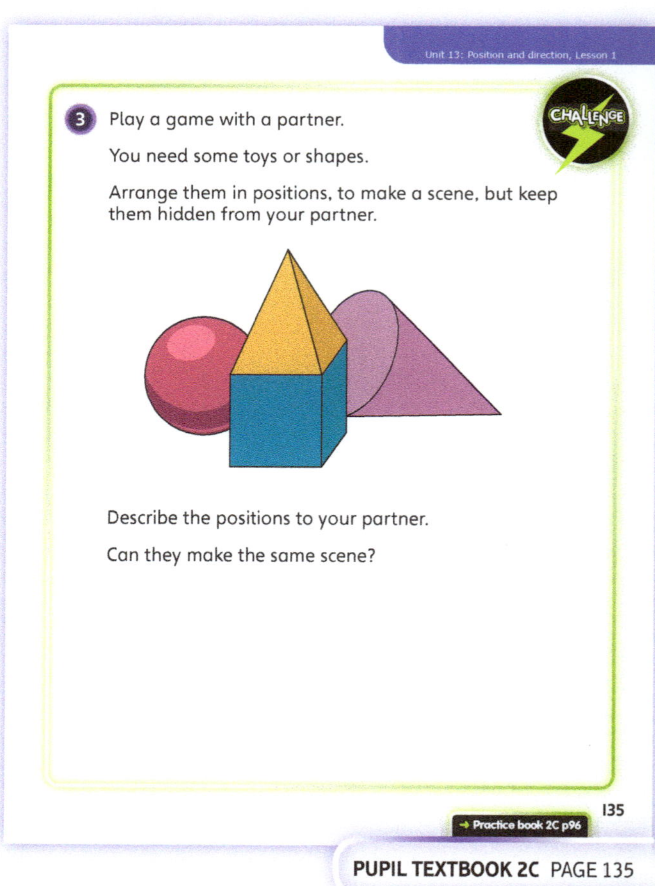

PUPIL TEXTBOOK 2C PAGE 135

Practice

WAYS OF WORKING Independent work

IN FOCUS Question **1** requires children to draw on their understanding of positional vocabulary to complete sentences that describe a given scene. Children are also provided with some key position words in question **2**. They can use these as prompts to write their own sentences describing the position of items that are shown in a grid arrangement. In question **3**, children choose their own vocabulary to best describe the position of an object in a real-life context. They should try to use words that they have already learnt in this lesson

STRENGTHEN Use small items to play games that use positional vocabulary.

DEEPEN Play position guessing games. For example, ask: *I am thinking of an object that is to the left of the door. What is it?*

ASSESSMENT CHECKPOINT Use question **1** to assess whether children can understand and describe relational positions.

ANSWERS Answers for the **Practice** part of the lesson can be found in the *Power Maths* online subscription.

PUPIL PRACTICE BOOK 2C PAGE 96

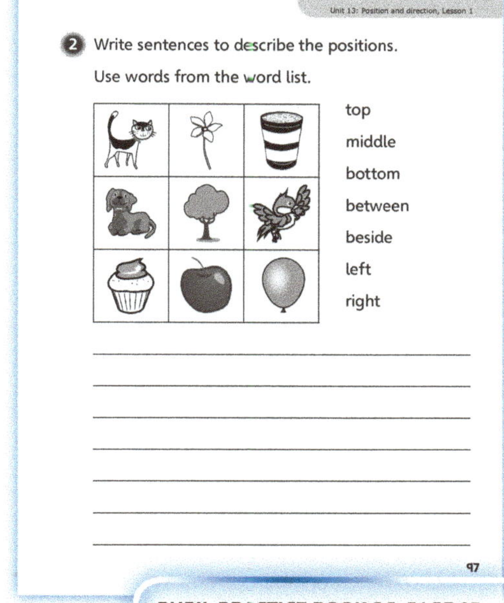

PUPIL PRACTICE BOOK 2C PAGE 97

Reflect

WAYS OF WORKING Pair work

IN FOCUS The **Reflect** part of the lesson prompts children to demonstrate their knowledge of positional vocabulary by listing key words and drawing corresponding pictures.

ASSESSMENT CHECKPOINT Assess whether children can correctly use position words and can draw pictures that accurately represent these.

ANSWERS Answers for the **Reflect** part of the lesson can be found in the *Power Maths* online subscription.

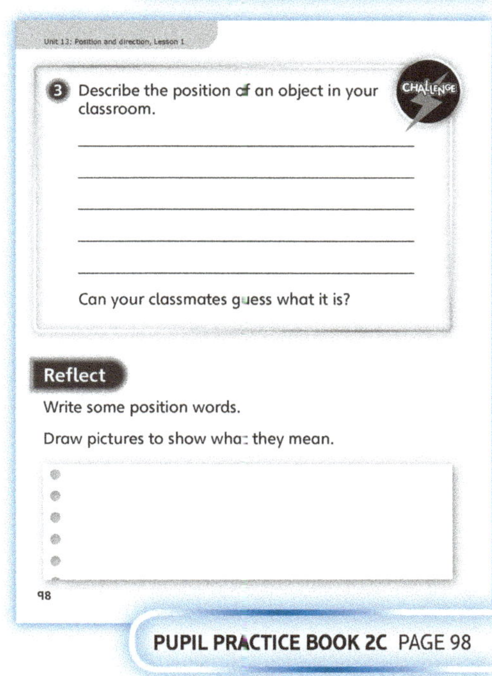

PUPIL PRACTICE BOOK 2C PAGE 98

After the lesson ⏸

- Were children able to use left and right accurately?
- Did children understand that the position of one item can be used to describe the position of another item?

Describe movement

Learning focus

In this lesson, children will describe quarter, half and three-quarter turns around a point using the terms 'clockwise' and 'anticlockwise'.

Before you teach 

- What experience have children had describing position and movement?
- What practical opportunities can you provide to support children's understanding of movement?
- Can you make links to previous work in other curriculum areas such as PE and computing?

NATIONAL CURRICULUM LINKS

Year 2 Geometry – position and direction

Use mathematical vocabulary to describe position, direction and movement, including movement in a straight line and distinguishing between rotation as a turn and in terms of right angles for quarter, half and three-quarter turns (clockwise and anticlockwise).

ASSESSING MASTERY

Children can correctly describe position and movement using the words 'left', 'right', 'forwards' and 'backwards'. Children can systematically follow a series of instructions involving movement.

COMMON MISCONCEPTIONS

Children may confuse which way is forwards when viewed on a 2D grid.

Children may confuse where a start point is, for example, counting the square on which they start.

Children may confuse left and right. Ask questions such as:
- *Who can raise their right hand? Who is sitting to the left of you?*

STRENGTHENING UNDERSTANDING

Ask children to follow instructions practically. This could start off as a simple game of 'Simon says' and build towards replicating the problems in the books using themselves or toys on a grid.

GOING DEEPER

Challenge children to find more than one possible solution to a problem, then ask them to create their own versions of the problems.

KEY LANGUAGE

In lesson: movement, forwards, backwards, left, right, middle, upwards

Other language to be used by the teacher: position, direction, above, below, between

STRUCTURES AND REPRESENTATIONS

Arrangements of objects in a grid or linear formation

RESOURCES

Optional: arrow cards, laminated copies of the problems, dry-wipe pens and a selection of grids and objects (including 2D and 3D shapes) to model the problems

 In the eTextbook of this lesson, you will find interactive links to a selection of teaching tools.

Quick recap 

Ask children to use each of the following words to describe the position of a different object in the classroom:

above   below   left   right

Discover

 Pair work

- Question **1** a): *Have you seen this arrangement of signs anywhere else?*
- Question **1** b): *What would be the 7th movement?*

 Question **1** b) requires children to apply vocabulary to the instructions as opposed to following and matching the arrow signs. Encourage children to be clear and concise in their descriptions.

 Mark a grid on the floor and hold up arrows like those on the screen, for a child to recreate the dance movements. Ask the rest of the class to call out *'forwards'*, *'backwards'*, *'left'* or *'right'* as you hold up each arrow.

Question **1**: See dance moves below:
    Step 1: forwards
    Step 2: backwards
    Step 3: left
    Step 4: forwards
    Step 5: right
    Step 6: forwards

PUPIL TEXTBOOK 2C PAGE 136

Share

 Whole class teacher led

- Questions **1** a) and b): *How many of each movement were there?*
- Questions **1** a) and b): *What would the 10th movement be? How do you know?* [Assume the dance repeats itself, so step 7 is the same as step 1. So, the 10th movement would be forwards.]
- Questions **1** a) and b): *Do the instructions change if the boy is facing the other way? Can you describe the movements of the dance in that case?*

 Question **1** secures understanding of backwards, forwards, left and right. If children consider what the instructions would be when the boy is facing towards them, make sure children understand that the boy's left and right hands would be opposite to an observer's. Children can test this by standing in front of a partner and both moving their right hands. They should see that the partner moves the hand diagonally opposite to their own.

PUPIL TEXTBOOK 2C PAGE 137

Think together

 Whole class teacher led (I do, We do, You do)

ASK

- Question **1**: *What different ways can you think of to describe the position of the green triangle flag?*
- Question **2** b): *What other routes can you describe?*
- Question **3**: *How many different solutions can you find?*
- Question **3**: *Where can the mouse not start? Which way is forwards?* (Away from the entrance.)

IN FOCUS Question **2** b) introduces the idea that there can be more than one way of describing a route. Discuss with children why they chose the route they did. Prompt children to see more complex routes, not simply the most direct.

STRENGTHEN Strengthen understanding of questions **2** and **3** by providing children with laminated copies of the questions and a dry-wipe pen so they can try out different possibilities. Children can use their laminations to justify why the mouse cannot start at certain points in question **3**.

DEEPEN In question **2** b), ask children if they can get from the star to the sun by entering each empty square only once. Ask them to explain how this is possible. In question **3**, Jake the mouse can only start in four of the nine squares and travels in a square shape. Ask children if they can adapt the instructions so that the mouse still travels in a square shape but can now start at different squares.

ASSESSMENT CHECKPOINT Use questions **1** and **2** to assess whether children are secure with left and right and are able to create a simple set of instructions for movement. Their responses to question **3** will determine if they can reason about different possibilities based on the instructions given.

ANSWERS

Question **1**: The green triangle flag is in the middle. The green triangle flag is to the right of the red square flag. The blue rectangle flag is to the right of the green triangle flag.

Question **2** a): To move from the heart to the sun, move 1 square up and 1 square left.

Question **2** b): To move from the star to the sun, move 2 squares up and 1 square right. You could also move 1 square right and 2 squares up.

Question **3**: Jake finishes where he started.

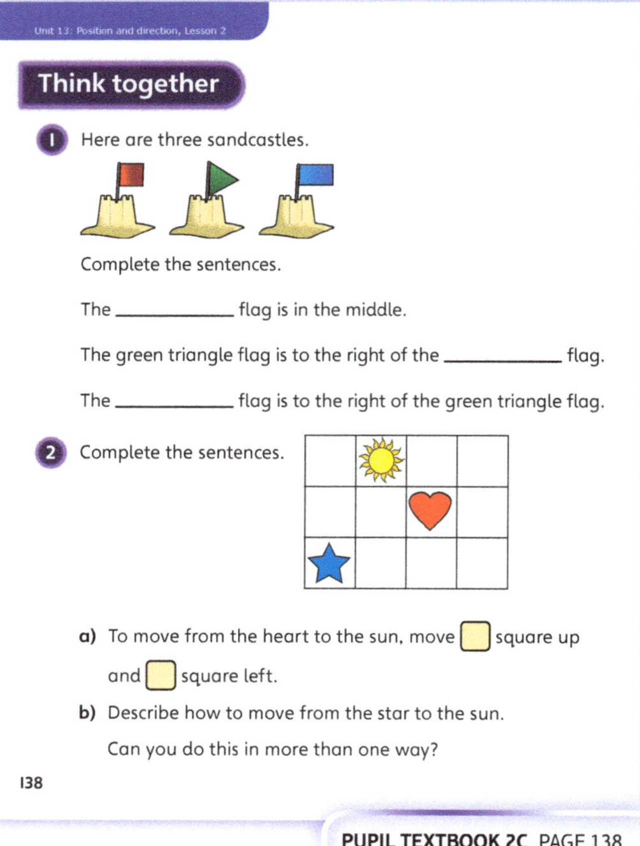

PUPIL TEXTBOOK 2C PAGE 138

PUPIL TEXTBOOK 2C PAGE 139

Practice

WAYS OF WORKING Independent work

IN FOCUS Question ⑤ requires children to follow the instructions to complete the grid. However, they cannot determine the exact position of any shape from the first instruction. Encourage children to look at the first instruction and discuss the possible positions of the square and circle, perhaps using a copy of the grid and 2D shapes. The second instruction enables them to place the square. Ask children how they know where each shape goes and how they went about solving the problem.

STRENGTHEN Provide children with the shapes represented in the problems in question ③. Ask children to build a tower by giving them instructions. They could then give instructions to each other.

For question ④, provide children with a laminated version of the problem and dry-wipe markers. This can encourage them to tackle the problem, knowing that they can easily start again.

DEEPEN Provide children with a blank copy of the grid from question ④. Can they make their own version of the problem, ensuring that they land in every square only once? They can then test it out on a partner.

THINK DIFFERENTLY Question ④ asks children to follow a series of instructions in order to move around a grid in sequence. If children do this correctly, they will find that they have shaded the whole grid.

ASSESSMENT CHECKPOINT Use questions ① to ③ to assess whether children are confident with the vocabulary of position and movement, and can identify the position of an object in relation to one or two other objects.

ANSWERS Answers for the **Practice** part of the lesson can be found in the *Power Maths* online subscription.

Reflect

WAYS OF WORKING Pair work

IN FOCUS The **Reflect** activity has more than one possible answer. It highlights the fact that an object's position can be described in relation to more than one other object. When children are secure in describing the position of objects relative to one another, ask them to describe the path of movement from one object to another. For example, Ask: *'Can you describe how to move from the watering can to the star?'*

ASSESSMENT CHECKPOINT Assess whether children can give clear and concise descriptions of positions using the correct vocabulary. Can they describe position in different ways?

ANSWERS Answers for the **Reflect** part of the lesson can be found in the *Power Maths* online subscription.

After the lesson ⏸

- How secure are children using the vocabulary of position and movement, in particular 'left', 'right', 'forwards' and 'backwards'?
- Were children able to reason about the decisions they made when identifying and describing position and movement?

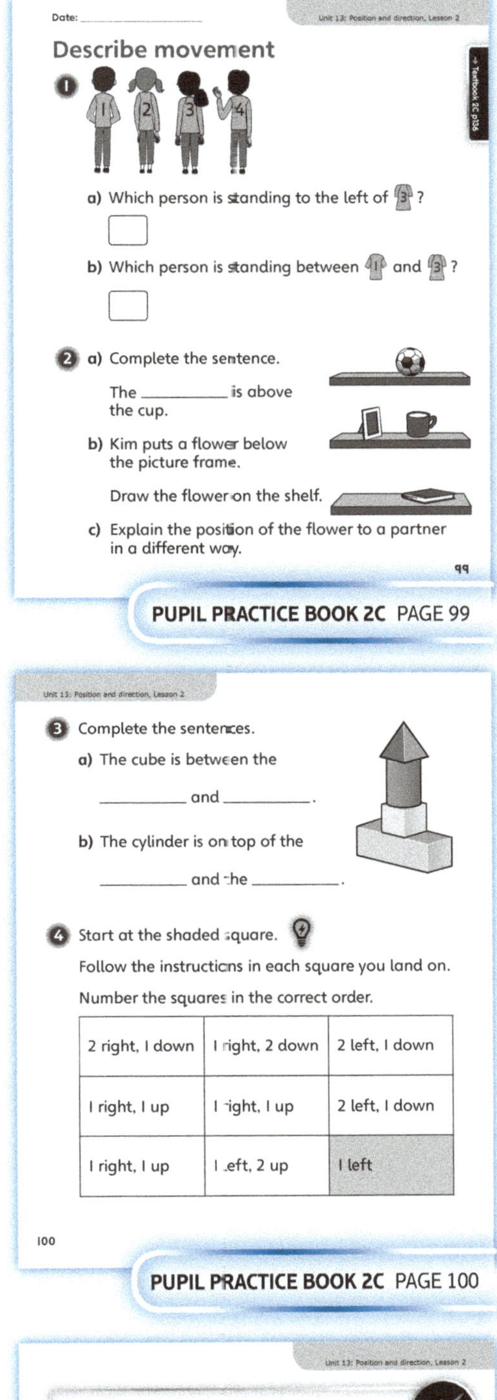

PUPIL PRACTICE BOOK 2C PAGE 99

PUPIL PRACTICE BOOK 2C PAGE 100

PUPIL PRACTICE BOOK 2C PAGE 101

Describe turns

Learning focus

In this lesson, children will describe quarter, half and three-quarter turns around a point using the terms 'clockwise' and 'anticlockwise'.

Before you teach 

- How will you make links to the previous lesson on movement?
- What resources could you provide to support children's understanding of turning clockwise and anticlockwise?
- How will you make the link to children's learning on fractions from previous units?

NATIONAL CURRICULUM LINKS

Year 2 Geometry – position and direction

Use mathematical vocabulary to describe position, direction and movement, including movement in a straight line and distinguishing between rotation as a turn and in terms of right angles for quarter, half and three-quarter turns (clockwise and anticlockwise).

ASSESSING MASTERY

Children can describe a turn around a point, using fractions to describe part turns and 'clockwise' and 'anticlockwise' to describe the direction of the turn. Children can describe a pair of turns in opposite directions that result in the same finishing position; for example, a quarter turn clockwise and a three-quarter turn anticlockwise.

COMMON MISCONCEPTIONS

Children may not understand the terms 'clockwise' and 'anticlockwise'. Use clocks to demonstrate and ask:
- *Can you show a clockwise turn? What about an anticlockwise turn?*

Children may also fail to recognise quarter, half and three-quarter turns. Show a whole turn and then focus on fractions of turns. Ask:
- *Can you show a quarter turn clockwise? What about a three-quarter turn anticlockwise?*

STRENGTHENING UNDERSTANDING

Ask children to carry out different turns whilst holding a clock in order to reinforce their understanding of clockwise and anticlockwise. Having objects to physically turn or programmable toys will support them in tackling the problems.

GOING DEEPER

Challenge children to predict the finishing position after two turns; for example, a half turn clockwise followed by a quarter turn anticlockwise. Children could check their prediction using programmable toys.

KEY LANGUAGE

In lesson: turn, half turn, quarter turn, **anticlockwise**, **clockwise**, **three-quarter turn**, whole turn

Other language to be used by the teacher: fraction, part, direction, position, rotation, left, right

STRUCTURES AND REPRESENTATIONS

Sketches of clock faces, arrows showing the direction and angle of turn

RESOURCES

Optional: clocks, objects to carry out rotations, programmable toys, animal pictures or toy animals

 In the eTextbook of this lesson, you will find interactive links to a selection of teaching tools.

Quick recap 

Look together at an analogue clock face. Show the hands and model how they move all the way around in a clockwise direction.

Discover

 Pair work

ASK

- Question **1** a): *Does it matter in which direction Sam turns?*
- Question **1** b): *Is there another way Sam could turn and still end up facing the same animal?*

IN FOCUS Question **1** requires children to visualise the turns. In question **1** a), no direction is given as a half turn will produce the same result regardless of the direction. In contrast, question **1** b) requires children to consider the direction of the turn as well. Use this opportunity to discuss direction of rotation and when it is important.

PRACTICAL TIPS Place animal pictures or toy animals to recreate the picture in the scenario. Ask a child to play the role of Sam and make the turns as instructed in the questions.

ANSWERS

Question **1** a): Sam turned a half turn and is now facing the cow.

Question **1** b): Sam made a quarter turn left, or anticlockwise, and is now facing the horse.

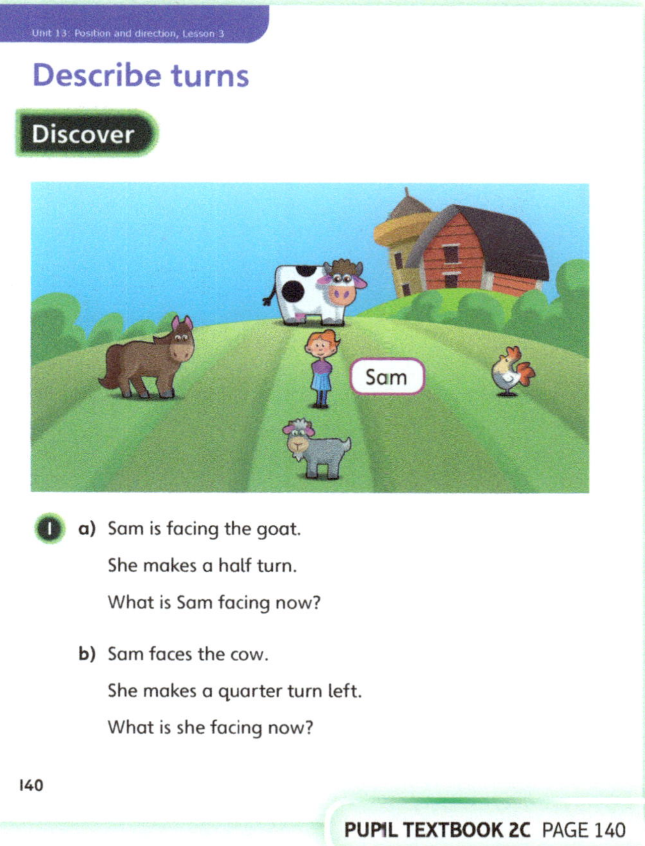

PUPIL TEXTBOOK 2C PAGE 140

Share

 Whole class teacher led

ASK

- Questions **1** a) and b): *If the horse is at 12 on a clock face, what numbers would the other animals be at?*
- Questions **1** a) and b): *Why do you think we use the terms 'clockwise' and 'anticlockwise'?*
- Questions **1** a) and b): *For what size turns does it not matter whether you turn clockwise or anticlockwise?*

IN FOCUS This part of the lesson introduces the terms 'clockwise' and 'anticlockwise', relating them to a pictorial representation of a clock. Provide children with clocks that they can manipulate in order to support them in exploring clockwise and anticlockwise turns. You can then relate what children are doing on the clocks to the picture.

PUPIL TEXTBOOK 2C PAGE 141

Think together

WAYS OF WORKING Whole class teacher led (I do, We do, You do)

ASK

- Question **2**: *Can you describe the turns in more than one way?*
- Question **3**: *How do you know Dai is incorrect?*
- Question **3**: *How else can you describe two quarter turns?*

IN FOCUS Question **2** shows rotations from two different viewpoints. The car is viewed from above, while the frog and bee are viewed from the side. Provide children with concrete representations to support their understanding of the change in viewpoint.

STRENGTHEN Give children objects so that they can carry out the turns practically. Having a clock face as a reference will support their understanding of clockwise and anticlockwise.

DEEPEN Question **3** gives two correct ways to describe a turn from the cow to the horse. Ask children to find combinations of turns with the same result. Challenge them to use both clockwise and anticlockwise; for example, a quarter turn clockwise and a half turn anticlockwise.

ASSESSMENT CHECKPOINT Assess whether children are able to describe fractions of turns and the direction of turns in questions **1** and **2**. Use question **3** to assess whether children understand that the finishing position can be reached by turning in either direction.

ANSWERS

Question **1** a): anticlockwise

Question **1** b): anticlockwise

Question **1** c): clockwise or anticlockwise

Question **2**:

Three-quarter turn clockwise

Quarter turn clockwise

Half turn anticlockwise

Question **3**: Harry and Amelia are both correct. If a quarter turn anticlockwise is made, this is the same as turning a three-quarter turn clockwise.

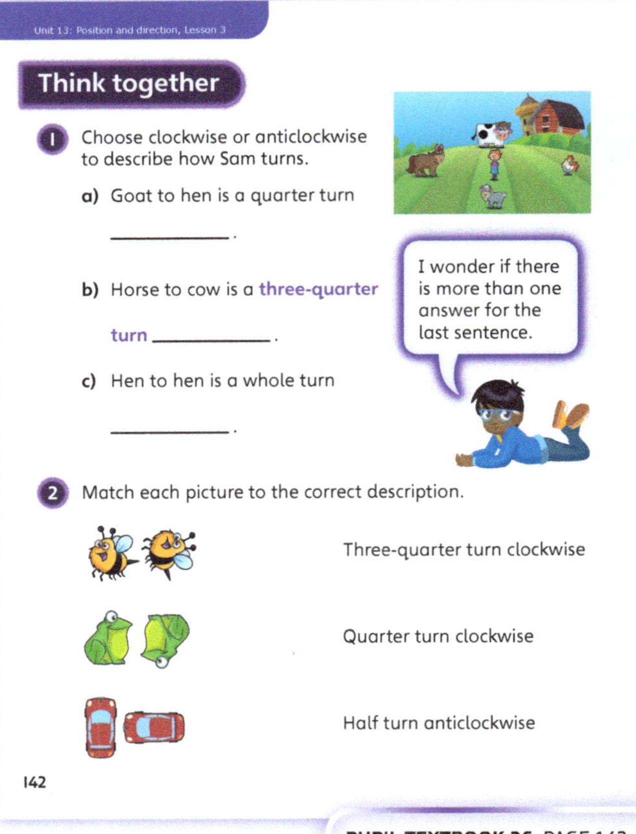

PUPIL TEXTBOOK 2C PAGE 142

PUPIL TEXTBOOK 2C PAGE 143

Practice

WAYS OF WORKING Pair work

IN FOCUS Question ① requires children to consider clockwise and anticlockwise turns. This is supported with clock face images. In question ②, children explore how to use fractions to describe turns and part turns.

Question ④ illustrates that it is not possible to determine the direction of the turn simply by observing the start and end position. Although clockwise would be a smaller turn, the fly could have rotated anticlockwise.

STRENGTHEN Provide children with clock faces as a reference and physical objects they can rotate. Ask children to justify their answers to prompt discussion and deepen their understanding of rotation.

DEEPEN Give a pair of children a programmable toy. Ask one child to write a series of turns (three or more) and give them to a partner. The partner should then try to determine which direction the programmable toy will be facing at the end, before trying the instructions out using the toy.

THINK DIFFERENTLY Question ③ highlights the fact that an object can be turned in either direction to reach a given position. 'Quarter turn clockwise' and 'three-quarter turn anticlockwise' are equivalent and each can be used to describe the ladybird. 'Half turn clockwise' and 'half turn anticlockwise' can each be used to describe the bird.

ASSESSMENT CHECKPOINT Questions ① and ② will confirm whether children are secure in their understanding of clockwise and anticlockwise, and fractions of turns. Questions ③ and ④ will determine whether children are able to recognise that the finishing position can be reached by turning in either direction.

ANSWERS Answers for the **Practice** part of the lesson can be found in the *Power Maths* online subscription.

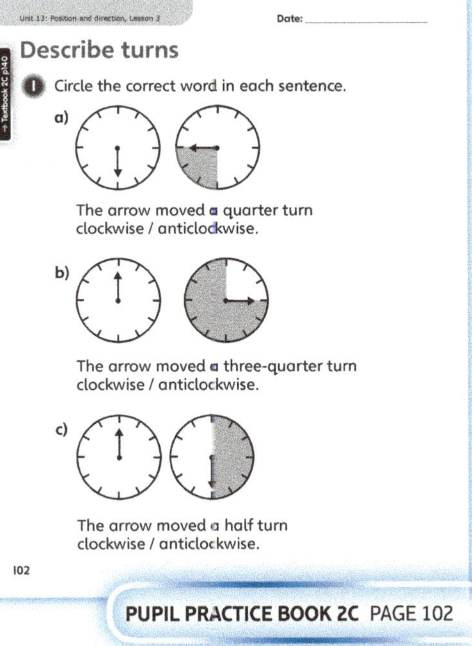

PUPIL PRACTICE BOOK 2C PAGE 102

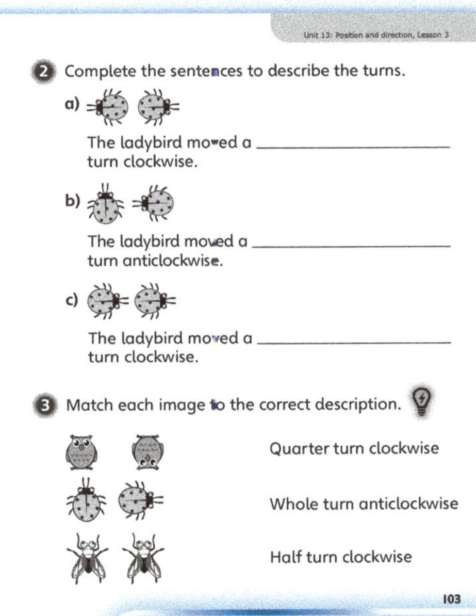

PUPIL PRACTICE BOOK 2C PAGE 103

Reflect

WAYS OF WORKING Independent thinking

IN FOCUS This part of the lesson requires children to visualise the final position of the house following the turns described. The final two images will be the same, reinforcing the idea that an object can turn either clockwise or anticlockwise to achieve the same final position.

ASSESSMENT CHECKPOINT Assess whether children are secure in their understanding of clockwise and anticlockwise. Do children understand how the same final position was achieved by rotating either a quarter turn clockwise or a three-quarter turn anticlockwise?

ANSWERS Answers for the **Reflect** part of the lesson can be found in the *Power Maths* online subscription.

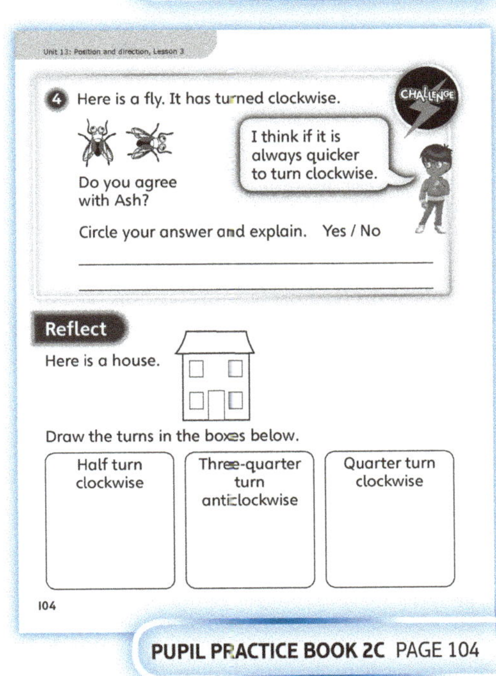

PUPIL PRACTICE BOOK 2C PAGE 104

After the lesson ⏸

- Are children confident in their understanding of the key vocabulary used in this lesson? How can you incorporate the learning into other areas of the curriculum (for example, PE, geography, computing)?
- Did children justify why there was more than one possible answer to some of the questions?

Describe movement and turns

Learning focus

In this lesson, children will combine rotation and linear movement in order to follow or describe a designated path.

Before you teach 

- How will you provide practical experiences for children to explore rotational and linear movement?
- How could you extend the activities to deepen children's thinking around position and direction?
- What references and prompts can you provide for children to ensure the correct use of vocabulary?

NATIONAL CURRICULUM LINKS

Year 2 Geometry – position and direction

Use mathematical vocabulary to describe position, direction and movement, including movement in a straight line and distinguishing between rotation as a turn and in terms of right angles for quarter, half and three-quarter turns (clockwise and anticlockwise).

ASSESSING MASTERY

Children can follow instructions involving linear movement and rotation in order to follow a desired route. Children can identify a variety of routes to a given goal and describe them using the correct vocabulary for rotation and linear movement.

COMMON MISCONCEPTIONS

Children may not realise that once an object is rotated to face the desired location, it then moves forward. They may still use left and right to describe direction in relation to their perspective as opposed to the object's. Ask:
- *Now the object is facing where it needs to go, what is the next instruction? Does it go left, right or forwards?*

STRENGTHENING UNDERSTANDING

Give children practical opportunities to explore the problems. They could carry out the instructions themselves or with the help of a friend on a grid drawn on the playground. The use of programmable toys will enable children to test and adapt their instructions immediately.

GOING DEEPER

Children could explore finding the most complex routes. They could also create their own versions of the problems to test on a partner. They could design a 'treasure hunt' in the playground, creating their own grid and instructions for how to find the 'treasure'.

KEY LANGUAGE

In lesson: movement, turn, forwards, backwards, clockwise, anticlockwise, quarter turn, three-quarter turn

Other language to be used by the teacher: position, direction, rotation, route, location, left, right, half, whole

STRUCTURES AND REPRESENTATIONS

Square grids

RESOURCES

Optional: counters, laminated copies of grids, dry-wipe pens, physical objects to use on the grids, programmable toys

 In the eTextbook of this lesson, you will find interactive links to a selection of teaching tools.

Quick recap 

Look together at an analogue clock face. Ask children to show how the hands make a clockwise turn. Then ask them to show an anticlockwise turn.

Discover

WAYS OF WORKING Pair work

ASK

- Question ① a): *Which way is the pirate facing? If he moves forward, in which direction will he go?*
- Question ① a): *Is your answer different from anyone else's?*

IN FOCUS Question ① a) is the first time children are asked to combine turns with moving forwards and backwards. Ensure children understand that in order for the pirate to move in a desired direction, he has to first turn so that he can go forwards. Ensure children understand that the pirate's hat points in his direction of travel.

DEEPEN There are several different routes to the treasure. If children find a route other than the fastest, use this as a prompt for a discussion about multiple solutions. Should the correct solution to question ① a) be the fastest route?

PRACTICAL TIPS Provide a large copy of the grid or mark it on the floor. Children can instruct a partner to move about the grid using commands such as 'forwards, backwards, clockwise, anticlockwise, quarter turn'. Both partners need to think very carefully about what one is saying and how the other is moving.

ANSWERS

Question ① a): To get to the treasure, the pirate needs to move: forwards 2, quarter turn clockwise, then forwards 1.

Question ① b): He is in the middle square on the top row.

Share

WAYS OF WORKING Whole class teacher led

ASK

- Question ① a): *Is the route you found the same as the one in the book?*
- Question ① a): *What is the longest route you can create to get the pirate to the treasure without entering a square more than once?*
- Question ① b): *Can you describe the most direct route back to the start from the pirate's final position?*
- Question ① b): *What route could the snake take to get to the pirate?*
- Question ① b): *Can the snake get to the treasure in fewer moves than the pirate?*

IN FOCUS The diagrams allow children to see the results of each step. Highlight how the pirate needs to turn in order to move in the desired direction. Encourage children to explore alternative routes that the pirate could take.

PUPIL TEXTBOOK 2C PAGE 144

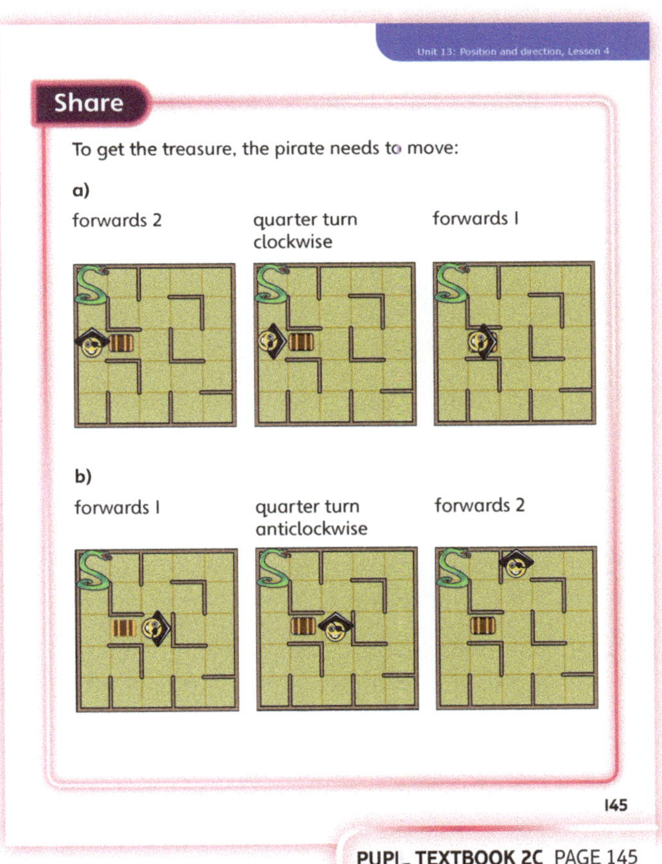

PUPIL TEXTBOOK 2C PAGE 145

Think together

WAYS OF WORKING Whole class teacher led (I do, We do, You do)

ASK

- Questions **1** and **2**: *Can you describe an alternative route?*
- Question **2**: *What happens if you do the instructions in a different order?*
- Question **3**: *What is the fastest route to the boat, treasure and cannon?*

IN FOCUS In question **2**, children should realise that the order in which you follow the instructions is important. If you make a turn before you are told to, the pirate will end up going off in the wrong direction. Having said that, there are different paths which will end up at the treasure, but each path has its own unique set of instructions which must be completed in the order they are given. Sparks prompts children to come up with their own instructions for question **3**. Children may fail to realise that a backward movement is possible and may give the unnecessary instruction to rotate a half turn. Asking children to find the fastest route (the one with the fewest instructions) can expose this misconception.

STRENGTHEN Give children laminated copies of each grid and an object such as a toy car that they can physically rotate and move. Children can carry out each movement before describing what they did.

DEEPEN Extend question **3** by asking children to describe a route where the pirate will pass all three objects. What is the fastest route they can find?

ASSESSMENT CHECKPOINT Use question **2** to assess whether children can link rotational and linear movement to get to a desired goal.

Question **3** will determine if children are able to follow a series of instructions involving rotational and linear movement.

ANSWERS

Question **1**: 1st: go forwards 1 space. 2nd: make a quarter turn anticlockwise. 3rd: go forwards 2 spaces.

Question **2**: Go forwards 2 spaces. Make a quarter turn clockwise. Go forwards 1 space.

Question **3**: The pirate will be at the boat.

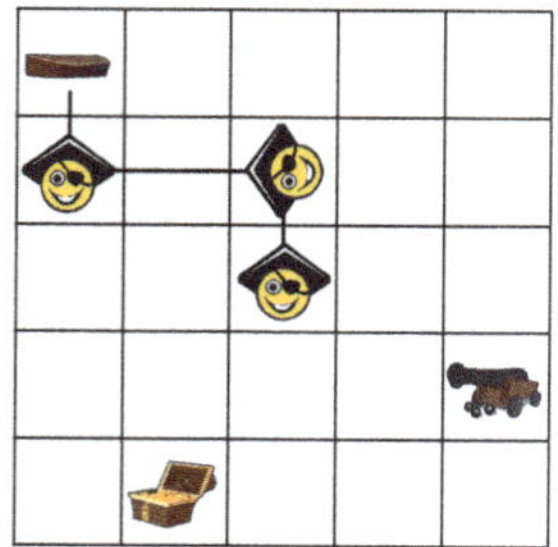

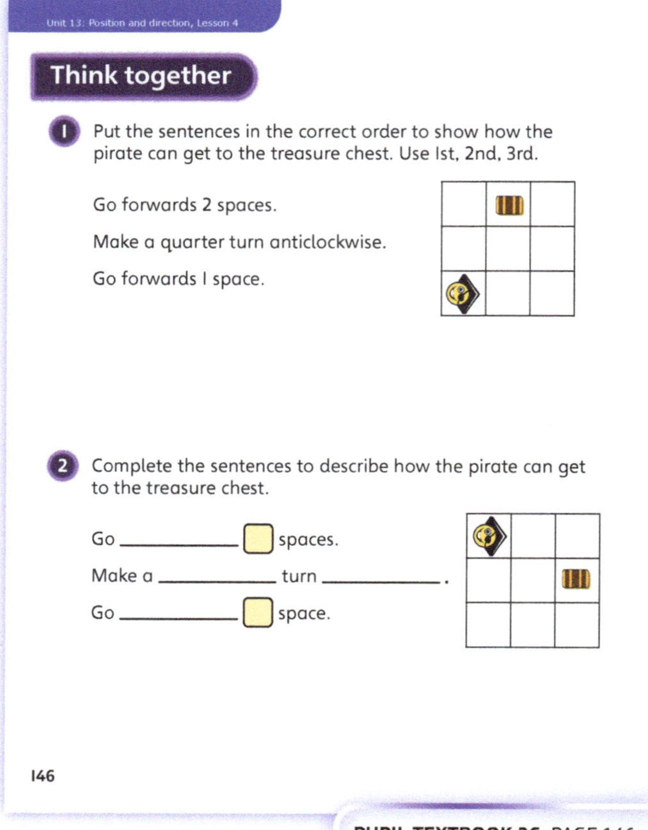

PUPIL TEXTBOOK 2C PAGE 146

PUPIL TEXTBOOK 2C PAGE 147

Practice

WAYS OF WORKING Pair work

IN FOCUS Question 2 asks children to get the bee to the beehive. Ask them to think of a similar but alternative route to the beehive. Changing the first instruction means children need to think about what else they need to change in the remaining instructions. Encourage them to think about how they can get the bee to the beehive by changing as few instructions as possible.

Question 3 a) requires children to match a set of instructions to a given start and end point. Encourage children to work systematically, trying the instructions on each grid in turn until they find the correct match. Question 3 b) then asks children to describe a simple route using the correct vocabulary. They may identify more than one possible solution.

STRENGTHEN Children could use programmable toys to test their ideas in questions 2 and 3.

DEEPEN Ask children to create their own challenge using a programmable toy or a grid drawn in the playground. They can place objects to collect, obstacles to avoid, and an end position. Encourage them to plan their entire route before testing it out.

ASSESSMENT CHECKPOINT Use questions 1 and 2 to identify any areas of weakness in the children's ability to follow or describe a simple route.

Question 3 a) will determine if children can match a set of instructions to a start and end point.

Use question 4 to check whether children remember that a quarter turn in one direction will have the same result as a three-quarter turn in the opposite direction. Can children justify their answer in the context of this lesson?

ANSWERS Answers for the **Practice** part of the lesson can be found in the *Power Maths* online subscription.

Reflect

WAYS OF WORKING Pair work

IN FOCUS This part of the lesson secures children's understanding of describing turns in both clockwise and anticlockwise directions. Children need to determine the arrow's movement from the end position, based on their knowledge of its starting position.

ASSESSMENT CHECKPOINT Assess whether children can accurately describe rotation using the correct vocabulary.

ANSWERS Answers for the **Reflect** part of the lesson can be found in the *Power Maths* online subscription.

After the lesson ⏸

- Are children secure in describing the direction of linear movement based on the orientation of an object?
- How secure are children in using the correct vocabulary in order to describe rotational and linear movement?

Describe movement and turns

1 The pirate moves to the treasure.

Put an X where the treasure is.

Instructions to find the treasure:
Forwards 2
Quarter turn clockwise
Forwards 1

2 Complete the sentences to get the bee to the nest.

clockwise
anticlockwise
forwards
backwards
quarter

Go ☐ spaces ____________.

Make a ____________ turn

____________.

Go ☐ spaces ____________.

105

PUPIL PRACTICE BOOK 2C PAGE 105

---

3 Match each image to the correct instructions to get from Start to the tortoise.

a)

Forwards 2, quarter turn anticlockwise, forwards 1

Forwards 2, quarter turn clockwise, forwards 2

Forwards 1, quarter turn clockwise, forwards 1

b) Create your own instructions for the grid below.

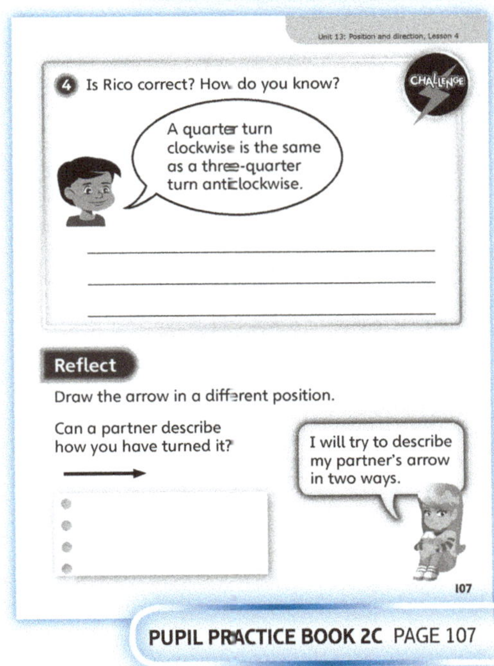

____
____

106

PUPIL PRACTICE BOOK 2C PAGE 106

---

4 Is Rico correct? How do you know?

____
____
____

Reflect

Draw the arrow in a different position.

Can a partner describe how you have turned it?

107

PUPIL PRACTICE BOOK 2C PAGE 107

185

Make patterns by turning shapes

Learning focus

In this lesson, children will apply what they have learnt about rotation and position in order to complete and describe patterns.

Before you teach 

- How will you display the key vocabulary for children to refer to?
- Do children have a secure understanding of 'clockwise' and 'anticlockwise'?
- What questions can you ask to prompt discussion about position and movement?

NATIONAL CURRICULUM LINKS

Year 2 Geometry – position and direction

Use mathematical vocabulary to describe position, direction and movement, including movement in a straight line and distinguishing between rotation as a turn and in terms of right angles for quarter, half and three-quarter turns (clockwise and anticlockwise).

Order and arrange combinations of mathematical objects in patterns and sequences.

ASSESSING MASTERY

Children can use correct mathematical vocabulary to describe position and rotation within a repeating pattern. Children can identify the repeating core of a pattern in order to continue it or identify missing terms.

COMMON MISCONCEPTIONS

When there are more than two shapes within the core of a pattern, children may not look at enough of the pattern in order to determine the missing shapes. Ask:

- *What part of the pattern is being repeated? Can you see where this shape is repeated again?*

Children may fail to draw on their knowledge from previous lessons to describe the rotations. They may use terms such as 'upside down' or 'on its side.' Ask:

- *How has the shape changed?*

STRENGTHENING UNDERSTANDING

Provide children with 2D shapes that are represented in the problems. Children can print or draw round the shapes in order to replicate the patterns. Discuss what they should do to the shape in order to create the next term.

GOING DEEPER

Children could create their own patterns and describe the cores of these patterns using vocabulary related to rotation and position. They could create patterns with more than one line.

KEY LANGUAGE

In lesson: pattern, triangle, half turn, turn, clockwise, anticlockwise

Other language to be used by the teacher: position, rotation, repeating, core pattern, term, quarter turn, semicircle, three-quarter turn, rectangle, circle, square

RESOURCES

Mandatory: variety of 2D shapes

 In the eTextbook of this lesson, you will find interactive links to a selection of teaching tools.

Quick recap 

Ask children to rotate a variety of different 2D shapes to show clockwise and anticlockwise turns of different sizes (half, quarter, three-quarter).

Discover

WAYS OF WORKING Pair work

ASK

- Question **1** a): *What will the next two shapes look like from the girl's point of view? Is it different from your point of view?*
- Question **1** b): *Is there more than one way to describe the differences between the two shapes?*

IN FOCUS Question **1** b) requires children to apply their learning about rotation to describe the differences between the two shapes in the pattern core. Children may revert to using words and phrases such as 'upside down' or 'other way round.' Have the vocabulary related to rotation available for children to refer to. This will help prompt a more mathematical description.

PRACTICAL TIPS Provide children with 2D shapes which they can arrange and rotate to replicate and continue the pattern in the **Discover** scenario.

ANSWERS

Question **1** a): The next two shapes will be:

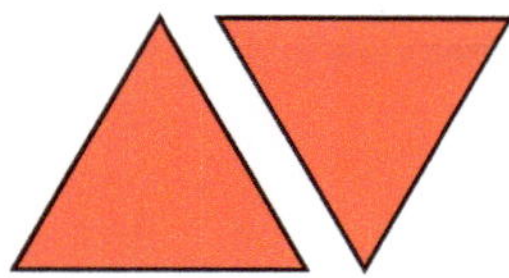

Question **1** b): All the shapes are triangles.
They all have three sides.
They are all the same size and colour.
The triangles are in a different position.
They have made a half turn after each shape in the pattern.

Share

WAYS OF WORKING Whole class teacher led

ASK

- Question **1** a): *Do your triangles look different?*
- Question **1** b): *Does the direction of the turn matter?*
- *What would the pattern look like if the triangle turned a quarter turn clockwise each time?*

IN FOCUS Question **1** b) allows children to use their understanding of rotation to describe how the orientation of the triangle changes in the pattern. Ash's comment reinforces the concept that the direction of rotation does not change the end position when the shape goes through a half turn.

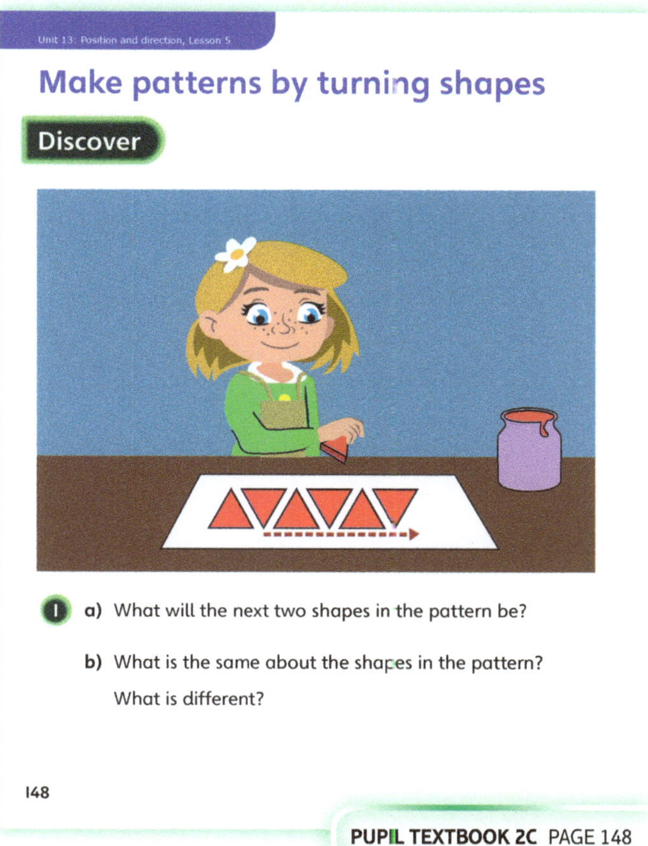

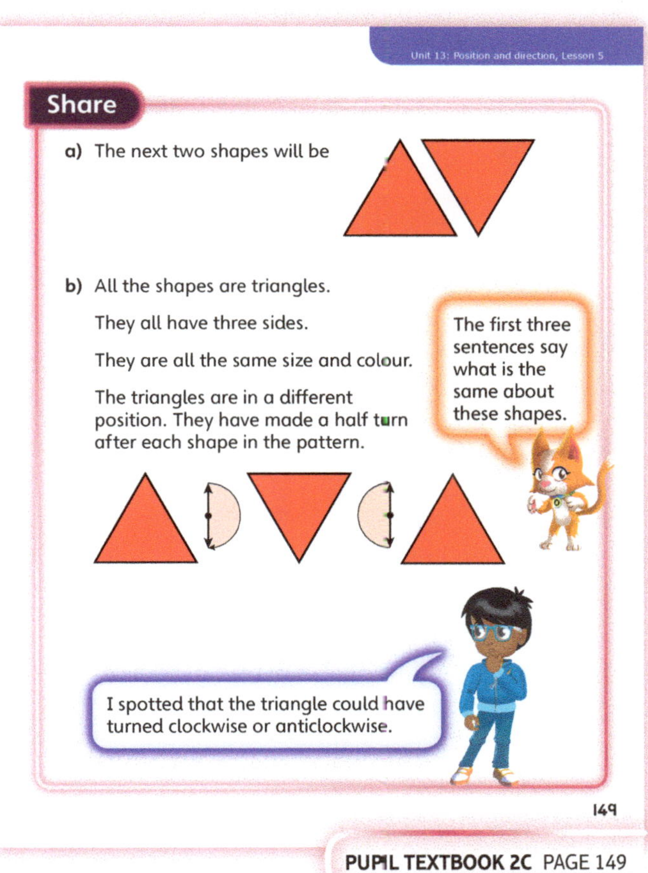

187

Think together

WAYS OF WORKING Whole class teacher led (I do, We do, You do)

ASK

- Questions **1** a) and **2**: *What is the repeating part of the pattern?*
- Questions **1** a) and **2**: *How do you know which shape comes next?*
- Questions **1** b) and **2**: *Can you describe how the shapes change in the patterns?*

IN FOCUS Question **3** has two different shapes and each shape has a different type of rotation. Children may be tempted to repeat the first three shapes as they look like a unit due to their orientation. Encourage them to look carefully so that they realise that the pattern's core has four terms. This problem requires children to apply their understanding of rotation in order to describe the pattern.

STRENGTHEN Provide corresponding shapes for children to arrange or draw around in order to replicate the patterns. Discuss when the shapes start to repeat and how the shapes turn and in which direction.

DEEPEN Challenge children to make a new pattern similar to the one in question **3**. Can children use the same shapes in the same order but change how they turn in order to create a different pattern?

ASSESSMENT CHECKPOINT Use question **2** to assess whether children can identify the pattern's core. Question **3** will determine not only whether children can identify the pattern's core but also whether they can use the correct vocabulary of rotation in order to describe the pattern.

ANSWERS

Question **1** a): Children should point to:

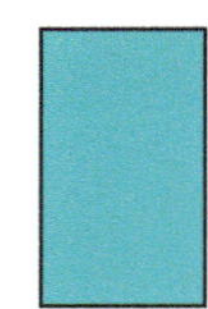

Question **1** b): The shape makes a quarter turn, either clockwise or anticlockwise.

Question **2**: The missing shape is the small triangle

Question **3**: The next three shapes in the pattern will be:

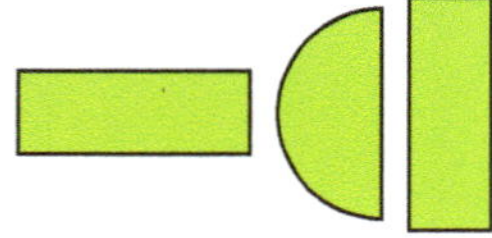

Check children are able to make their own repeating pattern. Can they successfully continue a pattern their partner creates for them?

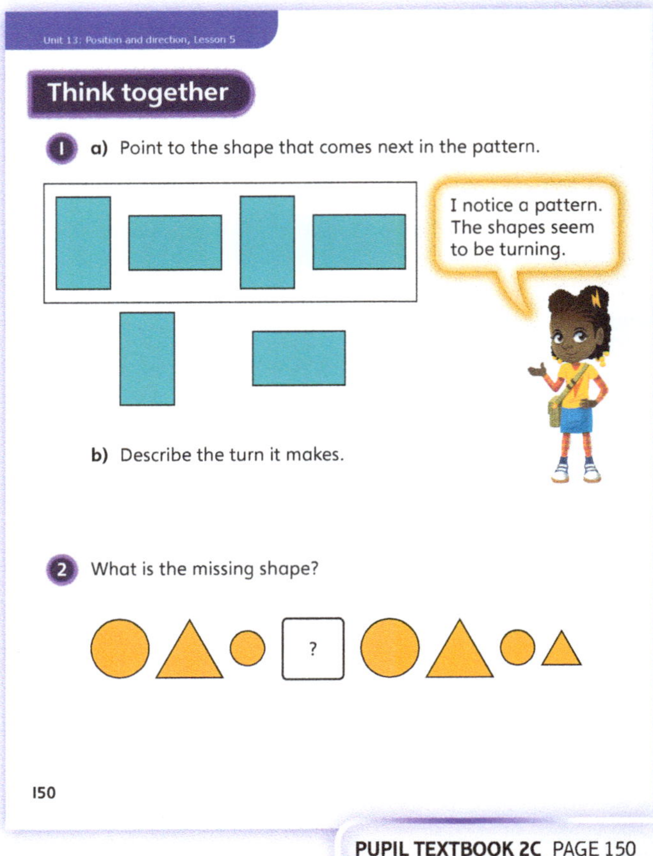

PUPIL TEXTBOOK 2C PAGE 150

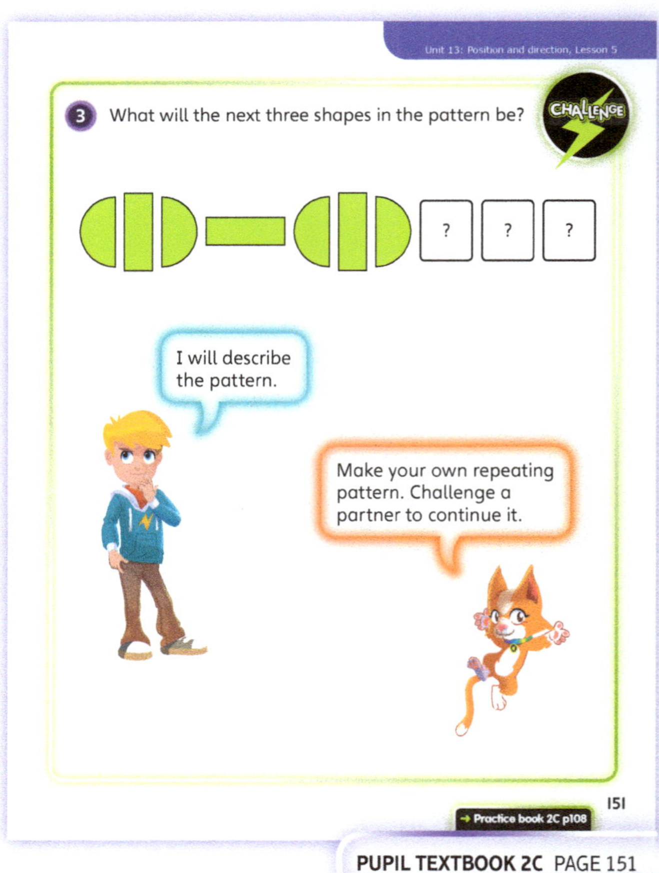

PUPIL TEXTBOOK 2C PAGE 151

Practice

WAYS OF WORKING Pair work

IN FOCUS Question **5** requires children to identify the shape that does not fit the pattern. In order to do this, children need to recognise how each shape has been rotated. They then need to justify their decision by describing how the rotation of their chosen shape does not fit the pattern. Provide a copy of the shapes for children to physically rotate to support their understanding.

STRENGTHEN Where possible, provide children with 2D shapes so that they can replicate the patterns in the problems. As children are physically rotating the shapes, talk to them about what they are doing. Have the key vocabulary available for children to refer to, to support their descriptions of the patterns. Ask them to create their own simple patterns by drawing around or printing with 2D shapes, carrying out a rotation for each term.

DEEPEN Ask children to create a pattern with two or more lines. The shapes should rotate as they go from left to right and from top to bottom. Can they write a description of their pattern? Can they describe a partner's pattern?

THINK DIFFERENTLY Question **4** requires children to describe the size and direction of the turns that a 2D shape makes in order to generate a given pattern. Look for children drawing on learning from the previous lesson to use the key vocabulary correctly.

ASSESSMENT CHECKPOINT Use questions **1** to **3** to check whether children can identify a pattern's core and how its shapes have been rotated. Use question **4** to assess their confidence in using the vocabulary of rotation to describe a pattern.

Question **5** will determine whether children can identify a shape that does not fit the rules of a pattern, and if they can justify this by applying their understanding of rotation.

ANSWERS Answers for the **Practice** part of the lesson can be found in the *Power Maths* online subscription.

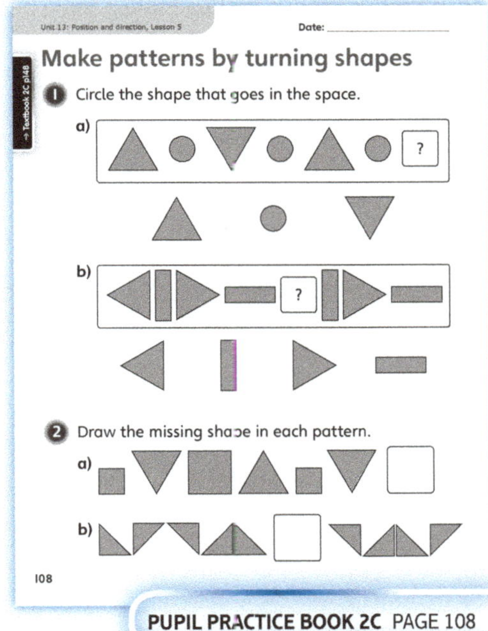

PUPIL PRACTICE BOOK 2C PAGE 108

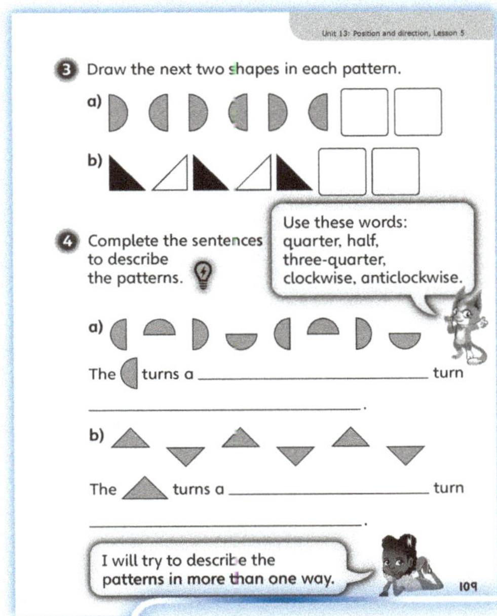

PUPIL PRACTICE BOOK 2C PAGE 109

Reflect

WAYS OF WORKING Pair work

IN FOCUS In this part of the lesson, children need to think carefully about what they have learnt in order to create their own pattern. The complexity of their pattern will indicate how secure children are in understanding and describing rotation.

ASSESSMENT CHECKPOINT This problem will determine whether children are able to apply their knowledge to create their own patterns.

ANSWERS Answers for the **Reflect** part of the lesson can be found in the *Power Maths* online subscription.

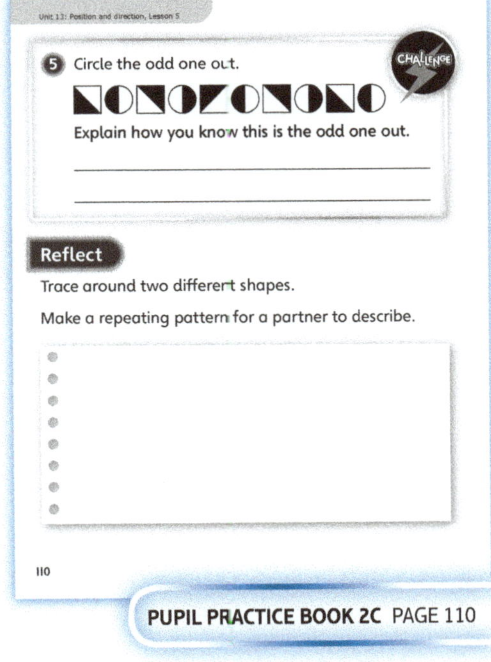

PUPIL PRACTICE BOOK 2C PAGE 110

After the lesson ⏸

- Were children consistent in their use of the correct mathematical language to describe rotation?
- Did children make links between this lesson and previous lessons?

End of unit check

Don't forget the unit assessment grid in your _Power Maths_ online subscription.

WAYS OF WORKING Group work adult led

IN FOCUS This **End of unit check** asks children to correctly use key language associated with position and movement. The questions require children to understand the amount of turns (in quarters) and direction of turn, and to be able to identify repeating cores within a pattern.

Think!

WAYS OF WORKING Pair work

IN FOCUS This question requires children to be able to describe the position of an object in relation to other objects. It assesses whether children can apply the correct vocabulary while giving enough information to avoid ambiguity.

Key vocabulary in this question includes: above, below, beside, next to, top, bottom, left, right and between.

Encourage children to think through or discuss the questions they would like to ask before writing in **My journal**.

ANSWERS AND COMMENTARY Children who have mastered this unit will be able to use the correct mathematical language to describe position and lateral and rotational movement. Children will be able to follow a series of instructions related to movement and they will be able to plan a series of instructions to arrive at a desired goal.

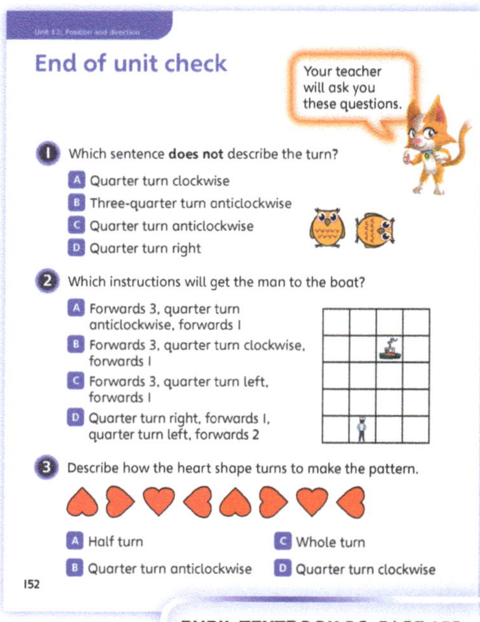

PUPIL TEXTBOOK 2C PAGE 152

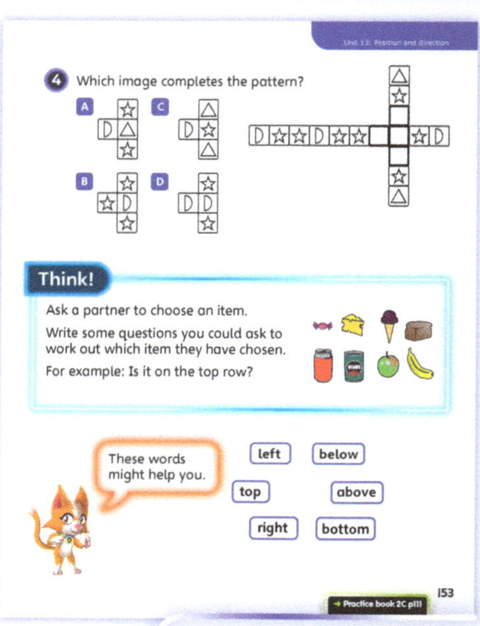

PUPIL TEXTBOOK 2C PAGE 153

Q	A	WRONG ANSWERS AND MISCONCEPTIONS	STRENGTHENING UNDERSTANDING
1	C	Incorrect answers may indicate that children do not understand the terms clockwise and anticlockwise or have confused left and right.	Provide practical opportunities to practise turning and movement, such as moving around a grid drawn on the playground, so that children can practise giving and following instructions. Provide children with a circular piece of card or paper with clockwise/right and anticlockwise/left drawn on it, which they can refer to in the playground.
2	B	A and C suggest children do not know the difference between clockwise and anticlockwise or left and right. D suggests miscounting.	
3	D	Wrong answers may indicate that children's ability to describe rotation is not secure.	Throughout the **End of unit check**, encourage children to describe each movement that is made so that they become secure in using the key language correctly.
4	C	Choosing A suggests that children have concentrated solely on the vertical pattern, whereas choosing B or D suggests that children have not identified the repeating core.	

My journal

WAYS OF WORKING Independent thinking

ANSWERS AND COMMENTARY

Encourage children to eliminate as many objects as possible with their first question. For example, ask: *Is it in the top row?* This allows children to rule out four objects straight away. Ask: *Is it between two other objects?* This will tell children whether the object is at the end of a row. Children then only need to ask one final question to solve the problem.

Provide children with a laminated copy of the objects and a dry-wipe pen so that they can cross out objects as they are eliminated. Ask: *How can you rule out as many objects as possible? How many objects did your question rule out? Could you ask a different question to rule out even more?* To ensure that children are using and applying the key language, ask them to include at least one of the words presented by Sparks in each of their questions. Challenge children to use a different word for each question that they ask.

PUPIL PRACTICE BOOK 2C PAGE 111

Power check

WAYS OF WORKING Independent thinking

ASK

- *How confident do you feel when you describe the position of an object?*
- *Do you now find it easier to follow a set of instructions related to movement?*
- *Do you still feel unsure about any of the key language?*
- *Has your ability to identify and describe repeating patterns improved?*

Power play

WAYS OF WORKING Pair work

IN FOCUS Use this game to see whether children can follow instructions relating to movement and to check that they are secure in their understanding of the direction and extent of a turn. Use counters and demonstrate the game first by playing against the whole class. To keep track of which direction the player is facing, draw an arrow on each counter to indicate the front of the counter.

If children are confident playing the game, they could roll two or three dice at a time to give them a series of instructions to follow. Ensure that children check that they can carry out all the instructions before moving their counter for the first time.

ANSWERS AND COMMENTARY If children struggle to make turns correctly, provide them with a paper circle divided into quarters with a quarter turn clockwise and a quarter turn anticlockwise drawn on it. Encourage them to remember that the direction they are facing makes a difference as to which way 'forwards' and 'backwards' is.

PUPIL PRACTICE BOOK 2C PAGE 112

After the unit ⏸

- Are there opportunities in other curriculum areas for children to apply their learning from this unit?
- Were all children confident using the key language in this unit?

Strengthen and **Deepen** activities for this unit can be found in the *Power Maths* online subscription.

Unit 14
Statistics

Don't forget to watch the Unit 14 video!

WHY THIS UNIT IS IMPORTANT

This unit is important because it is the first time children will have been introduced to statistics. The unit shows children how data can be collected effectively and then represented in a number of different ways. The unit will require children to use a range of different skills such as calculating and problem solving. These are great ways to consolidate prior learning. Children will be introduced to several different representations in the form of charts and diagrams and to some new mathematical language.

WHERE THIS UNIT FITS

→ Unit 13: Position and direction

→ **Unit 14: Statistics**

In this unit, children will build on their learning from a number of previous units. To interpret charts and diagrams, children must use their knowledge of addition and subtraction, counting and multiplication involving 2s, 5s and 10s. They will be introduced to symbols representing one or more pieces of data and to tally marks, which they will need to be able to count. Finally, previous units on problem solving will need to be called upon.

Before they start this unit, it is expected that children:
- can count in 2s, 5s and 10s
- can add and subtract 2-digit numbers
- can compare numbers to 100
- understand the language associated with problem solving.

ASSESSING MASTERY

In this unit, children will show mastery by being able to read charts and diagrams with ease and come to accurate conclusions. They will be able to explain both their methodology and their answers confidently. Children will be able to collect data independently, construct an appropriate chart or diagram and then interpret it (writing down useful statements using the correct terminology). Finally, they will learn to solve problems and puzzles efficiently.

COMMON MISCONCEPTIONS	STRENGTHENING UNDERSTANDING	GOING DEEPER
Children may miscount symbols used in pictograms or tallies on a tally chart.	On a tally chart, look closely at the oblique marks. Encourage children that they should recognise straight away that a set of 4 tallies with an oblique mark through represents 5 tallies. They do not need to count each one separately. Then, count in 5s as a class, pointing to each set of 5 tallies as you do. Finally, count any individual tallies in 1s. Your count might sound something like this: 'five, ten, fifteen, sixteen, seventeen.'	Ask children to think of more questions they could ask about a chart or diagram.
Children may assume one symbol in a pictogram always represents one item.	Pay particular attention to pictograms that have a half symbol. Ask children to point to the half symbol and see if they can suggest why half symbols always come at the end of a category. Look carefully at the key together. If 1 symbol represents 2 apples, for example, use counters to represent apples. Lay 2 counters on top of each complete symbol, and see if children can tell you how many counters should be on top of each half symbol.	Challenge children to think about where each chart or diagram would be used in real-life situations.

UNIT STARTER PAGES

Give children a few minutes to read the unit starter pages of the Textbook. Then read it together as a class.
Ask:

- *Which chart is a tally chart? Can you explain what it shows?*
- *Which chart is a pictogram? Can you explain what it shows?*
- *Is there anything you didn't understand?*
- *Which of the key words did you understand?*
- *What do they mean?*

STRUCTURES AND REPRESENTATIONS

Tally charts

Tariq	Amy
ЈHT ЈHT ЈHT |||	ЈHT ЈHT ЈHT ЈHT ||||

ЈHT means 5

Pictograms

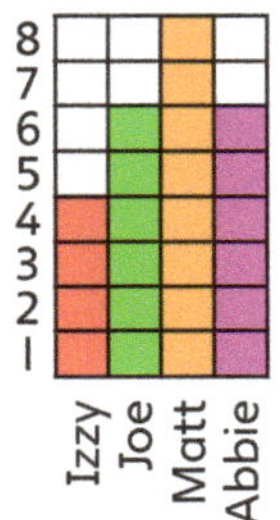

Name	Number
Tariq	🍎🍎🍎🍎🍎🍎🍎🍎🍎
Amy	🍎🍎🍎🍎🍎🍎

Block diagrams

Block diagram with values 1–8 on the vertical axis and names Izzy, Joe, Matt, Abbie on the horizontal axis.

Representations such as number lines, base 10 equipment and bar models may be used to support the interpretation of data or solving problems.

KEY LANGUAGE

There is some key language that children will need to know as part of the learning in this unit:

→ tally chart, tally
→ pictogram
→ block diagram
→ table
→ more, less, most, least
→ favourite, popular
→ equal
→ represent, symbol, key, information
→ total, altogether
→ compare

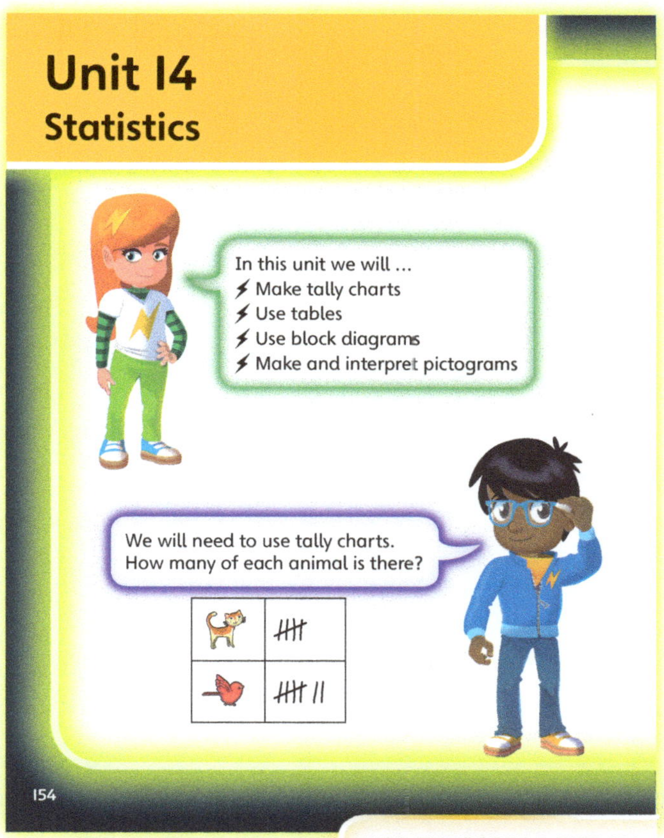

PUPIL TEXTBOOK 2C PAGE 154

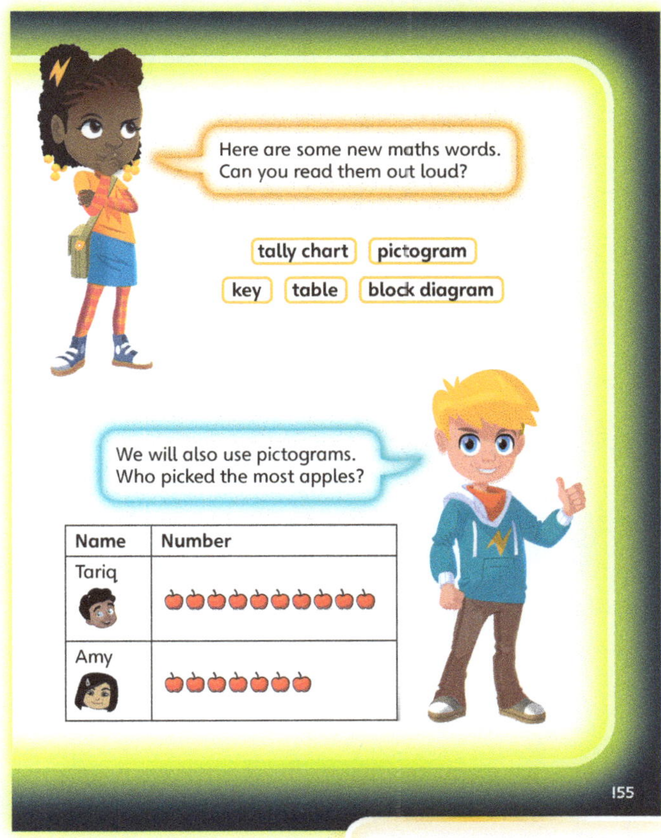

PUPIL TEXTBOOK 2C PAGE 155

Make tally charts

Learning focus

In this lesson, children will learn how to read and construct tally charts.

Before you teach 

- Do children need counting support?
- Can children count in 5s?
- Could you make a small display to reinforce tally charts and tallying?

NATIONAL CURRICULUM LINKS

Year 2 Statistics

Interpret and construct simple pictograms, tally charts, block diagrams and simple tables.

ASSESSING MASTERY

Children can confidently read a tally chart and explain it clearly, using the correct vocabulary. They can construct tally charts independently from a given set of data. Children understand why tally charts are useful and can say where they could be used in real-life situations.

COMMON MISCONCEPTIONS

Children may count a tally of 5 as 4 (due to not understanding that the oblique mark counts as 1). Ask:
- *What does the sloping mark mean? Why is it useful to tally like this?*

Be careful of children counting two tally marks as 11. Ask:
- *Is it eleven? If I put three marks, would it mean one hundred and eleven?*

STRENGTHENING UNDERSTANDING

This lesson is a great opportunity to do some extra counting practice, including counting in 5s and grouping in 5s (you could start with counters and move on to tallies). Children could practise recording numbers up to 20 as a tally and reading numbers to 40 given as a tally before working with actual tally charts.

GOING DEEPER

To deepen learning, ask children why tally charts are used. Ask: *Why are tally charts useful? Where could they be used in real-life situations? Why is it useful to record in 5s instead of 1s?*

KEY LANGUAGE

In lesson: tally chart, marks, table, count, fifth, total, tallied, groups of 5, choice, how many?, more, most, greatest, least, create, results

Other language to be used by the teacher: favourite, record

STRUCTURES AND REPRESENTATIONS

Tally charts and tables, number lines

RESOURCES

Mandatory: number lines, counters, ruler

Optional: number line counting up in 5s

 In the eTextbook of this lesson, you will find interactive links to a selection of teaching tools.

Quick recap

As a class, count up together in 5s. See if children can count on in 5s from different multiples of 5.

Discover

WAYS OF WORKING Pair work

ASK

- Question ❶ a): *What does the key mean? How does this make it easier to count?*
- Question ❶ b): *Do you need to count up all the tallies to find out who won? Why or why not?*

IN FOCUS In question ❶ a), children may be unfamiliar with tallies of 5, so let children practise recording numbers up to 20 as a 5 tally and reading numbers to 40 given as a tally before moving on to look at this question closely.

PRACTICAL TIPS Children can make groups of counters or towers of cubes to represent each tally of 5, which they then count in 5s to find the total.

ANSWERS

Question ❶ a): Tariq won 18 games; Amy won 24 games.

Question ❶ b): 24 is greater than 18. Amy won overall.

PUPIL TEXTBOOK 2C PAGE 156

Share

WAYS OF WORKING Whole class teacher led

ASK

- Question ❶ a): *Why is it useful to arrange the marks in groups of 5?*
- Question ❶ b): *Without counting, how can you see from the tally marks that Amy scored more than Tariq?*
- Question ❶ b): *Which mathematical sign could you use to explain your answer?*

IN FOCUS In question ❶ b), link back to prior learning of comparing numbers. Explain the vocabulary to children and remind them of the signs they can use. Can children use the signs in their answers? Children could work in pairs, one child recording a score on a number line showing jumps of 5 and some extra 1s, and one child recording a score as tally. Children should swap roles so that they each get a chance to record the number line score as a tally and to show the tally on a number line.

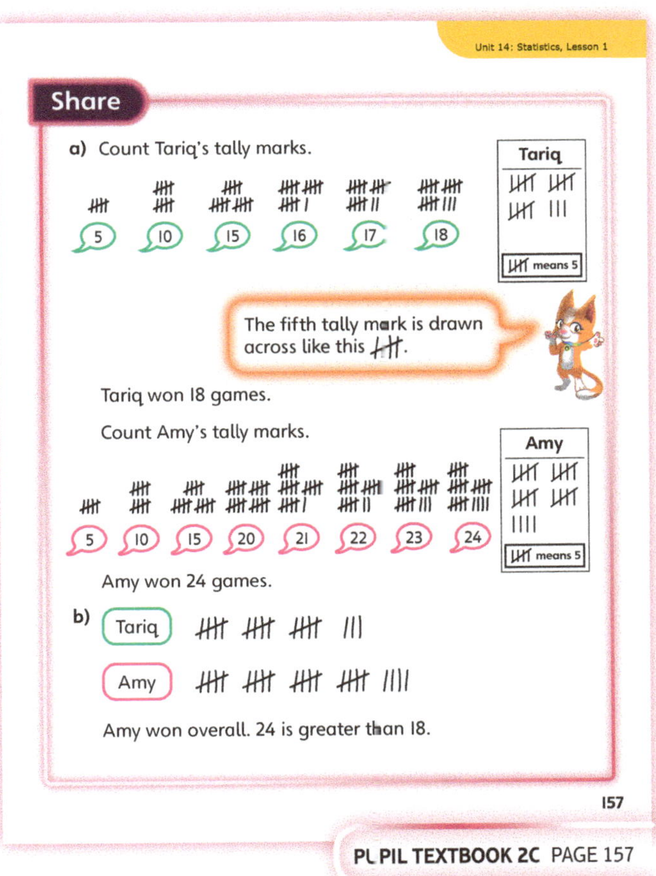

PUPIL TEXTBOOK 2C PAGE 157

Think together

WAYS OF WORKING Whole class teacher led (I do, We do, You do)

ASK

- Question **2** a): *Look at the question. What should you do first?*
- Question **2** a): *How many rows and columns should your chart have?*
- Question **2** a): *Should you use any equipment to help you create your tally chart?*

IN FOCUS Question **2** requires children to create a tally chart from the given data. Some children may need a table template and adult support with the layout. Others may need to be reminded of key features such as headings and using a ruler.

Question **3** is best suited for children working in groups of three, so that each child within a group can be responsible for collecting one type of object. You might want to have some counters, pencils and rulers handy so that the emphasis is on retrieving the correct number of objects. Children can check that the others in their group have found the correct number of objects too.

STRENGTHEN Check if any children are still counting the tallies of 5 individually. Run some intervention groups in which children are shown that it is more efficient to count in 5s. Afterwards, practise counting in 5s to make sure children are secure with this. Furthermore, you could run some extra intervention sessions in which children practise constructing tally charts from data.

DEEPEN Children could create more tally charts from their own collected data, such as tallying the colours of pencils in their pencil case. Children could also create a simple survey on, for example, eye colour or their favourite snack, sandwich or fruit from three or four choices.

ASSESSMENT CHECKPOINT Question **1** will allow you to assess whether children can count tally marks and find the total numerical value. Question **2** is an excellent opportunity to see whether children can apply what they have learnt and construct a tally chart from data.

ANSWERS

Question **1**: 13 rocks, 14 pieces of paper and 12 scissors.

Question **2** a): Children should complete the tally chart to show the following tallies and frequencies:

Object	Tally	Number
red button	卌 卌 I	11
yellow cube	卌	5
blue counter	卌 III	8

Question **2** b): Red buttons

Question **2** c): Yellow cubes

Question **3**: 12 red counters, 7 pencils, 4 rulers

Think together

1 These choices have been tallied in a **tally chart**.

Write the tally number for each choice.

Choice	Tally	Number
	卌 卌 III	
	卌 卌 IIII	
	卌 卌 II	

2 Here are some red buttons, yellow cubes and blue counters.

a) Create a tally chart to show how many of each colour.

b) Which item is there the greatest number of?

c) Which item is there the least number of?

158

PUPIL TEXTBOOK 2C PAGE 158

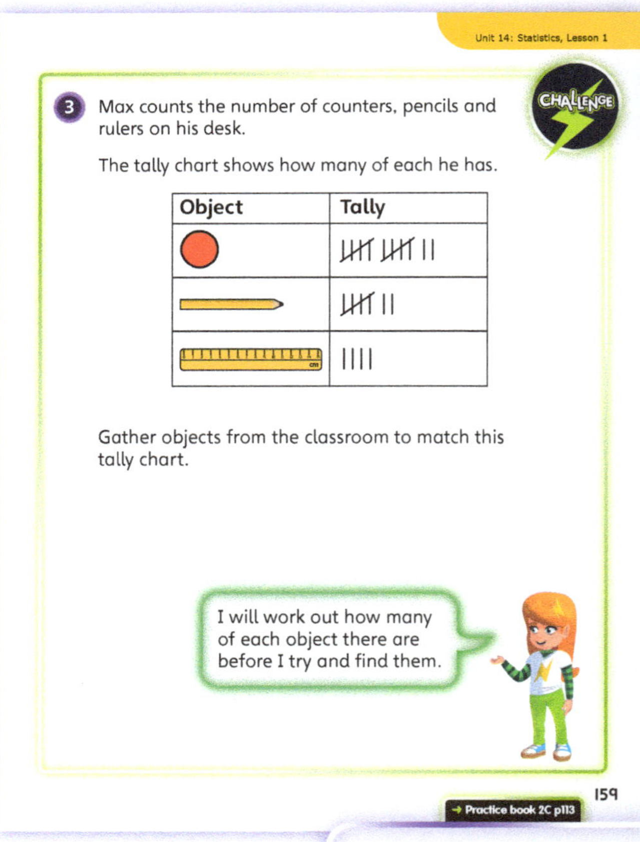

3 Max counts the number of counters, pencils and rulers on his desk.

CHALLENGE

The tally chart shows how many of each he has.

Object	Tally
	卌 卌 II
	卌 II
	IIII

Gather objects from the classroom to match this tally chart.

159

→ Practice book 2C p113

PUPIL TEXTBOOK 2C PAGE 159

Practice

WAYS OF WORKING Pair work

IN FOCUS Question **3** requires children to complete a tally chart where they have already been given some of the information. They are then asked to interpret the results. Children will need to write a tally as a number and a number as a tally. Then they will need to use the information to make statements about what the table shows.

STRENGTHEN Give children extra support when interpreting a tally chart. Discuss how you can tell if something is more popular than something else just by looking. There will certainly be a discussion about how one of them 'looks more'. This is a good opportunity to show the importance of making tally marks of similar size – so everyone can see results clearly.

DEEPEN After making some statements about what the chart shows in question **3**, ask children to write questions relating to the chart for a partner to answer. Ask them to see if they can write questions that involve addition or subtraction.

ASSESSMENT CHECKPOINT Question **1** a) will assess whether children can count in 5s to complete the tally chart. Questions **1** b) and c) will allow you to assess which children can interpret a tally chart to identify the most and the least popular animal.

ANSWERS Answers for the **Practice** part of the lesson can be found in the *Power Maths* online subscription.

Reflect

WAYS OF WORKING Whole class

IN FOCUS Create a group tally chart with five school subjects to choose from. You may want to have a template on the board and ask a child to be the scribe. Start by discussing the best way to create the tally chart – a useful way to assess how children will go about collecting data as a whole class. Complete the chart and ask children to think of some questions they might ask, such as: *Which subject was the most popular? Which subject was the least popular? Can you think of a question that will need subtraction to answer it? How many more children preferred maths to PE?*

ASSESSMENT CHECKPOINT This activity will allow you to assess which children can use efficient strategies to collect data and construct a tally chart. You will also be able to assess their interpretation of the tally chart by the questions they ask.

ANSWERS Answers for the **Reflect** part of the lesson can be found in the *Power Maths* online subscription.

After the lesson

- Do children need support with counting or recording scores in 5s?
- Can children construct a tally chart independently and discuss what the chart shows?

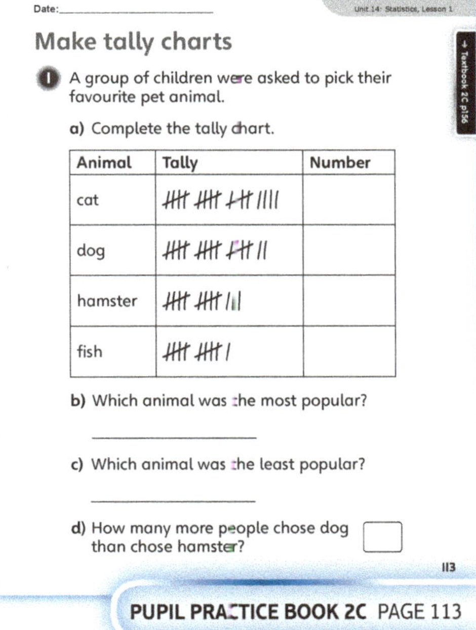

Animal	Tally	Number
cat	ⳊⳊ ⳊⳊ ⳊⳊ IIII	
dog	ⳊⳊ ⳊⳊ ⳊⳊ II	
hamster	ⳊⳊ ⳊⳊ III	
fish	ⳊⳊ ⳊⳊ I	

PUPIL PRACTICE BOOK 2C PAGE 113

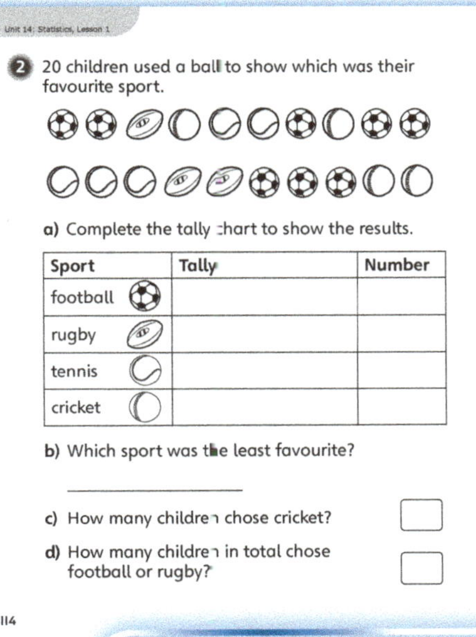

Sport	Tally	Number
football		
rugby		
tennis		
cricket		

PUPIL PRACTICE BOOK 2C PAGE 114

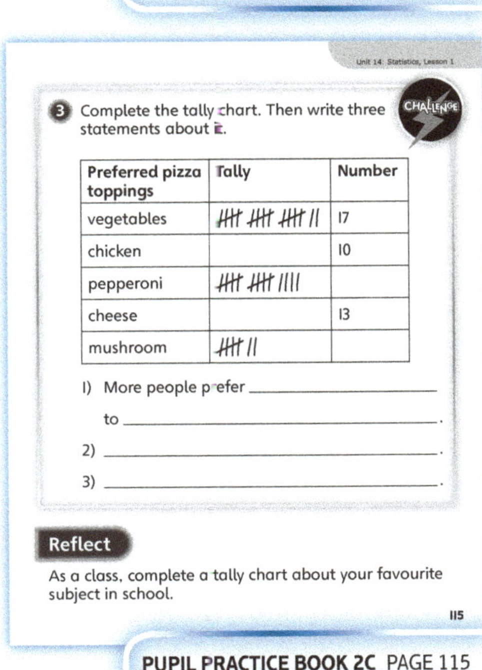

Preferred pizza toppings	Tally	Number
vegetables	ⳊⳊ ⳊⳊ ⳊⳊ II	17
chicken		10
pepperoni	ⳊⳊ ⳊⳊ IIII	
cheese		13
mushroom	ⳊⳊ II	

PUPIL PRACTICE BOOK 2C PAGE 115

Tables

Learning focus

In this lesson, children will interpret data which is presented in tables, and use this data to answer 1- and 2-step problems.

Before you teach 

- How will you help children to understand that a table lets you compare data systematically?
- Are you able to use real data, such as from a PE lesson, to complement some of the examples given in this lesson?

NATIONAL CURRICULUM LINKS

Year 2 Statistics

Interpret and construct simple pictograms, tally charts, block diagrams and simple tables.

ASSESSING MASTERY

Children can interpret data that is presented in tables. They can identify the correct row and column needed to find the required data. They can use this data to answer a range of 1- and 2-step problems, including questions involving comparison and ordering.

COMMON MISCONCEPTIONS

Children may wrongly identify the row and column needed to answer a question. Encourage children to make a decision about the column first, identifying this with a star, and then to look at the row. Ask:
- *Which column will you find the information in? Which row will you find the information in?*

STRENGTHENING UNDERSTANDING

Support children with correctly identifying the row and column that they need to refer to by encouraging the use of two rulers, one laid to the right of the column they need and the other laid under the row that they need, so that the data they require is where the rulers cross. Support children with the calculations required in this lesson by using the structures and representations from their earlier work on addition and subtraction.

GOING DEEPER

Encourage children to answer questions that involve more than one data source. For example, ask them to use linked data presented in a tally chart and a table to make comparisons and draw conclusions from the data.

KEY LANGUAGE

In lesson: table, how many?, total, most, more

Other language to be used by the teacher: row, column, data, smallest, greatest

STRUCTURES AND REPRESENTATIONS

Tables

RESOURCES

Optional: 3D shapes, ruler, toys for counting, a storage box

 In the eTextbook of this lesson, you will find interactive links to a selection of teaching tools.

Quick recap 

Ask children to order these numbers from smallest to greatest:

13   3   16   10   20   7

Discover

WAYS OF WORKING Pair work

ASK

- Question ❶ a): *Where is the information you need?*
- Question ❶ b): *Which row do you need to look at to find out how many cubes are in the box to start with?*
- Question ❶ b): *What do you need to do to find out how many cubes there are now?*

IN FOCUS In this activity, children are introduced to reading data presented in tables where they need to identify the correct row in order to find a particular piece of data. Encourage children to explore the structure of the table, and to identify where information for each shape is found.

PRACTICAL TIPS Use a box of real 3D shapes to recreate the scenario. Children sort and count the shapes and discuss how this information is captured in the table. Ask: *Why is the table a useful way to show this information?*

ANSWERS

Question ❶ a): There are 9 cones in the box.

Question ❶ b): There are 12 cubes in the box now.

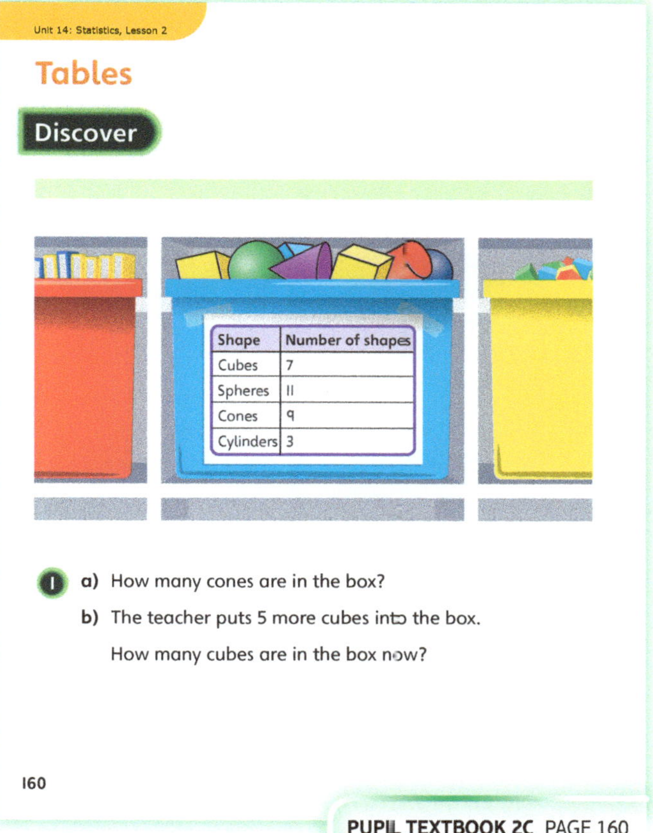

Shape	Number of shapes
Cubes	7
Spheres	11
Cones	9
Cylinders	3

PUPIL TEXTBOOK 2C PAGE 160

Share

WAYS OF WORKING Whole class teacher led

ASK

- Question ❶ a): *How did you work out the answer? What steps did you take?*
- Question ❶ a): *Which row did you look at?*
- Question ❶ b): *Where did you find the information about the number of cubes in the box to start with?*
- Question ❶ b): *How did you work out the total number of cubes after 5 more were added?*

IN FOCUS Question ❶ b) requires children to first identify the necessary data in the table and then to carry out a calculation with that number. Look for children who recognise that an addition is needed to find the new total. If children make any errors, consider whether this is due to misreading the table, misunderstanding what operation is required, or a calculation error.

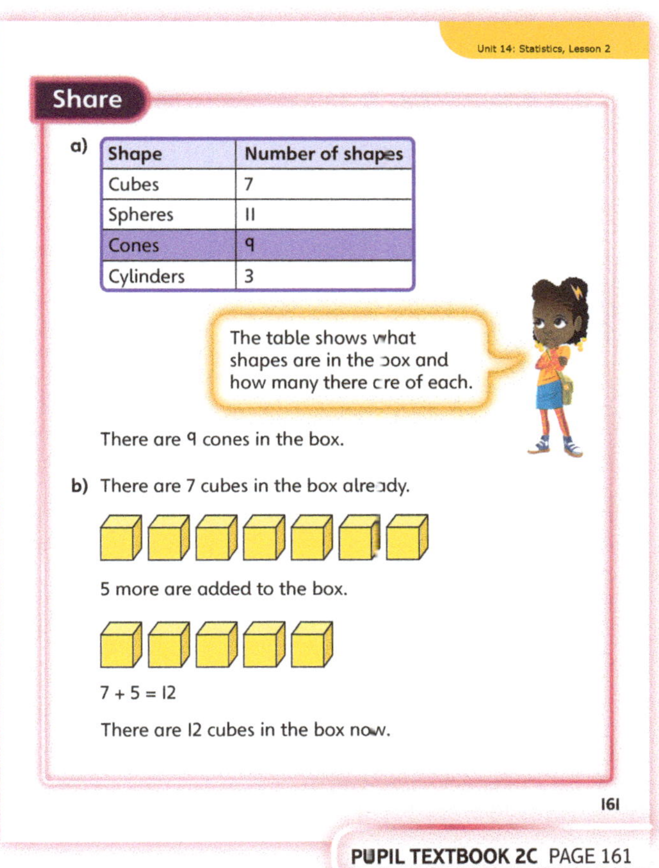

Shape	Number of shapes
Cubes	7
Spheres	11
Cones	9
Cylinders	3

PUPIL TEXTBOOK 2C PAGE 161

Think together

WAYS OF WORKING Whole class teacher led (I do, We do, You do)

ASK

- Question **1**: *What is there most of? What is there least of? Which two items have the same number?*
- Question **2**: *What calculation method are you going to use to find the total? Is there a strategy that might make it easier?*
- Question **3**: *Can Harry make the word STARS? How do you know?*

IN FOCUS Question **1** involves reading each row of a table correctly in order to interpret the given data. This is taken further in question **2**, where children are required to first read each row of the table and then complete an addition with these numbers in order to find the total.

In question **3**, children read and interpret a table and use the data to solve problems, including an open-ended question with more than one possible answer.

STRENGTHEN To help support children in linking and adding the values in question **2**, provide real toys or similar items in a box to match the information in the table. Children can identify, group and count the toys to find the total.

DEEPEN Question **3** gives children an opportunity to practise interpreting a table in a variety of different ways. Deepen understanding of the data available by further exploring question **3** d) which is an open question. Share all the words that children have made and check together that each word is possible based on the number of letters available. Ask: *Are there any other words you could make? What's the longest word you can make?*

ASSESSMENT CHECKPOINT Use question **1** to assess if children can independently identify the correct row in which to find the data they require.

ANSWERS

Question **1**: There are 6 footballs, 10 bats, 6 cones and 5 ropes in the box.

Question **2**: There are 15 toys in total in the box.

Question **3** a): There are 2 As and 3 Es.

Question **3** b): The letter Harry has most of is E.

Question **3** c): He only has one each of letters D, R and S.

Question **3** d): Children's answers will vary.

Think together

1 Here is a box of sports equipment.

Name	Number
Footballs	6
Bats	10
Cones	6
Ropes	5

How many of each item are in the box?

2 Here is another box of toys.

Name	Number
Cars	5
Teddies	7
Planes	3

How many toys are in the box in total?

162

PUPIL TEXTBOOK 2C PAGE 162

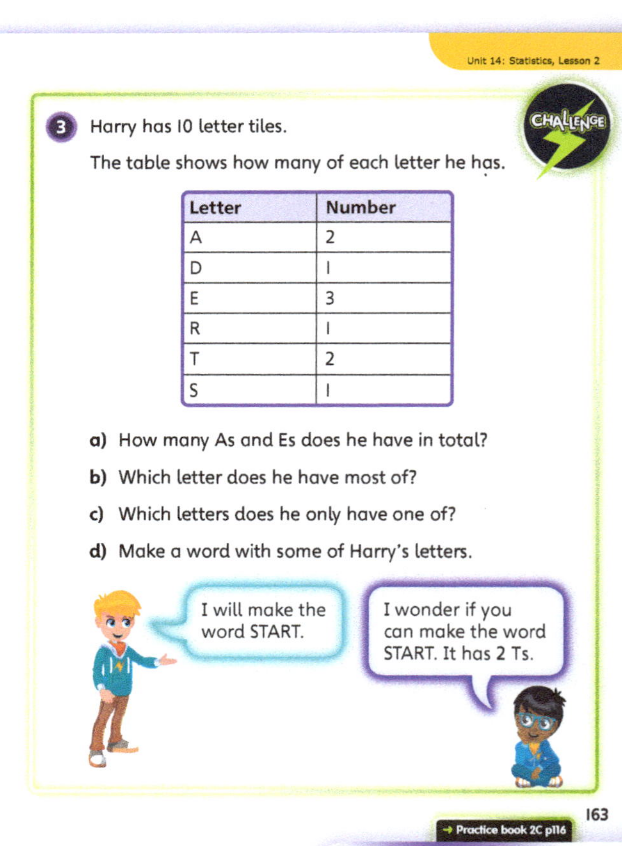

3 Harry has 10 letter tiles.

The table shows how many of each letter he has.

Letter	Number
A	2
D	1
E	3
R	1
T	2
S	1

a) How many As and Es does he have in total?

b) Which letter does he have most of?

c) Which letters does he only have one of?

d) Make a word with some of Harry's letters.

163

→ Practice book 2C p116

PUPIL TEXTBOOK 2C PAGE 163

Practice

WAYS OF WORKING Independent thinking

IN FOCUS Question ② involves children interpreting the data given in a table and using it to draw each pattern on the correct number of balloons.

Question ③ involves children using the data they have been given and interpreting what they are being asked in order to derive other information by using addition and subtraction.

DEEPEN Question ④ is a 2-step problem. Ensure children read the question carefully and describe what operations will be needed to solve the word problem. Can they draw a bar model to help them find the answer?

ASSESSMENT CHECKPOINT Use question ③ to assess if children are able to accurately carry out calculations based on the data presented in a table.

ANSWERS Answers for the **Practice** part of the lesson can be found in the *Power Maths* online subscription.

Reflect

WAYS OF WORKING Pair work

IN FOCUS This question encourages children to interpret a table and explore what information they can find from the data they have been given.

ASSESSMENT CHECKPOINT Use this activity to assess if children understand the features of tables, and can describe how to use these to find information. This should include facts that can be derived from the numbers in the table by using one or more operations to calculate a value.

ANSWERS Answers for the **Reflect** part of the lesson can be found in the *Power Maths* online subscription.

After the lesson ⏸

- Are children secure when answering questions based on tables?
- Can you incorporate tables into your day-to-day classroom or school environment?

Unit 14: Statistics, Lesson 2   Date:__________

Tables

① The table shows the number of pens, pencils and rubbers on a table.

Item	Number
Pen	6
Pencil	7
Rubber	6

a) How many pens are there? ☐

b) How many pencils are there? ☐

c) There are 4 rulers. Write this in the table.

② Holly has 10 balloons.

The table shows the patterns of the balloons.

Colour	Number
Spotty	3
Stripy	5
Plain	2

Draw the patterns on the correct number of balloons.

116

PUPIL PRACTICE BOOK 2C PAGE 116

③ The table shows the number of animals in a pet show.

Animal	Number
Dog	8
Cat	5
Rabbit	5
Mouse	2

a) How many dogs are in the show? ☐

b) How many rabbits and mice are in the show in total? ☐

c) What is the total number of dogs and cats in the show? ☐

d) What is the total number of animals in the show? ☐

e) How many more dogs than mice are there in the show? ☐

117

PUPIL PRACTICE BOOK 2C PAGE 117

④ Ambika is reading a 60 page book. **CHALLENGE**

The table shows how many pages she reads each day.

Day	Number of pages
Monday	12
Tuesday	20
Wednesday	15
Thursday	

By the end of Thursday she had finished the book. How many pages did Ambika read on Thursday?

Reflect

Pick a table from your lesson today.
Share with a partner some information from the table.

118

PUPIL PRACTICE BOOK 2C PAGE 118

Block diagrams

Learning focus

In this lesson, children will read, construct and interpret block diagrams.

Before you teach 

- Do you have equipment (cubes, squared paper) ready to reinforce learning in this lesson?
- How will you link block diagrams to pictograms?
- How will you link block diagrams to tally charts?

NATIONAL CURRICULUM LINKS

Year 2 Statistics

Interpret and construct simple pictograms, tally charts, block diagrams and simple tables.

ASSESSING MASTERY

Children can accurately construct a block diagram and then interpret it effectively. Children can spot mistakes and recognise when it is *not* useful to use a block diagram.

COMMON MISCONCEPTIONS

Children may assume that a block could represent more than one thing. Ask:
- *Look through the Textbook – are there any block diagrams where each block represents more than one block?*

Children may forget to label all parts of the block diagram when constructing it. Ask:
- *Did you remember the scale? Have you labelled the scale correctly? Have you missed any numbers?*

STRENGTHENING UNDERSTANDING

Using cubes to make 3D block diagrams (graphs) will strengthen understanding. Children could then lay the towers on squared paper with squares matching the size of the cubes to make a 3D block diagram before labelling it and then shading squares to replace the cubes. A checklist may be useful for them. Ask questions such as: *Did you use a ruler? Have you included a scale? Did you label what the blocks or cubes represent?*

GOING DEEPER

Children could think about when it is appropriate to use a block diagram, and when it is not. Ask them if it would represent very large numbers effectively (such as numbers greater than 100).

KEY LANGUAGE

In lesson: block diagram, most, least, same, amount, count, compare, scale, height, score, more, less, fewer, popular, prefer

Other language to be used by the teacher: block graph, label, labelled, construct, represent, strategy

STRUCTURES AND REPRESENTATIONS

Block diagrams, cubes

RESOURCES

Mandatory: interlocking cubes, ruler

Optional: 1 cm or 2 cm squared paper (same size as interlocking cubes)

 In the eTextbook of this lesson, you will find interactive links to a selection of teaching tools.

Quick recap 

Work together as a class to create a table showing lunch information for your class. Ask questions such as: *How many children have a hot meal? How many have a sandwich? How many have something else?*

Discover

WAYS OF WORKING Pair work

ASK

- Question ❶ b): *How can you see which columns have the same number of points?*
- Questions ❶ a) and b): *What strategies did you use to work out the answers?*

IN FOCUS Counting the blocks out loud is an effective strategy; it will make the questions easier to answer. Children may also be able to see, as with the length of pictograms, that the tallest tower shows the greatest amount; the shortest tower shows the least amount.

PRACTICAL TIPS Children can use cubes in different colours to recreate the block diagram.

ANSWERS

Question ❶ a): Jack has the most points as he has the highest tower.

Question ❶ b): Joe and Kara have the same number of points as the towers are the same height.

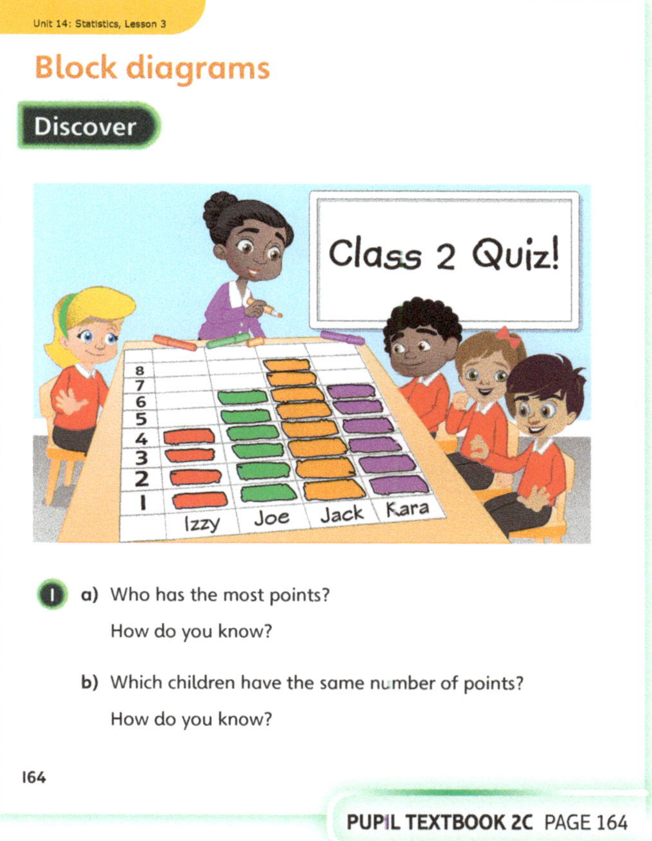

PUPIL TEXTBOOK 2C PAGE 164

Share

WAYS OF WORKING Whole class teacher led

ASK

- Question ❶ a): *Can you explain what you did to answer the question?*
- Question ❶ a): *Did anyone use a different way?*
- Question ❶ a): *Can you show me how to make the block diagram with cubes?*

IN FOCUS For question ❶, have cubes ready for your class to make the block diagrams. This will reinforce learning and understanding. Encourage children to line up the towers of cubes next to each other vertically so that children can easily see which tower is the tallest and therefore who has the most. Then lay the towers out horizontally so that it looks like a block diagram, ensuring children understand that the bottom of each tower needs to be level with the bottom of all the other towers.

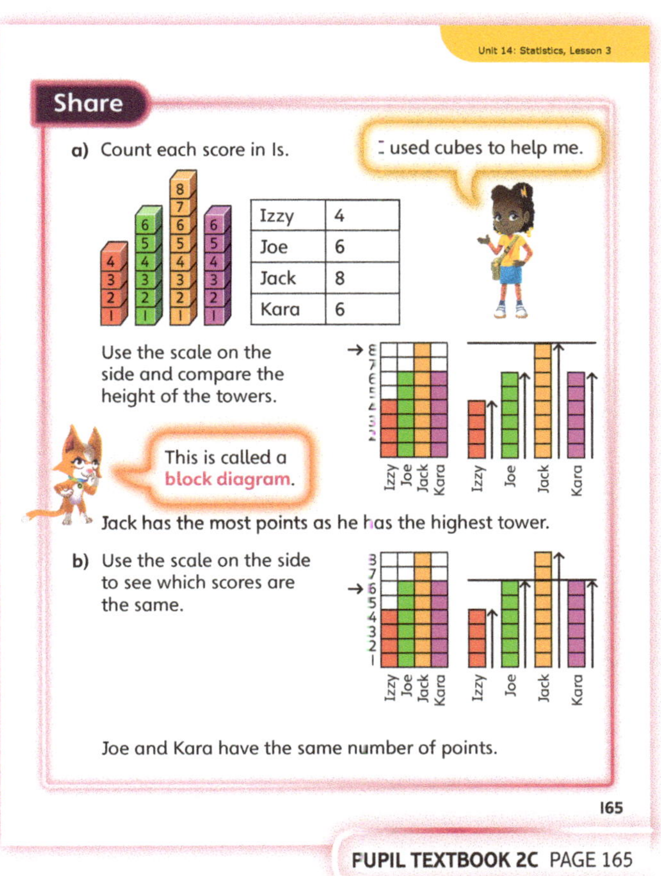

PUPIL TEXTBOOK 2C PAGE 165

Think together

WAYS OF WORKING Whole class teacher led (I do, We do, You do)

ASK

- Question **1**: *How does having a scale help?*
- Question **3**: *Why are the towers different colours?*

IN FOCUS In question **1**, encourage children to use the scale to find the answers more quickly. You may find that children are counting the blocks individually.

STRENGTHEN Continue to support children by providing equipment (blocks). When reading the scale, a ruler can be useful to ensure the correct number is read.

DEEPEN Can children construct a block diagram of the vowels in five of their friends' first names or surnames? Ask them to write three statements about what they found and write these under the block diagram. Children could compare results for first names and surnames if they do them separately. Ask: *Which vowels are used the most and which are used the least?*

ASSESSMENT CHECKPOINT Question **3** will give you an insight into which children are mastering interpretation of block diagrams, as they have to use the diagrams to check the statements but also use reasoning skills to generate their own facts.

ANSWERS

Question **1** a): Team D scored the most points as they have the tallest tower.

Question **1** b): Team E scored the least points as they have the shortest tower.

Question **2**: Children should create a block diagram with cubes. Check that the correct number of cubes have been used.

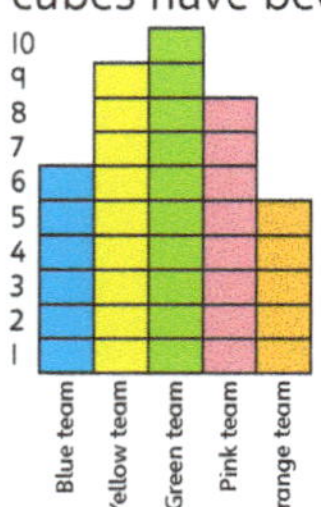

Question **3** a): The most popular animal is the tiger – correct.
5 fewer children prefer the elephant to the tiger – incorrect (3 fewer children prefer the elephant to the tiger).
8 more children prefer the monkey to the rhino – correct.

Question **3** b): Children's answers will vary. Look for accurate use of mathematical vocabulary and correct sense in the context of the block diagram.

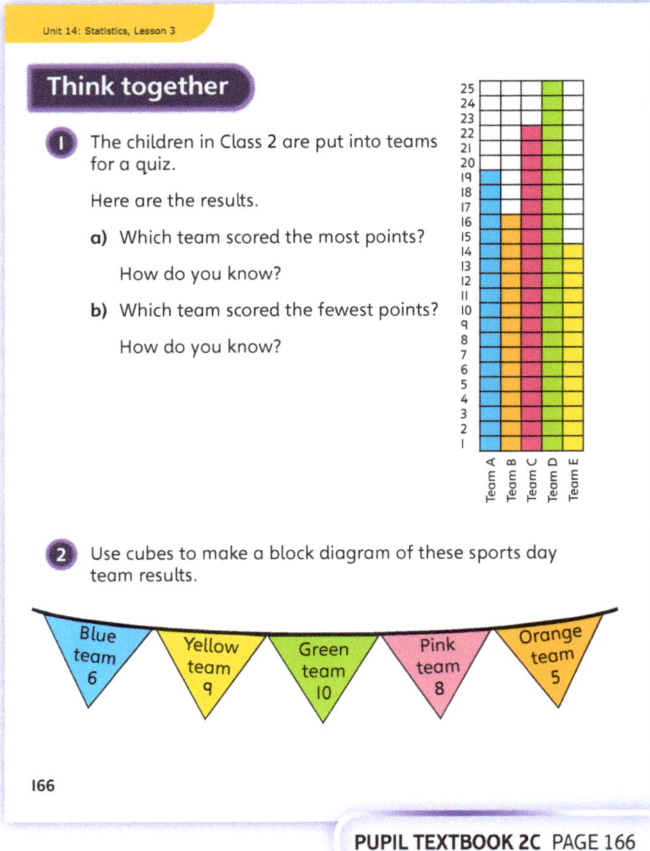

PUPIL TEXTBOOK 2C PAGE 166

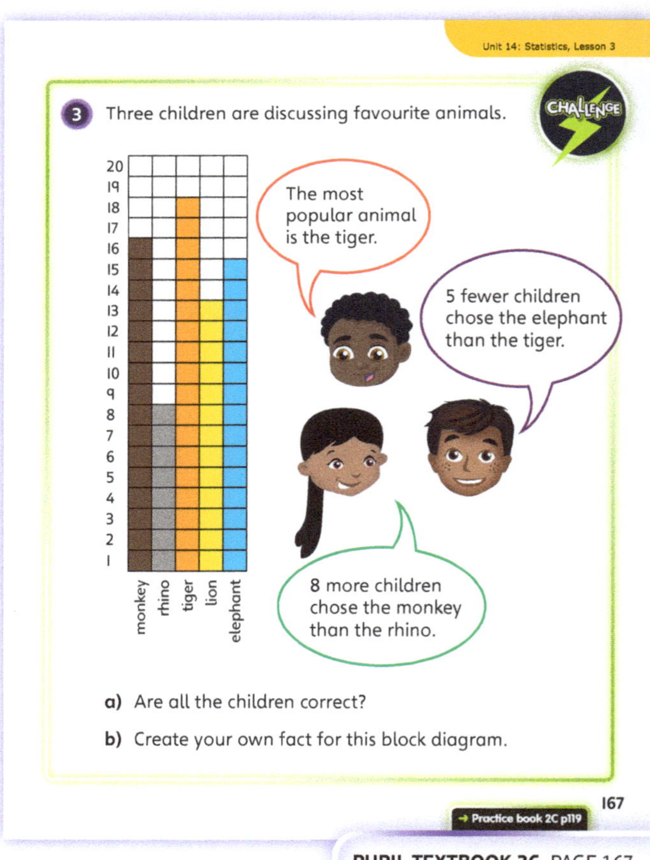

PUPIL TEXTBOOK 2C PAGE 167

Practice

WAYS OF WORKING Independent thinking

IN FOCUS Question **2** requires children to use a given set of data to complete their own block diagram. In question **3**, the block diagram is already completed and children use this to find information. They will need to solve 1- and 2-step problems in order to do this.

STRENGTHEN When comparing towers in questions **1** and **3**, ask children to place their finger on the top of the shorter tower and count on in 1s to the taller tower to find the difference.

DEEPEN Children have now covered tally charts and block diagrams. Ask children to discuss where they may be used in real life, and the pros and cons of each.

ASSESSMENT CHECKPOINT Question **2** will provide you with evidence of whether children can complete a block diagram. Question **3** will provide you with evidence of whether children can interpret a block diagram.

ANSWERS Answers for the **Practice** part of the lesson can be found in the *Power Maths* online subscription.

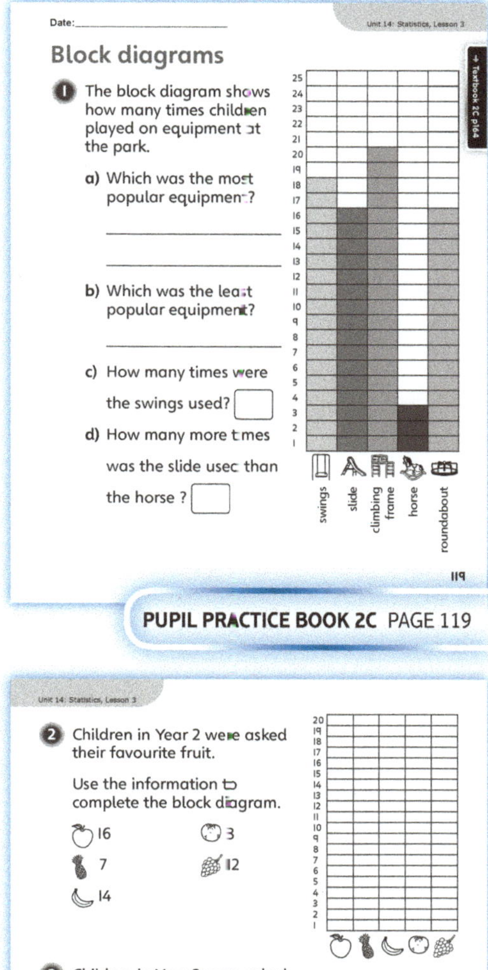

PUPIL PRACTICE BOOK 2C PAGE 119

PUPIL PRACTICE BOOK 2C PAGE 120

Reflect

WAYS OF WORKING Pair work

IN FOCUS This question will promote some deep thinking about the features of block diagrams. Children look at a block diagram which has not been drawn correctly and consider what mistakes they can see that would need to be improved for the block diagram to be more useful.

ASSESSMENT CHECKPOINT Assess whether children are able to explain their answers and reason about how a block diagram must be constructed. Children should spot that the block sizes are unequal, 11 is missing from the scale and the towers are not labelled.

ANSWERS Answers for the **Reflect** part of the lesson can be found in the *Power Maths* online subscription.

After the lesson ⏸

- Can you give children an opportunity to collect some data and construct a block diagram independently?
- Did children master block diagrams?

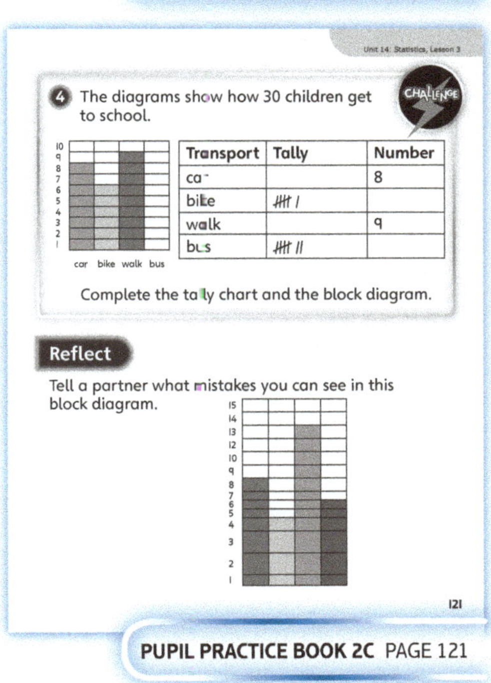

PUPIL PRACTICE BOOK 2C PAGE 121

Draw pictograms (1 to 1)

Learning focus

In this lesson, children will construct pictograms from given data, showing that one symbol represents one item. They will also link them to tally charts.

Before you teach 

- Have children mastered tally charts from Lesson 1?
- How can children be supported when converting a tally chart to a pictogram?
- Will you do a whole-class data collection and representation activity?

NATIONAL CURRICULUM LINKS

Year 2 Statistics

Interpret and construct simple pictograms, tally charts, block diagrams and simple tables.

ASSESSING MASTERY

Children can draw accurate pictograms with a key showing that one symbol is used to represent one item. They can explain their pictograms in detail using the correct mathematical vocabulary.

COMMON MISCONCEPTIONS

Children may not draw the symbols in a neat or organised manner. They may space them out or make some larger than others. This could mean that children think that the row that is more spaced out or the symbol that is larger is the greater in value. Sketch out some incorrectly drawn symbols and discuss why they are not right. Ask:

- *What is wrong with this row of symbols?* [spaced out more than the other rows]
- *What mistake have I made in this row?* [symbols are not the same size]
- *What are the 'rules' for drawing pictograms?* [same size, same spacing]

STRENGTHENING UNDERSTANDING

Children may need intervention activities in which pictograms are created from sets of data. They should focus on reading the data (this could be sets of objects or a tally chart), then creating symbols, a key, the outline table (using a ruler) and finally inputting the correct symbols.

If children create their own pictograms for a class survey, some children may need an outline template for support.

GOING DEEPER

Challenge children to collect data such as eye colour or favourite ice cream flavour, presenting it first as a tally chart and then as a pictogram. Children may need advice for choosing a suitable symbol relating to the data (simple oval for an eye and ice cream cone shape for ice cream) and they may also need reminding that there should be a limited choice of no more than five items. Alternatively, provide a class data list or quickly record choices on the board from a hand count.

KEY LANGUAGE

In lesson: represent, **pictogram**, symbol, tally chart, draw, make, count, **key**, result, compare

Other language to be used by the teacher: accurate, data, row, survey, favourite, rules, solve, strategy

STRUCTURES AND REPRESENTATIONS

Pictograms, tally charts

RESOURCES

Mandatory: rulers, some children may need pictogram outline templates

Optional: counters

 In the eTextbook of this lesson, you will find interactive links to a selection of teaching tools.

Quick recap

Play a simple game together as a class. Ask children to find a good way to keep score as the scores change during the game.

Discover

 Pair work

ASK

- Question **1** a): *Why is it important to arrange the symbols in a neat way?*
- Question **1** b): *How can you solve this? What strategy will you use?*

IN FOCUS Question **1** a) focuses children on thinking about the layout of the symbols in the pictogram. You could show neatly arranged counters next to erratically placed counters on children's desks. This will reinforce the fact that arrangements of the symbols are important for displaying information and interpreting this information easily.

PRACTICAL TIPS Use counters in different colours to recreate the pictogram. Start with the groups of counters arranged at random as shown, and then ask children to move the counters in order to line them up for each team.

ANSWERS

Question **1** a):

Team 1	⚪⚪⚪⚪⚪
Team 2	⚪⚪⚪⚪⚪⚪⚪⚪
Team 3	⚪⚪⚪⚪⚪
Team 4	⚪⚪⚪⚪⚪
Team 5	⚪⚪⚪

Each ⚪ represents 1 team point.

Question **1** b): Team 2 has the most points. They have 8 points.

Share

 Whole class teacher led

ASK

- Question **1** a): *Look back to the **Discover** section. Which pictogram is easier to read? Why?*
- Question **1** a): *Why has a different colour been used for each row?*

IN FOCUS Question **1** a) provides a good opportunity to talk about using a different colour for each row. Explain that the clearer you can make information, the easier it is to read. Ensure that children notice the equal sizing and spacing of the counters.

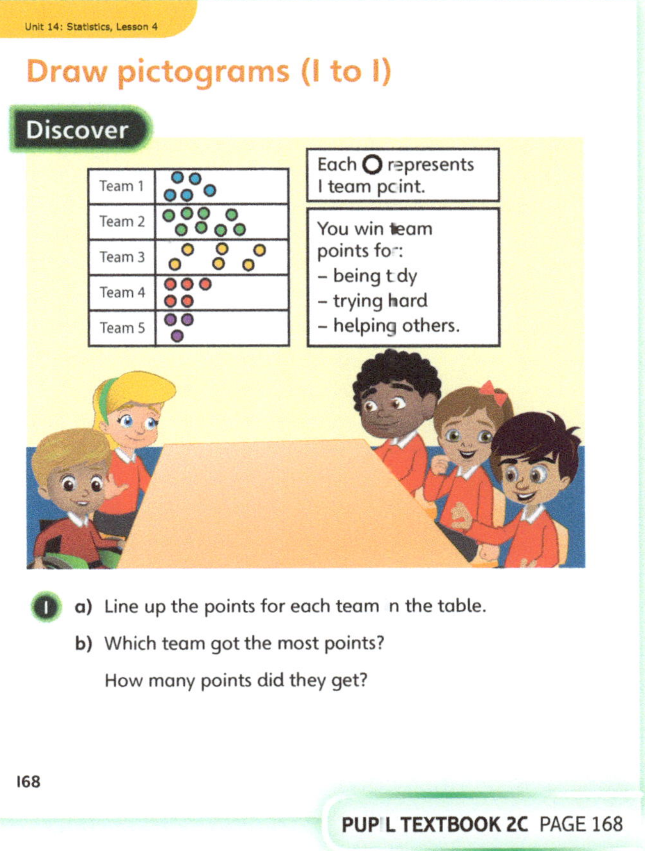

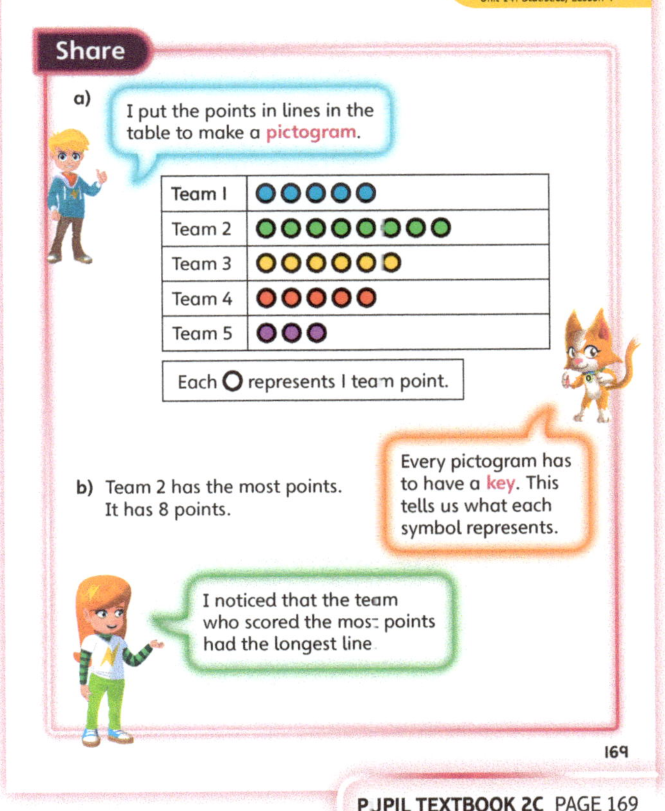

Think together

 Whole class teacher led (I do, We do, You do)

ASK

- Question **3**: *We need a symbol to represent an animal. What do lots of animals have?* (a paw, for example)
- Question **3**: *Could you draw a complicated symbol 20 times – could you draw it the same each time?*

IN FOCUS Question **1** asks children to use the information in a table to complete a pictogram with the key already provided. Question **2** explores several misconceptions about drawing a pictogram accurately. Question **3** asks children to choose animals when creating their pictograms. It is very important to explain that the symbol they choose to use must be kept simple but it should also reflect the subject of the data.

STRENGTHEN Question **3** requires children to think of a suitable symbol to use in the pictogram and in the key. If children find drawing difficult, suggest a very simple design for them to use. Check that their symbols are all roughly the same size and equally spaced.

DEEPEN Give children a choice of four 'favourite sports', then say: *12 children took part in a survey of their favourite sport. Make as many different pictograms as you can.* Can children work methodically?

Alternatively, get children to compare the effectiveness of tally charts and pictograms. Ask: *What is the same? What is different? Which one is easier to read? Which one is more interesting to look at?*

ASSESSMENT CHECKPOINT Question **1** will allow you to assess whether children can complete a 1 to 1 pictogram accurately. Question **2** will allow you to assess whether children can explain how to draw a meaningful pictogram with a sensible choice of symbol and a clear layout.

ANSWERS

Question **1**:

walk	☺☺☺☺☺☺☺☺☺
car	☺☺☺
bike	☺☺☺☺☺☺
other	☺☺

Each ☺ represents 1 child.

Question **2**: The symbols are not the same, they are not the same size and there is not the same spacing between the symbols.

Question **3** a): Answers will vary depending on children's choices.

Question **3** b): Check that children have chosen five animals, tallied accurately, and found the totals with appropriate symbols or names.

Question **3** c): Check for a carefully constructed pictogram – symbols should be all the same, ordered neatly and the same size.

Think together

1 20 children are asked how they travel to school.

Here are the results.

How child travels to school	Number of children
walk	9
car	3
bike	6
other	2

Complete the pictogram for this data.

walk	☺☺☺☺☺☺☺☺☺
car	
bike	
other	

Each ☺ represents 1 child.

170

PUPIL TEXTBOOK 2C PAGE 170

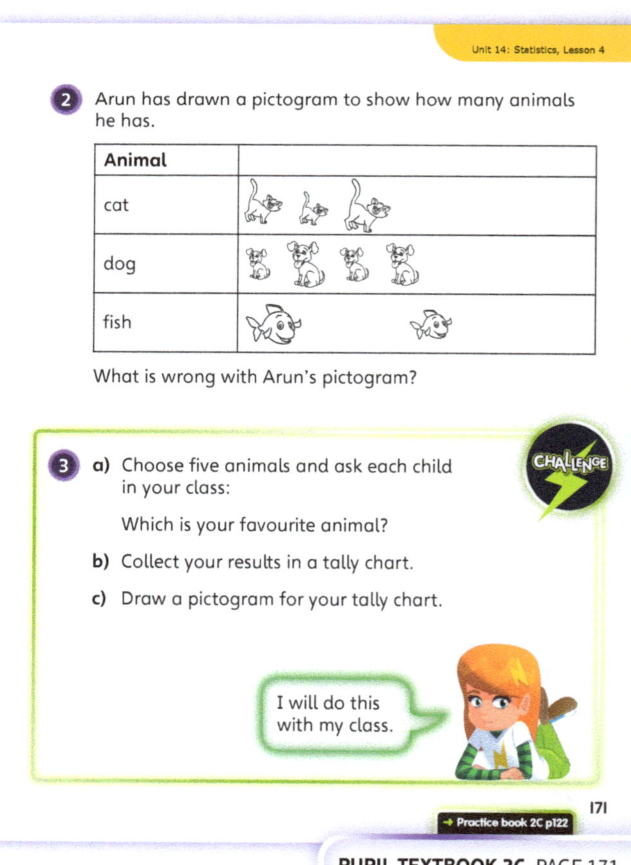

2 Arun has drawn a pictogram to show how many animals he has.

Animal	
cat	
dog	
fish	

What is wrong with Arun's pictogram?

3 a) Choose five animals and ask each child in your class:

Which is your favourite animal?

b) Collect your results in a tally chart.

c) Draw a pictogram for your tally chart.

→ Practice book 2C p122

171

PUPIL TEXTBOOK 2C PAGE 171

Practice

WAYS OF WORKING Pair work

IN FOCUS Question **4** will make children think carefully. They will first have to find the total number of children who like rugby and football. Then they will have to find the difference between that total and the overall pictogram total. This problem may need to be broken down into the two steps for some children.

STRENGTHEN In question **1**, children will find it easier to draw the pictogram if they work out the totals first and write these next to each row. If they do not, children might have to keep counting rows, which could mean they lose their train of thought.

DEEPEN Give children some data that has all even totals. See if they can use a key to represent each symbol as two items. Repeat with data with totals that are all multiples of 5.

Alternatively, give children a similar challenge to question **4**. Could they then make up their own problem relating to the number of children in the class?

THINK DIFFERENTLY Question **3** requires children to study two different pictograms and identify which one shows the information provided in a tally chart.

ASSESSMENT CHECKPOINT Question **4** will give you a clear indication of which children can confidently read and interpret pictograms. Question **1** will assess their understanding of the connection between tally charts and pictograms.

ANSWERS Answers for the **Practice** part of the lesson can be found in the *Power Maths* online subscription.

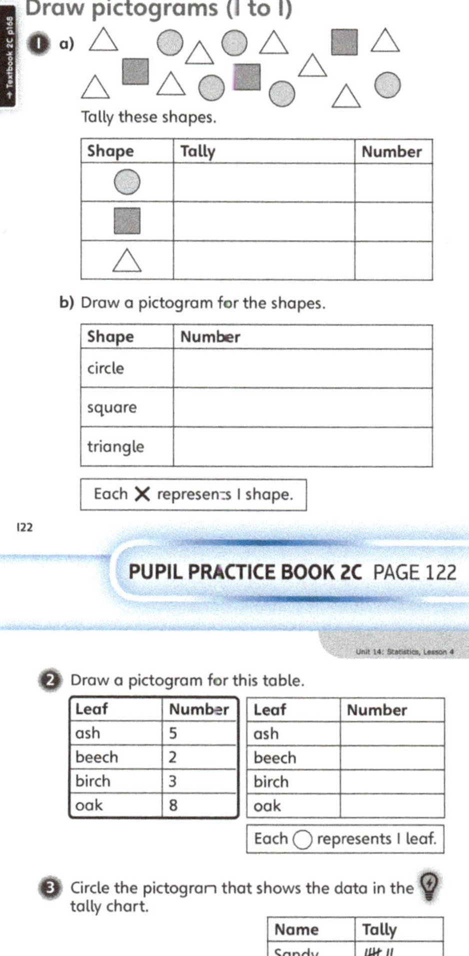

PUPIL PRACTICE BOOK 2C PAGE 122

PUPIL PRACTICE BOOK 2C PAGE 123

Reflect

WAYS OF WORKING Pair work

IN FOCUS This is an excellent opportunity to reinforce the vocabulary associated with tally charts and pictograms. Children should describe the key features of both tally charts and pictograms in order to compare them.

ASSESSMENT CHECKPOINT This reflective exercise will allow you to assess children's understanding of the vocabulary associated with tally charts and pictograms. Check if children can explain what each word means.

ANSWERS Answers for the **Reflect** part of the lesson can be found in the *Power Maths* online subscription.

After the lesson 🕚

- What went well in the lesson?
- Could you display the key words somewhere in the classroom?
- What home-learning activity would fit in well with this lesson?

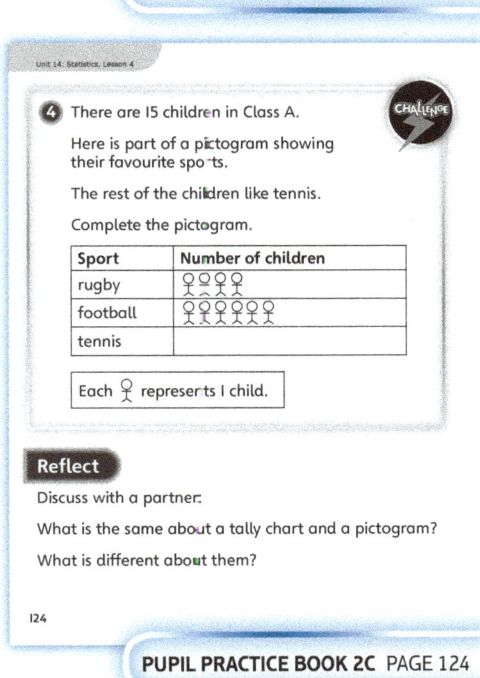

PUPIL PRACTICE BOOK 2C PAGE 124

Interpret pictograms (1 to 1)

Learning focus

In this lesson, children will read and interpret pictograms where one symbol represents one item. They will find totals and compare amounts.

Before you teach 

- Are children secure with drawing pictograms?
- Have you prepared resources for extension activities?

NATIONAL CURRICULUM LINKS

Year 2 Statistics

Ask and answer simple questions by counting the number of objects in each category and sorting the categories by quantity.

Ask and answer questions about totalling and comparing categorical data.

ASSESSING MASTERY

Children can understand and have a strong grasp of the vocabulary associated with interpreting data and pictograms. Children can answer questions relating to pictograms accurately and in detail, where one symbol represents one item.

COMMON MISCONCEPTIONS

When children see the word 'more' in a word problem, they often assume it means they need to add. Practise answering these type of questions using counters to illustrate the problem. Ask:

- *If you have 1 sweet and I have 4 sweets, how many more do I have? The picture shows there are 4 crabs and 8 sea snails. How many more sea snails than crabs are there?*

STRENGTHENING UNDERSTANDING

Understanding mathematical vocabulary will be very important in this lesson. Children may need extra support with key words – especially understanding the meaning when used in context. Make flashcards and highlight key words in problems to help children explain the problem, not just the answer.

GOING DEEPER

In this lesson, to deepen learning, challenge children to think of their own questions for a given pictogram.

KEY LANGUAGE

In lesson: pictogram, how many?, most, least, count, represent, fewer, more, amount, tally chart, altogether

Other language to be used by the teacher: table, diagram, check, answers, opposite, symbol

STRUCTURES AND REPRESENTATIONS

Pictograms

RESOURCES

Mandatory: counters, number lines, extra pictograms

 In the eTextbook of this lesson, you will find interactive links to a selection of teaching tools.

Quick recap 

Play a simple team game together as a class. Ask children to find a good way to keep score during the game. Then ask them to draw a simple block diagram showing the number of points for each team.

Discover

WAYS OF WORKING Pair work

ASK

- Question **1** a): *What is the name of this table? How do you know that it is a pictogram?*
- Question **1** a): *What do you need to do to answer this question? Could you count the creatures in 2s?*
- Question **1** b): *How does a pictogram show which creatures were found the most?*

IN FOCUS For question **1** a), encourage children to check their answers. Ask: *Which creatures are easiest to count first?* Draw attention to the fact that the information children need is in the large picture at the top of the page and that the pictogram shows clearly which creature there is most of or least of.

PRACTICAL TIPS Children can arrange rows of counters to recreate the pictogram. They count each counter as they place it down, in order to find each total.

ANSWERS

Question **1** a): The labels in order from top to bottom: crab, sea snail, shrimp, razor shell, starfish.

Question **1** b): There were 10 starfish. The creature the children found most of was the starfish.

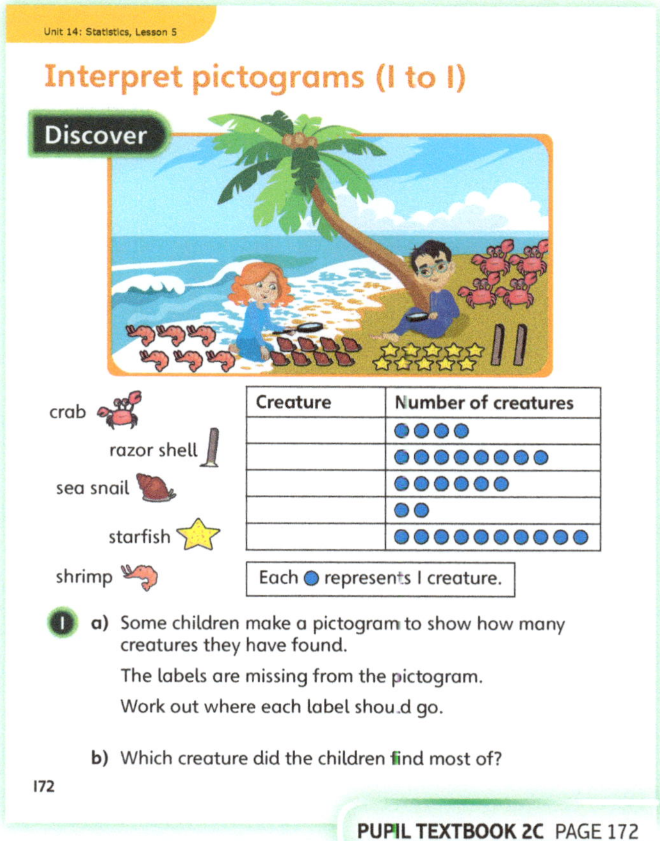

PUPIL TEXTBOOK 2C PAGE 172

Share

WAYS OF WORKING Whole class teacher led

ASK

- Question **1** a): *Can you repeat the steps to make the pictogram without looking at the page?*
- Question **1** b): *Was there the same amount of any creature? How do you know?*
- Questions **1** a) and b): *What makes this pictogram easy to read?*

IN FOCUS In question **1** b), focus on the word 'most'. Discuss what it means with children. Ask children if they know the opposite word ('least'). You could deepen learning here by asking: *If you found 10 crabs and 10 starfish, which would be the most? Neither? Both?*

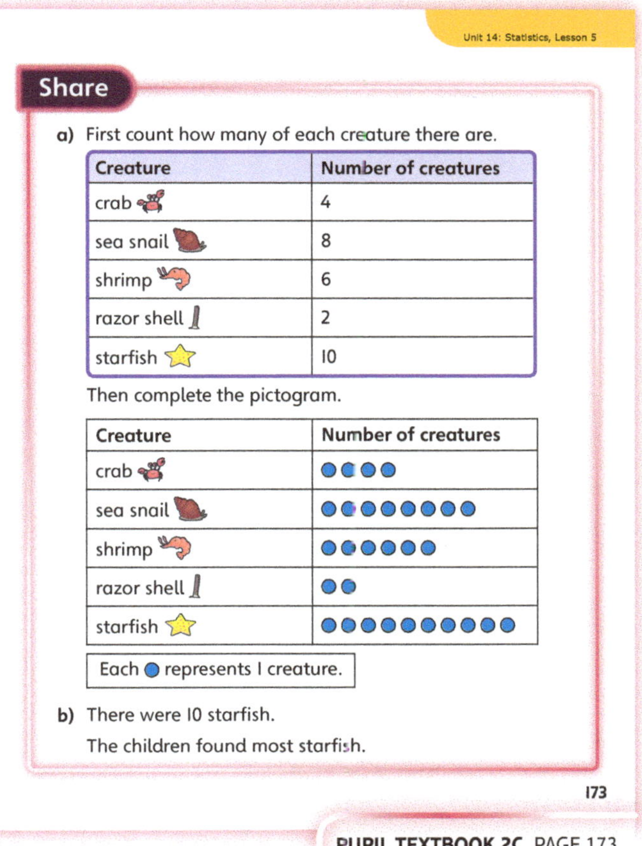

Creature	Number of creatures
crab	4
sea snail	8
shrimp	6
razor shell	2
starfish	10

Creature	Number of creatures
crab	●●●●
sea snail	●●●●●●●●
shrimp	●●●●●●
razor shell	●●
starfish	●●●●●●●●●●

PUPIL TEXTBOOK 2C PAGE 173

Think together

WAYS OF WORKING Whole class teacher led (I do, We do, You do)

ASK

- Question **1** b): *Do you have to count all the symbols or is there another way to tell which day had the most or least sunshine?*
- Question **2**: *What key words can you use to describe the pictogram?*
- Question **3**: *What facts do you know from this pictogram?*

IN FOCUS Question **2** requires children to draw on their understanding of drawing and interpreting pictograms to describe information from a given pictogram. They should consider the row headings, the number of symbols and the key.

STRENGTHEN Consider efficient strategies for comparing the rows of the pictogram in question **1**. Children may feel that they always need to count each symbol, but for questions **1** a) and b), they can make effective comparisons simply by identifying which row of symbols is longest and which is shortest. In contrast, discuss why they will need to count symbols to answer questions **1** c) and d).

DEEPEN Ask children to use the pictogram in question **3** to write three more questions using the words: more than, total, 4 less.

ASSESSMENT CHECKPOINT Question **1** will give you a useful insight into whether children can understand the word problems and subsequently interpret the data. Look carefully at any markings or strategies that children use.

ANSWERS

Question **1** a): Tuesday had the most sunshine.

Question **1** b): Wednesday had the least sunshine.

Question **1** c): There were 7 hours of sunshine on Tuesday.

Question **1** d): There were 3 more hours of sunshine on Monday compared to Wednesday.

Question **2**: Children's answers will vary.
There are the same number of boys as girls.
There are 6 boys and 6 girls.
There are 12 children in total.

Question **3**: Sam and Mo have the same number of stickers. Kim has the fewest stickers.
Children should create their own pictograms for a partner.

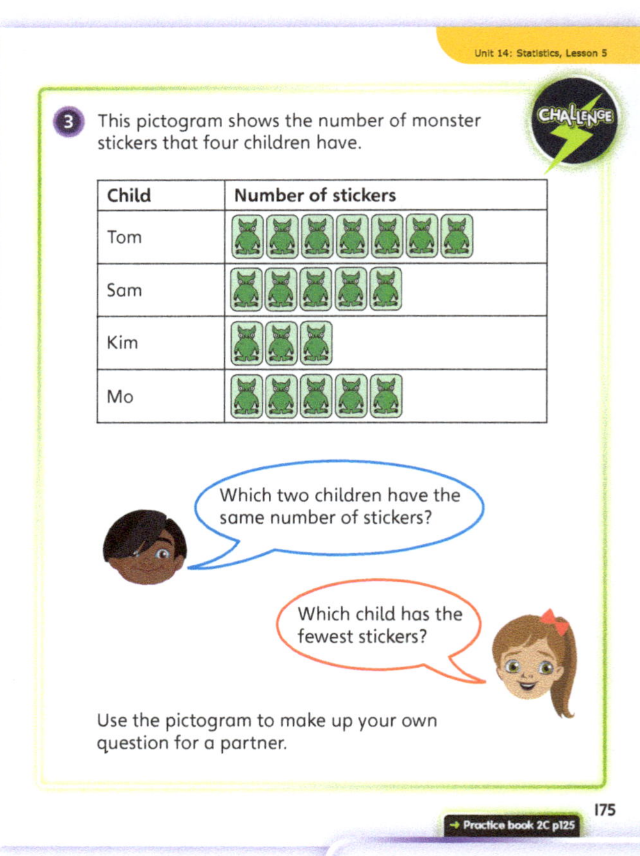

PUPIL TEXTBOOK 2C PAGE 174

PUPIL TEXTBOOK 2C PAGE 175

Practice

WAYS OF WORKING Independent thinking

IN FOCUS Question ① is a useful way to recap the features of a pictogram and how these can be used to find information and solve problems.

STRENGTHEN In question ②, encourage children *not* to count all of the flowers individually. Show children how to count on or back using the actual symbols rather than the total for each flower. Ask: *Do you need to know how many of each flower there are before you can work out how many more or fewer there are?* It may help some children to write totals of each flower on the pictogram.

DEEPEN Challenge children to create a pictogram based on the following information:
- There are 7 cows.
- There are more sheep than cows.
- The total number of animals is 24.
- There are 2 fewer cats than cows.
- There are fewer chickens than cows.
- There are more dogs than chickens but fewer dogs than cows.
- Then ask: *Are there any other solutions?*

ASSESSMENT CHECKPOINT Use question ③ to assess whether children can read the amounts on pictograms and use their knowledge of number bonds to 20 to solve the problem.

ANSWERS Answers for the **Practice** part of the lesson can be found in the *Power Maths* online subscription.

Reflect

WAYS OF WORKING Pair work

IN FOCUS This **Reflect** exercise will stimulate discussion about pictograms. You should find children discussing observations about the value of each symbol, the number of symbols in each row and comparisons or calculations that they can do to derive further facts.

ASSESSMENT CHECKPOINT Can children read pictograms accurately? Can children explain why the symbols should be the same size and equally spaced so that the pictogram can be read accurately? Can children explain why a key is always needed?

ANSWERS Answers for the **Reflect** part of the lesson can be found in the **Power Maths** online subscription.

After the lesson ⏸

- Do children understand the key vocabulary relating to pictograms?
- How well did the prompts and questions promote learning and what were children's responses to them?
- Could children be given an opportunity to create pictograms in other areas of the curriculum?

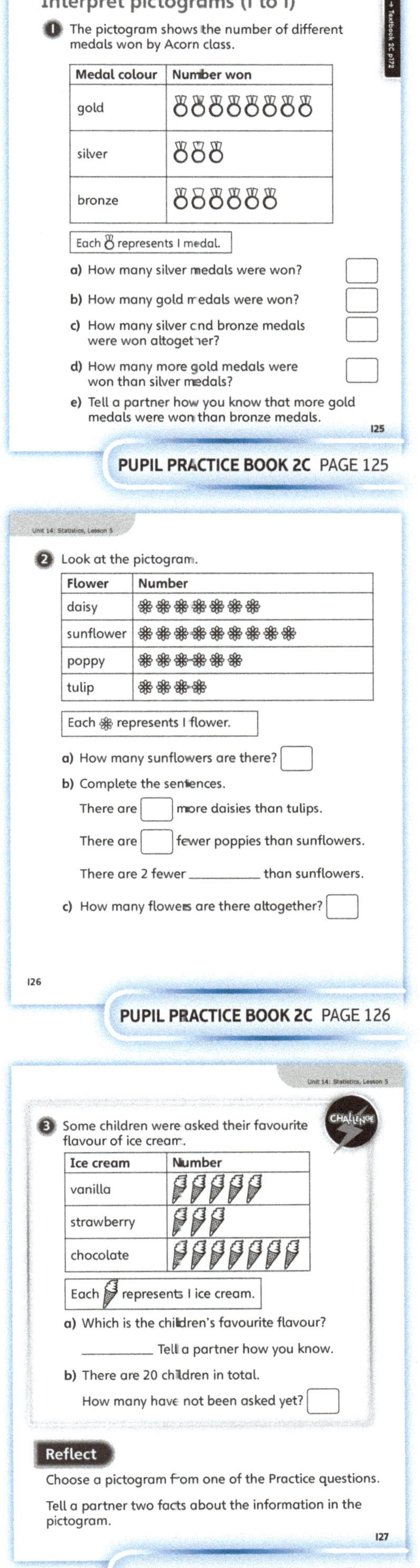

PUPIL PRACTICE BOOK 2C PAGE 125

PUPIL PRACTICE BOOK 2C PAGE 126

PUPIL PRACTICE BOOK 2C PAGE 127

Draw pictograms (1 to 2, 5 or 10)

Learning focus

In this lesson, children will read and construct pictograms in which symbols represent more than one item.

Before you teach

- Do children need to recap prior learning of halving shapes?
- How will you give children extra counting support?
- Can children construct a 1 to 1 pictogram?

NATIONAL CURRICULUM LINKS

Year 2 Statistics

Interpret and construct simple pictograms, tally charts, block diagrams and simple tables.

ASSESSING MASTERY

Children can confidently construct pictograms to display data effectively by choosing the most appropriate key (for example, choosing one symbol to represent 2, 5 or 10 items as appropriate).

COMMON MISCONCEPTIONS

Children may not realise that a symbol can represent more than one object. Ask:
- *What does the key say? How many objects does each symbol represent?*

STRENGTHENING UNDERSTANDING

There is a lot of counting in this lesson. Children may need a counting intervention, in particular practising counting on in 2s, 5s and 10s. Support children with printed number lines in intervals of 2, 5 or 10.

GOING DEEPER

Deepen learning by asking children to create keys for data sets. For instance, give children the following set of data: 35, 0, 5, 10, 30, 20, 40. Ask: *If you were to draw a pictogram to show this data, what would one symbol represent? How would you write a key to show this?* Then ask children to explain why they have chosen the key they have.

KEY LANGUAGE

In lesson: pictogram, symbol, how many?, key, equal, =, results, tally chart, represent, difference, most, least, popular, key, title, odd number

Other language to be used by the teacher: collect, survey, even number, multiple, number line, data, data set

STRUCTURES AND REPRESENTATIONS

Pictograms and tally charts, number lines

RESOURCES

Mandatory: pictogram templates, number lines

 In the eTextbook of this lesson, you will find interactive links to a selection of teaching tools.

Quick recap

As a class, count up together in 2s.

Discover

WAYS OF WORKING Pair work

ASK

- Question **1** a): *What information does this key give us?*
- Question **1** b): *How long would it take you to draw one symbol for each piece of fruit?*

IN FOCUS Question **1** a) requires children to consider carefully the key for a pictogram that they will draw. They should realise that, unlike in previous lessons, each symbol for this pictogram represents not 1 but 2 pieces of fruit. Consider why this might be and agree that it will make it easier for children to draw their pictograms as they will have fewer symbols to draw. Count up to 12 together, first in 1s and then in 2s. Agree that counting in 2s is quicker.

PRACTICAL TIPS Children can use counters in three colours to create their pictogram before they draw it. They can explore what would happen if each counter represented only one piece of fruit. Agree that it would make the pictogram very big and more difficult to count.

ANSWERS

Question **1** a):

Fruit	Pieces of fruit
apples	● ● ● ●
oranges	● ● ● ● ● ● ●
bananas	● ● ●

Question **1** b): When you use a red circle to represent 2 pieces of fruit, you do not need to draw as many circles.

Share

WAYS OF WORKING Whole class teacher led

ASK

- Question **1** a): *Why is it helpful to know that 2 × 4 = 8?*
- Question **1** a): *Why has Flo halved each number?*
- Question **1** b): *If you use 1 counter for each piece of fruit, do you think that is too many circles to draw or count?*

IN FOCUS For question **1** b), focus children on the 2-digit number of oranges. Children may find that if they had 1 circle representing 1 piece of fruit, they would be more likely to make mistakes when drawing or interpreting a pictogram with a number this big.

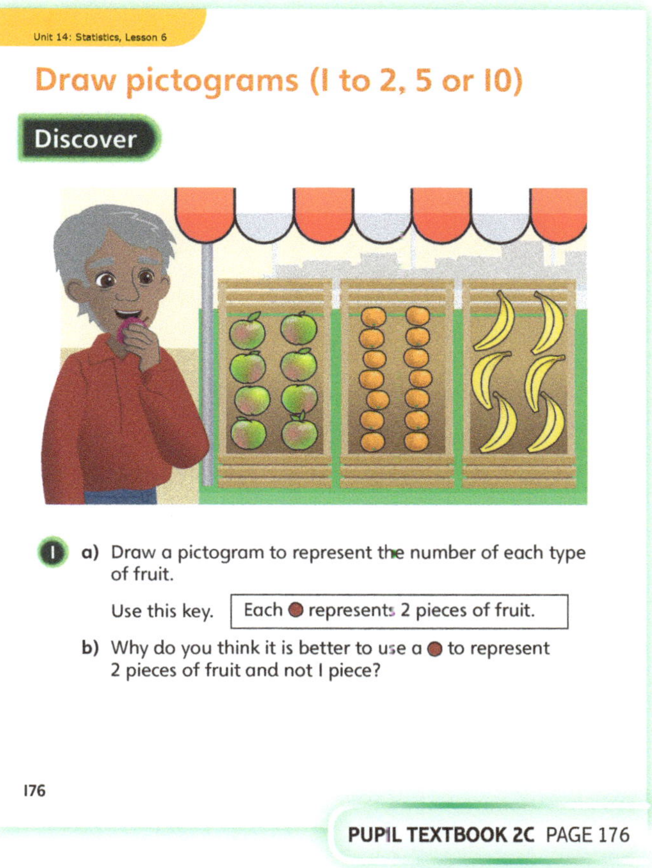

PUPIL TEXTBOOK 2C PAGE 176

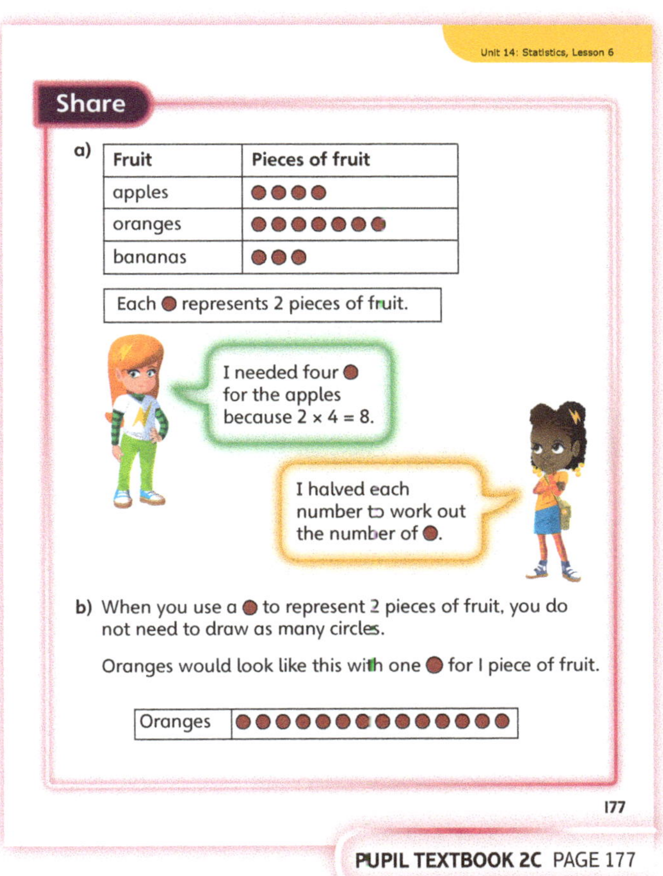

PUPIL TEXTBOOK 2C PAGE 177

Think together

WAYS OF WORKING Whole class teacher led (I do, We do, You do)

ASK

- Question **1**: *Why is this pictogram a good way to represent information from a tally chart?*
- Question **2**: *If a symbol represents 10 children, how might you represent 5 children?*
- Question **3**: *One of these is an odd number? How will you show that on your pictogram?*

IN FOCUS For question **3**, children may find it difficult to represent an odd number when each symbol represents 2 items. Ask children what they could do with a symbol to make it represent 1 item. Remind them of prior learning about halving shapes (because half of 2 is 1). Discuss how it is important that they choose a suitable symbol to represent 2 (or 4) items, so that the symbol can easily be halved (or quartered).

STRENGTHEN Some children may need intervention in halving regular shapes and in representing odd numbers. Discuss how to show odd numbers such as 3, 5 and 7 using whole symbols and one half symbol.

DEEPEN Set children the challenge of creating a pictogram using the same data collected in question **3**. This time, however, tell children that one symbol represents 4 sheets. How would they represent 1, 2 and 3 sheets here?

ASSESSMENT CHECKPOINT Questions **1** and **2** should help you decide whether children can create a pictogram in which a symbol represents 5 or 10. Look carefully to see if they managed to represent each number accurately.

ANSWERS

Question **1**:

Leo	⭐⭐⭐⭐
Jane	⭐⭐⭐⭐⭐⭐⭐
Milo	⭐⭐⭐
Amir	⭐⭐⭐⭐⭐

Each ⭐ represents 5 questions.

Question **2**:

Favourite flavour	Number of children
strawberry	☺☺☺☺
vanilla	☺☺☺☺☺☺
chocolate	☺☺☺☺☺☺☺

Each ☺ represents 10 children.

Question **3**:

Colour	Number of sheets
blue	○○○○○○○
red	○○○○○◖

Each ○ represents 2 sheets of paper.

Think together

1 Reena counts the number of questions that 4 people answer correctly in a quiz. The tally chart shows her results.

Name	Tally	Number of questions
Leo	卌 卌 卌 卌	20
Jane	卌 卌 卌 卌 卌 卌 卌	35
Milo	卌 卌 卌	15
Amir	卌 卌 卌 卌 卌	25

Complete the pictogram for Reena's data.

Leo	⭐⭐⭐⭐
Jane	
Milo	
Amir	

Each ⭐ represents 5 questions.

178

PUPIL TEXTBOOK 2C PAGE 178

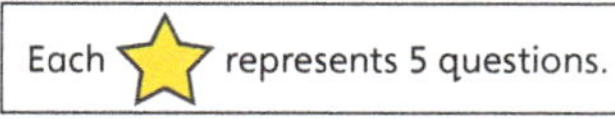

2 Complete a pictogram to represent this data.

Favourite flavour	Number of children
Strawberry	40
Vanilla	60
Chocolate	70

Use ☺ to represent 10 children.

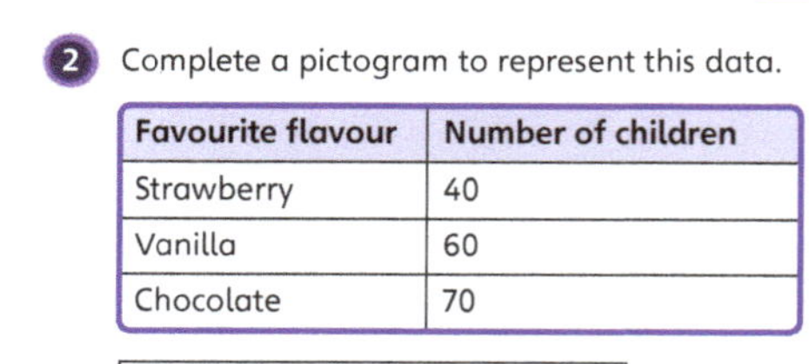

3 Josh counts the number of red sheets and blue sheets of paper.

CHALLENGE

Colour	Number of sheets
Blue	14
Red	11

Complete a pictogram to represent the data.

Use ○ to represent 2 sheets of paper.

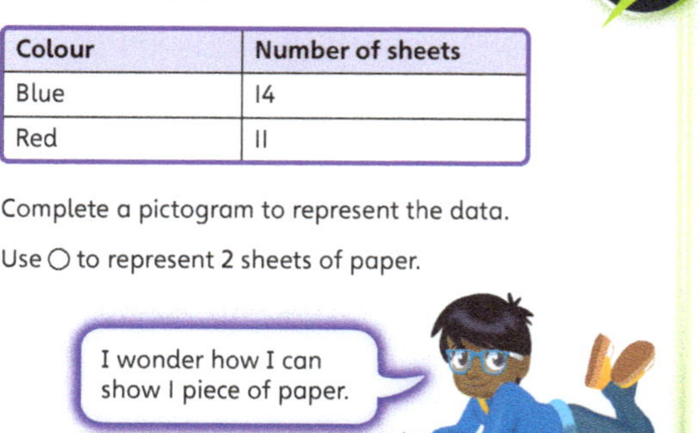

179

→ Practice book 2C p128

PUPIL TEXTBOOK 2C PAGE 179

Practice

WAYS OF WORKING Independent thinking

IN FOCUS In question **2**, children must look carefully at the key. They should notice that each symbol represents 5 days, as does each tally group. Children may also need support in understanding that the sun icon represents a way of counting all types of days (whatever the weather).

STRENGTHEN Work with children who require support with question **3**. The problem may need to be broken down into bitesize chunks. Start by asking children to read the statements aloud, then ask them what they already know. Finally, explore which clue helps first, second and last. Discuss why it is important to know that the symbol represents 10.

DEEPEN Present children with a pictogram without a key, including some half symbols. Ask them what the data could have been. Can they show it in a table? They will probably start by counting the symbols one-by-one. Ask: *But, what if each symbol represents 2?* Children will realise there are different options.

ASSESSMENT CHECKPOINT Question **2** will allow you to assess children's progress towards mastery of reading pictograms in which symbols represent more than one thing. Question **3** presents a good opportunity to assess children's reasoning skills when completing a pictogram from the clues given.

ANSWERS Answers for the **Practice** part of the lesson can be found in the *Power Maths* online subscription.

Reflect

WAYS OF WORKING Pair work, whole class

IN FOCUS This question will generate a discussion about symbols representing more than 1 object. Children may mention the benefits of using symbols that represent 2, 5 or 10 objects. They may also make the link to tally charts and to halving when representing an odd number.

ASSESSMENT CHECKPOINT This activity is an excellent way to observe language used and assess which children use effective reasoning when choosing which symbol to use on a pictogram.

ANSWERS Answers for the **Reflect** part of the lesson can be found in the *Power Maths* online subscription.

After the lesson

- Have children mastered reading and constructing pictograms?
- Can children represent more than 1 item with just 1 symbol?
- Are children ready for interpreting pictograms where 1 symbol represents 2, 5 or 10 objects?

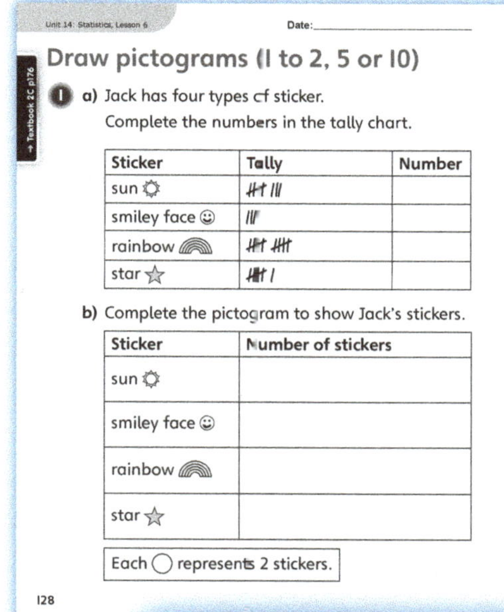

Sticker	Tally	Number
sun	Htt III	
smiley face	IIII	
rainbow	Htt Htt	
star	Htt I	

Sticker	Number of stickers
sun	
smiley face	
rainbow	
star	

PUPIL PRACTICE BOOK 2C PAGE 128

Weather	Tally	Number of days
sunny	Htt Htt Htt	15
cloudy	Htt Htt	
rain	Htt Htt Htt Htt Htt	

Weather	Number of days
sunny	
cloudy	
rain	

PUPIL PRACTICE BOOK 2C PAGE 129

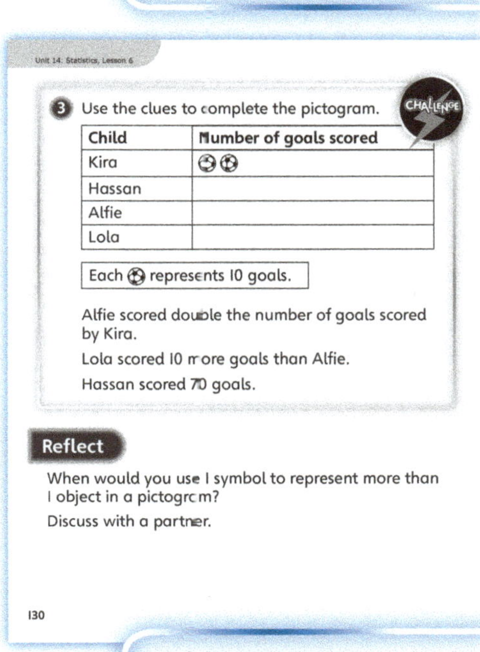

Child	Number of goals scored
Kira	☺ ☺
Hassan	
Alfie	
Lola	

PUPIL PRACTICE BOOK 2C PAGE 130

Interpret pictograms (1 to 2, 5 or 10)

Learning focus

In this lesson, children will read and interpret pictograms that have symbols representing more than one item.

Before you teach 

- How will you make links to the previous lesson?
- Are children confident with the language that will be used in this lesson?
- Are there any misconceptions that need addressing?

NATIONAL CURRICULUM LINKS

Year 2 Statistics

Ask and answer simple questions by counting the number of objects in each category and sorting the categories by quantity.

Ask and answer questions about totalling and comparing categorical data.

ASSESSING MASTERY

Children can answer questions about given pictograms confidently and in detail. Children can think of appropriate questions that show deeper interrogation of data.

COMMON MISCONCEPTIONS

Children may not read the key and may count each symbol as one item. Ask:
- *How can you find out what each symbol represents?*

Children may forget about the key midway through counting. For example, if each symbol represents two items and there are five symbols, they may count: 2, 4, 6, 7, 8. Ask:
- *Can you check your answer? What are you counting in? Are you counting in 1s, 2s or 5s?*

STRENGTHENING UNDERSTANDING

In this lesson, intervention may be needed in which children study a pictogram and then answer questions about it. Guiding their answers is important to ensure they use the correct terminology. Children may need support in counting in 2s, 5s or 10s and in interpreting half symbols for symbols representing 2 and 10.

GOING DEEPER

Arrange for a traffic survey with your class. Collect the data, create pictograms (emphasise the need for an appropriate key depending on the amount of traffic), then interpret the data. Children could write questions about the pictogram and answer them underneath the pictogram. Finally, ask children: *Who might find these statistics useful?*

If a traffic survey is not possible, an alternative might be to collect data about the books on the classroom bookshelf.

KEY LANGUAGE

In lesson: pictogram, represent, symbol, groups, multiplication, how many?, worth, equals, count, more, less

Other language to be used by the teacher: most, least, total, altogether, check, answer, data, statistics, survey

STRUCTURES AND REPRESENTATIONS

Pictograms

RESOURCES

Optional: Counting materials such as number lines, 100 squares, multiplication grids (2s, 5s, 10s)

 In the eTextbook of this lesson, you will find interactive links to a selection of teaching tools.

Quick recap 

As a class, count up together in 2s to 20, in 5s to 50 and in 10s to 100.

Discover

 Pair work

ASK

- Question **1** a): *Why are 5 hot dinners represented by 1 plate?*
- Question **1** b): *If you have 50 plates, why not just draw all 50?*
- Questions **1** a) and b): *Would you change the key if you needed to represent 100 hot dinners? What would each plate represent?*

IN FOCUS Question **1** a) requires children to read the key and realise that each plate represents 5 hot dinners. This may be difficult for some children to understand. Explain that when you are working with higher numbers, it would be difficult to draw so many plates. So instead we represent 5 items with just 1 symbol.

PRACTICAL TIPS Children can use counters to recreate the pictogram, counting up in 5s as they place each counter.

ANSWERS

Question **1** a): 25 hot dinners were made on Tuesday.

Question **1** b): You need to draw 7 plates.

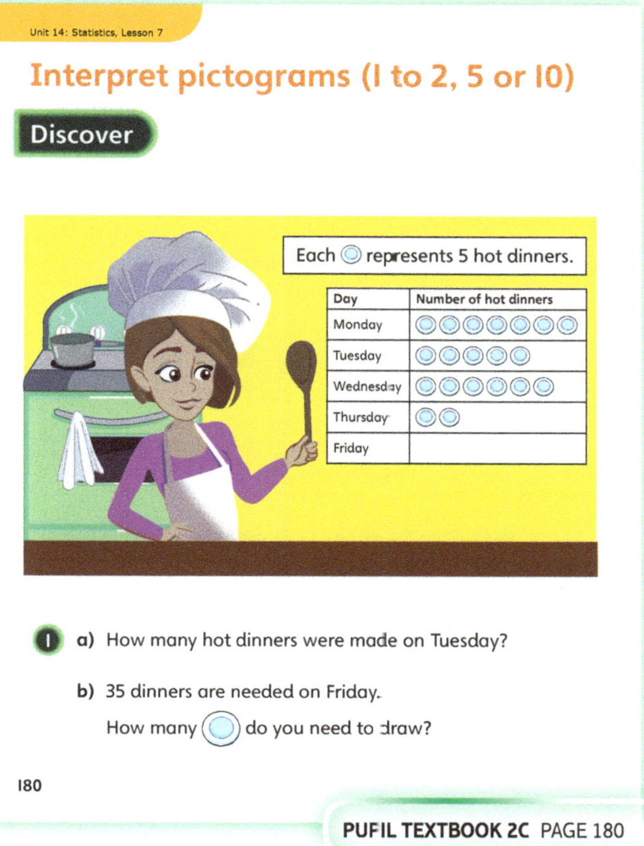

PUPIL TEXTBOOK 2C PAGE 180

Share

 Whole class teacher led

ASK

- Question **1** b): *What is an efficient strategy to work this out?*
- Question **1** b): *Could Monday's amount help you find the answer?*
- Questions **1** a) and b): *Can you explain your answer?*

IN FOCUS For question **1** b), focus children on the strategies they can use to find the answer. Some may draw the 35 dinners and then make 7 groups; others may use their knowledge of counting in 5s. It is good practice here to link the long-winded way with the efficient method. Making connections is important for achieving mastery.

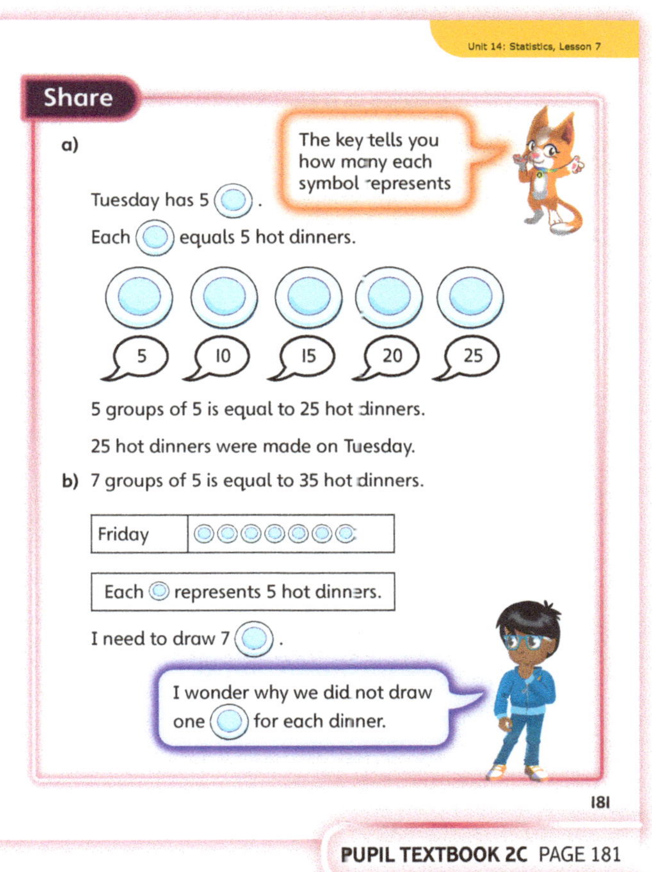

PUPIL TEXTBOOK 2C PAGE 181

Think together

WAYS OF WORKING Whole class teacher led (I do, We do, You do)

ASK

- Questions ❶ and ❷: *What do you need to do to answer a 'How many fewer?' question or a 'How many more?' question with pictograms?*
- Question ❸: *Why might you use half a symbol in pictograms? What is half a symbol worth in this pictogram?*

IN FOCUS For question ❸, children will have to think carefully about the key of this pictogram. Discuss that 1 symbol represents 2 books and use this to count the number of books for each child before exploring the three statements in more detail. Ask: *How do you know, without counting, that no one read an odd number of books?*

STRENGTHEN Counting in 2s, 5s and 10s is a good way to strengthen learning. If children cannot count fluently, they will find interpretation of data much more difficult.

Use counters to make physical representations of the pictograms to strengthen understanding. Kinaesthetic learners may find this easier to understand when interpreting.

DEEPEN Get children to create two pictograms from the same set of data but with different keys. Once complete, ask children to evaluate which one is more effective. You may need to suggest that one key is double the other so that a half symbol has a definite value.

ASSESSMENT CHECKPOINT Question ❷ will help you decide whether children can derive information from a pictogram where 1 symbol represents 10 objects.

ANSWERS

Question ❶ a): There were 15 blackbirds, 10 robins, 5 sparrows and 20 blue tits seen.

Question ❶ b): Blue tits were seen the most.

Question ❶ c): There were 5 fewer robins than blackbirds seen.

Question ❷ a): There are 40 red balloons.

Question ❷ b): There are 30 more blue balloons than red balloons.

Question ❸: Bella's statement is false.

Filip's statement is true.

Half a symbol would represent 1 book; so Milo's statement is also true.

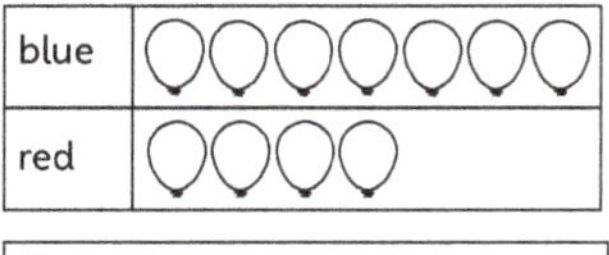

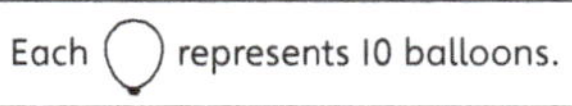

PUPIL TEXTBOOK 2C PAGE 182

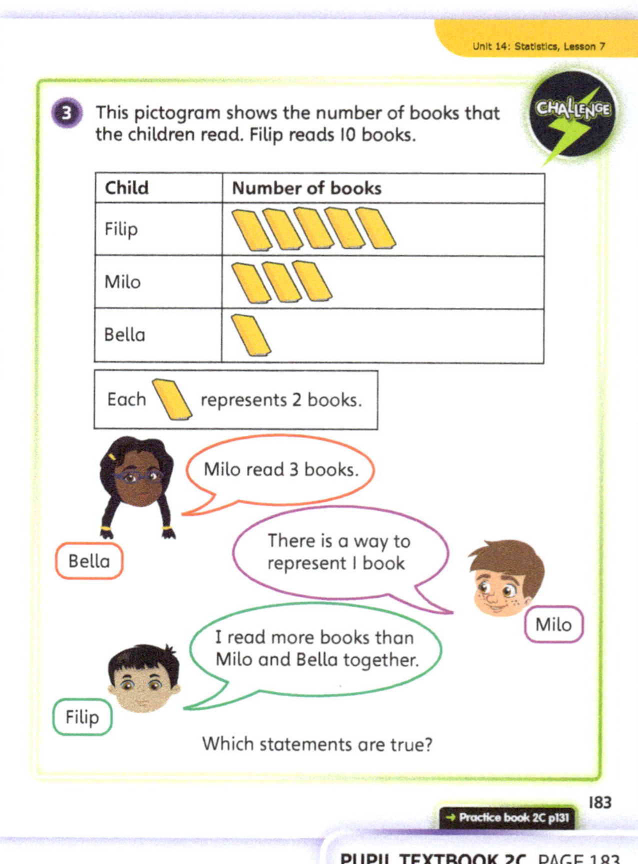

PUPIL TEXTBOOK 2C PAGE 183

Practice

WAYS OF WORKING Independent thinking

IN FOCUS Question **3** b) asks children to find the total number of children in both years. Children will have to use the information from both rows of the pictogram. Discuss useful strategies with them.

STRENGTHEN Can children spot any patterns that will help them count in 2s, 5s or 10s? Ask children to write the total amounts at the end of each row so they do not have to keep recounting.

DEEPEN Ask children to write some rules that help to decide what a key should be.

ASSESSMENT CHECKPOINT Question **1** will give you an opportunity to assess whether children can interpret data from a pictogram where one symbol represents two objects. Look for children using efficient strategies to count and calculate with the data provided.

ANSWERS Answers for the **Practice** part of the lesson can be found in the *Power Maths* online subscription.

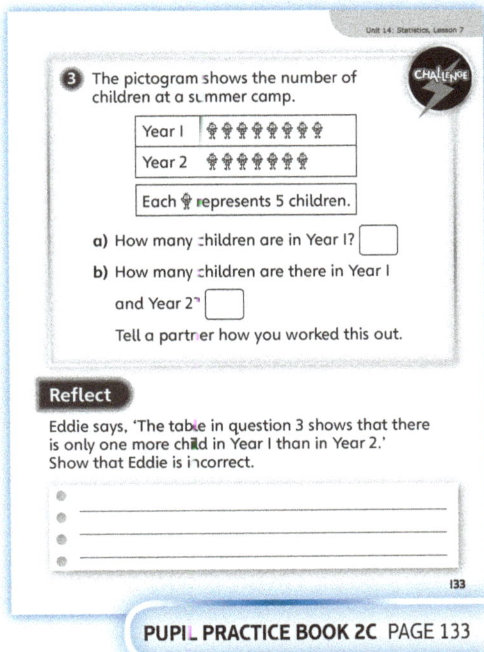

PUPIL PRACTICE BOOK 2C PAGE 131

PUPIL PRACTICE BOOK 2C PAGE 132

PUPIL PRACTICE BOOK 2C PAGE 133

Reflect

WAYS OF WORKING Independent thinking

IN FOCUS Children should come up with some clear, concise statements for this **Reflect** question such as, 'The symbol represents 5 children'.

ASSESSMENT CHECKPOINT This activity will let you see which children have fully understood the importance of reading the key and interpreting each symbol correctly.

ANSWERS Answers for the **Reflect** part of the lesson can be found in the *Power Maths* online subscription.

After the lesson

- What opportunities can you identify to reinforce and apply this lesson's learning?
- Did children understand the issues with using keys representing more than one item?

End of unit check

Don't forget the unit assessment grid in your *Power Maths* online subscription.

WAYS OF WORKING Group work adult led

IN FOCUS All questions in the **End of unit check** will give you a good insight into which children can read and interpret information from a range of charts and diagrams.

Think!

WAYS OF WORKING Pair work or small groups

IN FOCUS The purpose of the **Think!** question is to find out whether children are secure enough to spot Ola's mistake. See if children can explain where she went wrong and if their explanations are confident.

The second part of the question requires children to read the pictogram and come up with their own statements. This will involve some deep thinking.

Children should use words such as: same, equal, more than, less than, least, most, total and altogether.

ANSWERS AND COMMENTARY Children who have mastered this unit should be secure with interpreting data and explaining methods and answers. They will also be confident in solving problems involving charts and diagrams.

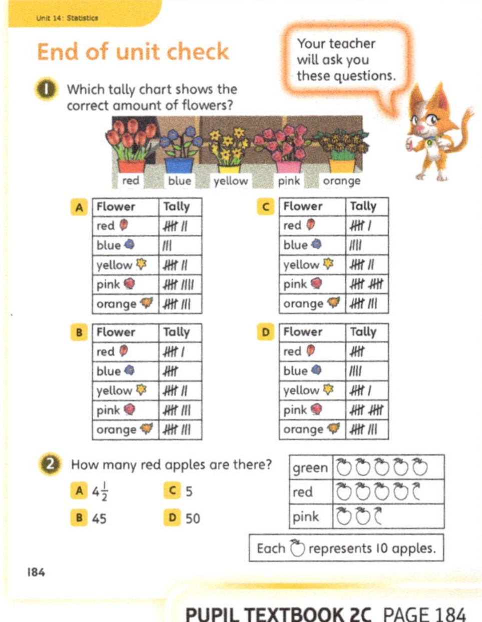

PUPIL TEXTBOOK 2C PAGE 184

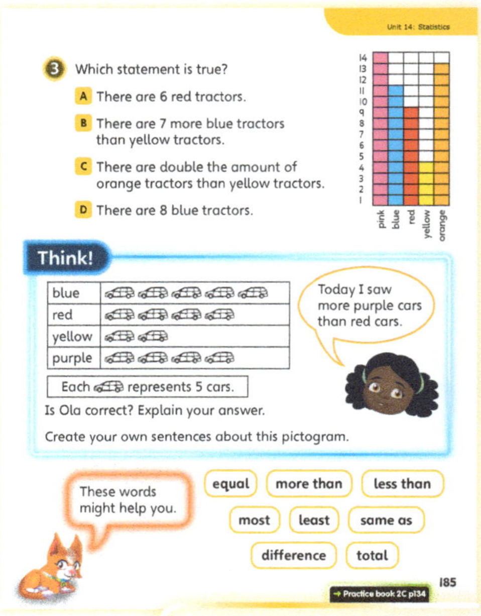

PUPIL TEXTBOOK 2C PAGE 185

Q	A	WRONG ANSWERS AND MISCONCEPTIONS	STRENGTHENING UNDERSTANDING
1	C	Any other answer suggests that children are not secure with counting the tally marks. They may be miscounting or not counting the oblique.	Run intervention activities in which children construct tally charts from data. Particularly provide support when reaching '5' – the oblique mark. Challenge children to make pictograms and block charts from tally charts by first writing the total given in each tally chart category.
2	B	A suggests children have not looked at the key. C or D suggest children have misunderstood that the final symbol is a half-symbol.	Practise counting tally chart and pictogram totals. Children could use counters or blocks to support them (and prevent miscounting).
3	B	A or D suggest that children have not counted each block as 1. C might tell you that children are not secure with counting and doubling numbers.	Move children on to pictograms in which a symbol represents 2 items or 10 items. This work will link closely with practising times-tables.

My journal

WAYS OF WORKING Independent thinking

ANSWERS AND COMMENTARY

Children will need to realise that Ola is incorrect as the amounts are equal. They should make use of the vocabulary that Sparks gives at the bottom of page 134 in the Practice Book to form their answer. An exemplary answer would be: *Ola is incorrect because there are 4 red cars and 4 purple cars. This means the amounts are equal.*

Children may need strengthening exercises such as reading the question in groups and discussing what it is asking.

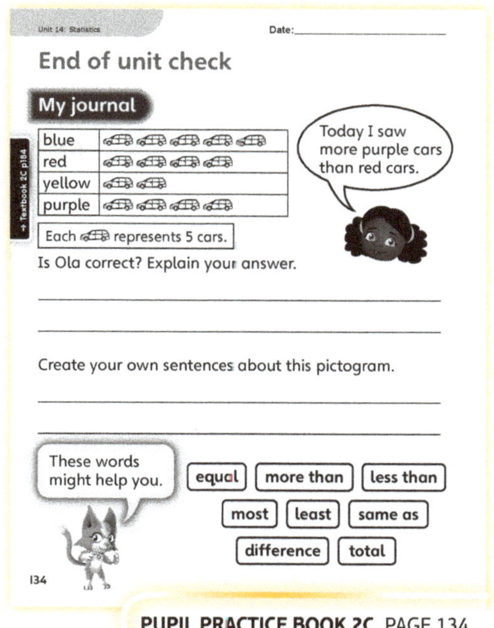

PUPIL PRACTICE BOOK 2C PAGE 134

Power check

WAYS OF WORKING Independent thinking

ASK

- *Why do you feel like that about pictograms?*
- *What was your main strength in this unit?*
- *Are there some charts or diagrams that you feel better about than others?*

Power puzzle

WAYS OF WORKING Pair work or small groups

IN FOCUS Children will need to use reasoning to work out how many of each fruit there are before they can complete the block diagram. If children do not know where to start, suggest they start with the two amounts they know (bananas and apples) and use them to work out how many pears there would be, before focusing on the oranges. Encourage children to add in order to check the total is correct, remembering to also check their answers against each sentence again.

ANSWERS AND COMMENTARY 5 pears, 7 oranges and 10 apples and 3 bananas = 25 pieces of fruit

Children could use counters or cubes of different colours to represent the fruits but should not try to complete the block graph before they know how many pieces of each fruit are needed. Children will need to use reasoning to reach the correct answers.

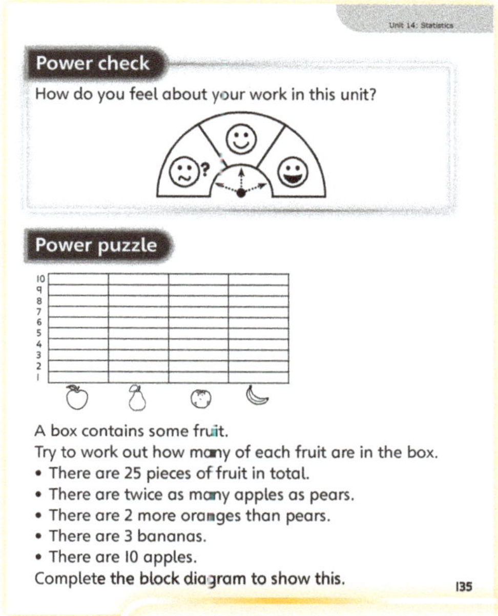

PUPIL PRACTICE BOOK 2C PAGE 135

After the unit

- How did the unit go?
- What lessons went well? Why did they go well?
- Did children depend on equipment or support? Or did they get more secure as the unit progressed?

Strengthen and **Deepen** activities for this unit can be found in the *Power Maths* online subscription.

Published by Pearson Education Limited, 80 Strand, London, WC2R 0RL.

www.pearsonschools.co.uk

Text © Pearson Education Limited 2017, 2023
Edited by Pearson and Florence Production Ltd
First edition edited by Pearson, Little Grey Cells Publishing Services and Haremi Ltd
Designed and typeset by Pearson and PDQ Digital Media Solutions Ltd
First edition designed and typeset by Kamae Design
Original illustrations © Pearson Education Limited 2017, 2023
Illustrated by Laura Arias, Fran and David Brylewski, Nigel Dobbyn and Nadene Naude at Beehive
Illustration; Emily Skinner at Graham-Cameron Illustration; Paul Higgins at Hunter Higgins
Illustrations; and Kamae Design
Back cover illustration © Will Overton at Advocate Art and Nadene Naude at Beehive Illustration
Images: The Royal Mint, 1971, 1982, 1990, 1992, 1997: 52, 53, 123, 156, 165
Images Power Up: The Royal Mint, 1971, 1982, 1990, 1992: Y2 Unit 12 Slides: Lesson 6
Cover design by Pearson Education Ltd

Series editor: Tony Staneff; Lead author: Josh Lury
Authors (first edition): Tony Staneff, David Board, Jonathan East, Tim Handley, Julia Hayes and
Timothy Weal
Consultants (first edition): Professor Liu Jian

The rights of Tony Staneff and Josh Lury to be identified as authors of this work have been
asserted by them in accordance with the Copyright, Designs and Patents Act 1988.

First published 2017
This edition first published 2023

27 26 25 24 23
10 9 8 7 6 5 4 3 2 1

British Library Cataloguing in Publication Data
A catalogue record for this book is available from the British Library

ISBN 978 1 292 45052 0

Printed in the UK by Ashford Press Ltd

For Power Maths online resources, go to:
www.activelearnprimary.co.uk

Note from the publisher
Pearson has robust editorial processes, including answer and fact checks, to ensure the
accuracy of the content in this publication, and every effort is made to ensure this publication
is free of errors. We are, however, only human, and occasionally errors do occur. Pearson is
not liable for any misunderstandings that arise as a result of errors in this publication, but it is
our priority to ensure that the content is accurate. If you spot an error, please do contact us at
resourcescorrections@pearson.com so we can make sure it is corrected.